Studies in Infrastructure and Control

The book series aims to publish top-quality state-of-the-art textbooks, research monographs, edited volumes and selected conference proceedings related to infrastructure, innovation, control, and related fields. Additionally, established and emerging applications related to applied areas like smart cities, internet of things, machine learning, artificial intelligence, etc., are developed and utilized in an effort to demonstrate recent innovations in infrastructure and the possible implications of control theory therein. The study also includes areas like transportation infrastructure, building infrastructure management and seismic vibration control, and also spans a gamut of areas from renewable energy infrastructure like solar parks, wind farms, biomass power plants and related technologies, to the associated policies and related innovations and control methodologies involved.

Mahmoud Elsisi · Noorman Rinanto ·
Chun-Lien Su
Editors

Maritime Infrastructure for Energy Management and Emission Reduction Using Digital Transformation

Springer

Editors
Mahmoud Elsisi
Department of Electrical Engineering
National Kaohsiung University of Science and Technology
Kaohsiung City, Taiwan

Department of Electrical Engineering
Faculty of Engineering (Shoubra)
Benha University
Cairo, Egypt

Noorman Rinanto
Department of Marine Electrical Engineering
Shipbuilding Institute of Polytechnic Surabaya
Surabaya, Indonesia

Chun-Lien Su
Department of Electrical Engineering
National Kaohsiung University of Science and Technology
Kaohsiung City, Taiwan

ISSN 2730-6453 ISSN 2730-6461 (electronic)
Studies in Infrastructure and Control
ISBN 978-981-96-4437-7 ISBN 978-981-96-4438-4 (eBook)
https://doi.org/10.1007/978-981-96-4438-4

This Springer imprint is published by the registered company Springer Nature Singapore Pte Ltd.
The registered company address is: 152 Beach Road, #21-01/04 Gateway East, Singapore 189721, Singapore

Preface

The book delves into the transformative potential of digital technologies in the maritime sector, specifically focusing on energy management and emission reduction. The maritime industry is undergoing a significant shift as it grapples with environmental challenges and regulatory demands. This book explores how digital transformation, particularly through the integration of advanced technologies like IoT, artificial intelligence (AI), and digital twins, can optimize maritime operations, reduce energy consumption, and lower emissions. It also addresses the need for new methodologies and the innovative application of these technologies in real-world scenarios.

The book is structured into 11 chapters, each contributing to a comprehensive understanding of the role digital transformation can play in the maritime industry. The chapters are as follows.

Chapter 1 provides an introduction to maritime infrastructure and its significance in the global trade network. It also highlights the challenges the maritime sector faces in terms of energy consumption and emissions, setting the stage for the discussion on how digital technologies can help address these issues. The chapter outlines the goals of the book and introduces the concept of digital transformation in maritime operations.

Chapter 2 explores the difficulties the maritime industry faces in managing energy consumption and reducing emissions. It examines the regulatory pressures, technological barriers, and environmental impacts associated with maritime operations. The chapter emphasizes the need for digital solutions to help meet sustainability goals and the increasing pressure to comply with environmental regulations.

Chapter 3 dives into the digital technologies available for monitoring and optimizing energy use in maritime operations. Technologies such as IoT, sensors, and data analytics are discussed in detail, focusing on how they can be used to gather real-time data, monitor energy consumption, and make operational adjustments to improve energy efficiency.

Chapter 4 focuses on the role of advanced data analytics, including machine learning and predictive modeling, in improving energy efficiency in maritime operations. The chapter explains how these tools can help analyze operational data, predict maintenance needs, and optimize resource allocation to enhance overall efficiency and reduce emissions.

Chapter 5 explores the concept of digital twins—virtual replicas of physical assets. The authors explain how digital twins are used to simulate real-world operations, optimize performance, and test energy management strategies. Through this technology, maritime operators can monitor systems in real time, predict failures, and improve operational efficiency without the need for physical adjustments.

Chapter 6 addresses the cybersecurity challenges faced by the maritime industry in the age of digital transformation. It explores strategies for safeguarding critical systems from cyber threats and ensuring the security of data and operations, particularly in environments that rely heavily on digital tools.

Chapter 7 examines the integration of renewable energy sources, such as wind, solar, and hydrogen fuel cells, into maritime infrastructure. It discusses the technological advancements in renewable energy systems, their feasibility, and their role in reducing emissions and improving energy efficiency in maritime operations. The chapter emphasizes the potential of hybrid systems that combine renewable and traditional energy sources.

Chapter 8 presents real-world case studies of maritime operations that have successfully implemented digital transformation strategies to improve energy efficiency and reduce emissions. These case studies highlight the practical applications of digital technologies in the maritime sector, showcasing the results of energy optimization efforts and the lessons learned.

Chapter 9 explores the application of Artificial Intelligence of Things (AIoT) in managing emissions at smart ports. The authors discuss how AIoT systems can optimize port operations, reduce emissions, and predict equipment failures. The chapter highlights how the combination of AI and IoT technologies can lead to cleaner, more efficient ports, offering a pathway for smarter, more sustainable maritime infrastructure.

Chapter 10 focuses on the regulatory landscape governing maritime energy management and emission reduction. It discusses international conventions, regional regulations, and national policies that shape the maritime industry's efforts to reduce its carbon footprint. The chapter examines the role of policy in driving digital transformation and incentivizing the adoption of green technologies.

Chapter 11 looks to the future, exploring emerging trends and innovations that could further transform the maritime sector's approach to energy management and emission reduction. The chapter discusses the potential for new technologies, such as fully autonomous vessels and zero-emission ships, as well as the role of continued digital innovation in achieving sustainability goals for maritime infrastructure.

The book aims to provide comprehensive resources for professionals, researchers, policymakers, and academics interested in understanding the critical role of digital technologies in shaping the future of the maritime industry, particularly in the areas of energy management and emission reduction.

Kaohsiung City, Taiwan — Mahmoud Elsisi
Surabaya, Indonesia — Noorman Rinanto
Kaohsiung City, Taiwan — Chun-Lien Su

Contents

About the Editors

Mahmoud Elsisi received the B.Sc., M.Sc., and Ph.D. degrees from the Electrical Engineering Department, Faculty of Engineering at Shoubra, Benha University, Cairo, Egypt, in 2011, 2014, and 2017, respectively. He worked as Assistant Professor at the Electrical Engineering Department, Faculty of Engineering at Shoubra, Benha University. From August 2019 to July 2022, he worked as Assistant Professor with the Industry 4.0 Implementation Center, Center for Cyber-Physical System Innovation, National Taiwan University of Science and Technology, Taipei, Taiwan. He is currently Associate Professor with the Electrical Engineering Department, National Kaohsiung University of Science and Technology, Kaohsiung, Taiwan. His research interests include studying the machine learning, deep learning, the Internet of Things (IoT), cybersecurity, model predictive control, neural networks, fuzzy logic, Kalman filter, observers, decentralized control of large-scale systems, robotic control, autonomous vehicle control, renewable energy, power system dynamics.

Noorman Rinanto received the B.E. and M.E. in electrical engineering from Sepuluh Nopember Institute of Surabaya (ITS), Surabaya, Indonesia, in 2006 and 2012, respectively. He completed the Ph.D. degree at the Department of Electrical Engineering, National Taiwan University of Science and Technology (NTUST), Taipei, Taiwan, in 2023. He is currently Assistant Professor with the Marine Electrical Engineering Department, Shipbuilding Institute of Polytechnic Surabaya, Surabaya, Indonesia. His research interests include sensor fusion, machine learning, artificial intelligence, intelligent robot and automation system, battery management system, smart control system, Industrial Internet of Things (IIoT), signal and image processing.

Chun-Lien Su received the diploma degree in electrical engineering from National Kaohsiung Institute of Technology, Taiwan, and the M.S. and Ph.D. degrees in electrical engineering from the National Sun Yat-Sen University, Taiwan, in 1992, 1997, and 2001, respectively. In 2002 and 2006, he was Assistant Professor and Associate Professor at the Department of Marine Engineering, National Kaohsiung Marine University, Taiwan, respectively. From 2012 to 2017, he was as Full Professor where he was Director at the Energy and Control Research Center. From August 2017 to January 2018, he was Visiting Professor at the Department of Energy Technology, Aalborg University, Denmark. He was Director at the Maritime Training Center, National Kaohsiung University of Science and Technology (NKUST) from February 2018 to July 2020. Since August 2020, he has been at the Department of Electrical Engineering in NKUST and Director at Center for Electrical Power and Energy, where he is now Distinguished Professor.

Chapter 1
Introduction to Maritime Infrastructure and Digital Transformation

Septaviola Dini Utami, Prativi Khilyatul Auliya, and Cries Avian

Abstract Maritime infrastructure has a vital role in global trade, connecting nations, enhancing economic growth, and ensuring sustainable development. To realize those roles, it is necessary to maintain and develop the maritime infrastructure by benefiting the digital transformation. This chapter covers four sections. The first section provides the overview of maritime infrastructure. The second section elaborates the significance of maritime infrastructure in global trade. The third section discusses the role of digital transformation. The last section focuses on challenges and opportunities in digital transformation.

Keywords Maritime infrastructure · Global trade · Digital transformation

1 Maritime Infrastructure

Maritime infrastructure is the sum of all facilities and systems necessary for maritime activities, such as ports, shipping lanes, and logistic networks that connect islands [1]. Maritime infrastructure has a vital role in global trade, connecting nations, enhancing economic growth, and ensuring sustainable development. It encompasses a wide range of physical structures and systems that support maritime transportation, from ports and harbors to navigational aids and shipping routes. This infrastructure is essential for the efficient operation of the maritime industry, which handles over 80%

S. D. Utami (✉) · P. K. Auliya
Department of Shipbuilding Engineering, Shipbuilding Institute of Polytechnic Surabaya, Surabaya, Indonesia
e-mail: septavioladini@ppns.ac.id

P. K. Auliya
e-mail: prativiauliya@ppns.ac.id

C. Avian
Department of Electronic and Computer Engineering, National Taiwan University of Science and Technology, Taipei, Taiwan
e-mail: cries.avian@ub.ac.id

M. Elsisi et al. (eds.), *Maritime Infrastructure for Energy Management and Emission Reduction Using Digital Transformation*, Studies in Infrastructure and Control,
https://doi.org/10.1007/978-981-96-4438-4_1

of global trade by volume [2]. Thus, investing in maritime infrastructure is important for achieving economic goals, particularly in further solidifying the standing of the country in the global trade situation.

Even though there are five types of maritime infrastructures proposed by [3], in early debates on critical infrastructure protection, the focus has been on maritime transport and energy, especially ports [4]. Ports are integral to the functioning of global trade and transportation. They serve as crucial hubs where ships dock to load and unload cargo, as well as where passengers embark or disembark. As one of the most vital components of maritime infrastructure, ports are not just physical locations but dynamic ecosystems that connect maritime and terrestrial transport networks.

In general, the function of ports is divided into two main functions: primary and secondary. The primary functions of a port are traffic function and transport function. As a traffic function, the port is a nodal point in the traffic, connecting water and various land modes. Ports are often connected to major road, rail, and sometimes air networks, allowing for the seamless transfer of goods between sea and land transport. These intermodal connections are vital for the continued success of global trade and the movement of goods over long distances. As transport functions, ports are turntables for various cargo flows. On secondary function, ports play an important role in industrial activities. It is often about the cargo flows, ship repair, shipbuilding, or offshore supply. Ports are also crucial for commercial and financial services, including banks [5].

2 The Significance of Maritime Infrastructure in Global Trade

Maritime infrastructure is the backbone of global trade. Port, the critical maritime infrastructure, and its development are vital components in increasing the competitiveness in the global maritime industry. It means that the quality of infrastructure is an important determinant of trade performance [6]. The improvement of port facilities and shipping logistics can enhance trade efficiency, boosting the volume of imports and exports. Efficient and well-developed ports drive greater trade activity, increase the availability of goods, improve foreign investments, and reduce the overall cost of commodities. These outcomes collectively strengthen the national economy by boosting both supply and demand across industries.

The national economy relies on maritime trade. Trade being carried out at ports and employing seaway transportation is one of the fundamental elements that have a direct effect on the economy of a country. It is also one of the factors affecting a country's economic development. Therefore, investing in maritime infrastructure not only facilitates international trade but also contributes to environmental sustainability and social inclusivity. In other words, great maritime infrastructure development can drive sustainable economic growth, create jobs, improve people's welfare, [7], and can serve as a significant tool in combating poverty [8].

Further, the development of good maritime infrastructure influences economic growth and increases the capital income of the surrounding population [9]. In increasing the capital income of the surrounding population, ports create job opportunities due to the high demand for labor [10] both in the construction phase and in the operational phase. As an example, in Panama, the maritime industry supports employment in port operations, logistics, ship registration, and other related services [8], while in Indonesia, ports have driven job creation in fisheries and tourism, directly contributing to poverty reduction. Additionally, the presence of seaports can transform undeveloped areas into bustling trade centers and industrial zones, raising the standard of living for nearby residents. This development often leads to the emergence of port cities and the overall growth of coastal communities, illustrating the wide-reaching effects of a well-integrated port industry on tropical coasts [11].

Therefore, as global trade expands and the demand for faster, more reliable shipping grows, the limitations of traditional port operations have become increasingly apparent. As mentioned before currently, it still relies upon manual processes, fragmented communication, and paper-based systems that struggle to keep pace with the scale and complexity of modern logistics. This is where digitalization and automation come into play to offer a much-needed solution to these inefficiencies. By integrating them with advanced technologies, connected devices, and data-driven operations, ports can eliminate bottlenecks and delays of manual operations and shift toward transitioning to faster, more accurate, and adaptable systems to pave the way for the future of global shipping.

3 The Role of Digital Transformation

3.1 Definition and Key Technologies

The "digital age" has dramatically changed competition in many industries, including maritime infrastructure. Digital transformation (DT) is reshaping the maritime sector by integrating advanced technologies and changing traditional business practices. In the maritime context, DT involves key changes that promote digitalization within organizations and their interactions, guided by a strategic digital approach. This shift creates significant opportunities to improve both environmental sustainability and the economic performance of those involved in maritime operations. At the core of this transformation are four pivotal technologies that dominantly transform traditional technology into a digital era. Those include the Internet of Things (IoT), artificial intelligence (AI), blockchain, and big data analytics. These technologies serve as catalysts for improved efficiency and collaboration in maritime infrastructure, one of which is the maritime supply chain. Before exploring how these technologies are revolutionizing and modernizing maritime infrastructure, it is crucial to define one by one to clearly understand these foundational concepts and then highlight their significance and roles in steering the maritime sector toward a more efficient, sustainable, and resilient future.

3.1.1 Internet of Things (IoT)

The Internet of Things (IoT) is a global infrastructure for the information society that enables applications and services by interconnecting physical devices referred to as "things" through the Internet. These devices are embedded with sensors, software, and other technologies, which allow them to communicate, exchange data, and interact with each other over the Internet connection. As it is known for connecting "things" through Internet connectivity, the Internet of Things (IoT) relies on several essential components to enable seamless integration and communication between devices. These components work together to collect data, process information, and then take actions based on insights which then makes IoT a powerful tool for automation, optimization, and decision-making across various sectors of application. These components include the following.

(a) **Sensors Devices and Actuators**:

At the heart of the Internet of Things (IoT) are devices, especially sensors, which capture a wide range of data for IoT systems. These sensors monitor key environmental factors like temperature, humidity, water quality, and maritime-specific conditions such as wave height and wind speed. GPS sensors are also crucial, providing real-time positioning data for tracking vessels and marine assets, essential for navigation and fleet management. The IoT ecosystem can involve hundreds or thousands of these sensors working together for comprehensive monitoring and data collection. Acting as the "eyes and ears" of IoT systems, sensors gather information that enables real-time analysis and informed decision-making, enhancing the intelligence and responsiveness of maritime operations. Actuators, the "hands and feet" of IoT, execute actions based on sensor data, adjusting operations in real time. For example, they can control engine systems to prevent collisions or activate safety measures like alarms or emergency shutdowns in response to hazardous conditions detected by sensors [12].

(b) **Connectivity**:

Another fundamental aspect of the Internet of Things (IoT) that enables seamless communication and data sharing is connectivity. IoT devices can be integrated with various communication protocols and technologies, with the choice depending on the specific tasks and situations they are designed to address. For instance, Wi-Fi is often used for high-speed data transfer in localized areas, while Bluetooth is favored for low-power applications requiring short-range communication. Zigbee is typically utilized in short–medium range communication due to its low energy consumption and ability to support multiple devices in a mesh network. For broader coverage, cellular networks like 4G, 5G, and 6G offer higher data transfer rates, making them suitable for real-time applications such as remote monitoring and fleet management.

Additionally, Long Range (LoRa) technology is ideal for long-distance communication with minimal power usage, particularly in remote monitoring scenarios. In maritime operations, satellite communication also plays a crucial role in maintaining connectivity in areas where traditional networks may be limited.

(c) **Data Processing**:

Another key element of the Internet of Things (IoT) is the infrastructure for data storage, processing, and analysis. IoT systems generate large amounts of data that need to be efficiently stored and processed. Cloud computing helps manage this by transmitting data from IoT devices to cloud servers, where scalable resources handle the information. This allows for complex tasks, like running machine learning models, to support data-driven decision-making. Edge computing complements cloud computing by processing data closer to the source, reducing latency and bandwidth usage through local connectivity. While edge computing enables real-time responses, its capacity is limited by the smaller devices used. In practice, combining cloud and edge computing creates a balanced IoT system. For example, in maritime infrastructure, integrating cloud computing into maritime networks (MNs), which cover vast areas and involve many vessels and sensors, improves global oceanic monitoring. This combined approach enhances low-latency applications and improves decision-making in managing multiple maritime operations.

(d) **User Interfaces**:

The other crucial component of IoT systems, typically manifested as mobile apps or web dashboards, allow users to monitor and control devices seamlessly and interact with IoT systems. In the context of maritime infrastructure, user interfaces facilitate real-time monitoring of vessel conditions, navigation systems, and environmental parameters. This information will give operators access to critical information and allow them to control maritime assets remotely, especially for enhancing situational awareness and decision-making.

3.1.2 Artificial Intelligence (AI)

Artificial Intelligence (AI) refers to the ability of digital computers or computer-controlled robots to perform tasks commonly associated with human intelligence. These tasks include reasoning, interpreting meaning, generalizing from data, and learning from past experiences. AI focuses on creating systems that can replicate human intellectual processes, and then allowing machines to handle complex decision-making based on specific information. This enables AI systems to adapt, learn, and improve their performance over time, simulating cognitive functions traditionally considered exclusive to humans. Over the years, AI has evolved into several branches, each with unique contributions and applications across various industries, including maritime infrastructure. Among the most prominent branches are Machine Learning (ML), Deep Learning, Reinforcement Learning (RL), and Metaheuristics, all of which have driven innovation in AI itself.

(a) **Machine Learning (ML)**

Machine Learning (ML) is a key branch of AI that focuses on creating algorithms that allow computers to automatically learn from and make decisions based on data, without needing explicit programming for each specific task. Instead of following predetermined rules, ML models recognize patterns within data, make predictions, and improve their performance through experience. The idea is that as these models are exposed to more data over time, they will become better at solving problems and making more accurate decisions. At the heart of Machine Learning is data; ML models require input data to learn and make decisions. The data can come from various sources, sensor readings, images, texts, financial records, etc., and is processed by the model to detect patterns, relationships, or trends. The type of task that the model performs depends largely on the nature of the data and the problem that needs solving.

In ML, the tasks a model undertakes can generally be categorized into two main types. Those are supervised learning and unsupervised learning. The differences between these two are based on whether the input data comes with labeled outcomes or not. In supervised learning, the model is built based on input data and its corresponding output. This approach is particularly useful in maritime infrastructure, where it can be applied to classification tasks. In this context, supervised learning is employed to learn the relationships between the input data and the predefined classes determined by the user. For example, consider the diagnosis of wear faults in marine diesel engines [13]. In this case, a machine learning-based model utilizes data generated during engine operation as input. The output of this model is the accurate identification of various fault modes, allowing for swift intervention to prevent severe accidents. By effectively classifying these faults, operators can take proactive measures to maintain engine performance and ensure safety. Then, in supervised learning, the regression paradigm is utilized when the goal is to predict a specific continuous value rather than assign a discrete label, as seen in classification tasks. For instance, the task is to predict fuel oil consumption (FOC) [14]. The input data comprises sensor readings from a Very Large Crude Oil Carrier (VLCC), including speed through water, relative wind direction, relative wind speed, mean draft, trim, days since last drydock, and the vessel's laden or ballast state. The output is the predicted FOC for the vessel.

(b) **Deep Learning (DL)**

Deep Learning (DL) is a specialized area of Machine Learning (ML) that utilizes artificial neural networks (ANNs) with multiple layers to model complex patterns in large datasets. While both DL and ML share similar tasks such as classification and regression, DL excels at handling unstructured data like images, audio, and text due to its ability to learn hierarchical feature representations automatically. Key architectures in DL include ANNs, Long Short-Term Memory (LSTM) networks, Bidirectional LSTM (biLSTM), Transformers, and other architectures that use backpropagation as their learning mechanism. Backpropagation is a crucial algorithm in training neural networks, involving a two-step process: the forward pass, where input data is fed into the network to produce an output and compute a loss based on the difference between the predicted and actual values, and the backward pass, where the algorithm calculates gradients of the loss concerning each weight in the network using the chain rule. These gradients inform weight adjustments to minimize the loss which will make the network to learn from the data effectively. The robustness of deep learning compared to traditional ML methods stems from its ability to automatically extract relevant features, scale effectively with large datasets, and model complex patterns without extensive feature engineering.

Since ML and DL have similar capabilities, there is no gold standard for choosing between them. However, several trade-offs must be considered to optimize the functionality of these approaches. Firstly, if the dataset is small or features are well understood and can be manually engineered, traditional ML methods like decision trees or support vector machines may be more efficient and interpretable. Conversely, if the dataset is large, complex, and unstructured, such as images or audio, deep learning models can provide superior performance despite requiring more computational resources and longer training times. Additionally, deep learning often demands more data to prevent overfitting, making it crucial to assess data availability before deciding. Another important consideration is the computational resources available; deep learning is notorious for its high computation costs but can sometimes handle complex data more successfully than traditional ML methods.

(c) **Reinforcement Learning (RL)**

Reinforcement Learning (RL) is a branch of Artificial Intelligence (AI) where an agent learns to make decisions by interacting with an environment. Unlike traditional Machine Learning (ML), where models learn from labeled data (supervised learning) or find patterns in unlabeled data (unsupervised learning), RL is based on a trial-and-error approach. The agent takes actions in an environment, observes the outcomes, and receives feedback in the form of rewards or penalties. The goal of the agent is to learn a policy that maximizes cumulative rewards over time. This learning process is driven by the concept of exploration (trying new actions) and exploitation (using known actions to get rewards). Common algorithms in RL include Q-Learning, Deep Q-Networks (DQN), and policy gradient methods.

The key difference between RL and other ML methods, including Deep Learning (DL), lies in the learning paradigm. In RL, learning is dynamic and sequential, as the agent learns from its actions and the consequences in an interactive environment, which involves time-based decision-making. In contrast, ML, especially for supervised learning, relies on static datasets where the relationship between input and output is learned in a single step. Deep Learning (DL), a subfield of ML, focuses on neural networks with many layers to learn complex patterns, but it usually operates on static data. However, in the recent development then RL can integrate DL for decision-making in complex environments, forming Deep Reinforcement Learning (DRL), where deep neural networks help the agent handle high-dimensional inputs (e.g., vision-based tasks). In short, it can be said that RL has a unique focus on interaction, learning from feedback, and a long-term strategy that differentiates it from ML and DL methods.

In more detail, Reinforcement Learning (RL) has several key points that make it distinguishable from other AI approaches. First, RL involves continuous interaction between an agent and an environment, where the agent acts, and the environment responds with new states and feedback. Second, the agent learns through trial and error and tries different actions to observe their outcomes. Third, feedback comes in the form of rewards or penalties, and the agent's goal is to maximize cumulative rewards over time. Unlike many other learning methods, RL involves sequential decision-making, where each action affects future states and makes it ideal for tasks that require long-term strategies. Another critical aspect of RL is policy learning, where the agent learns a strategy to decide the best actions based on the current state. Last, the agent must balance exploration and exploitation, trying new actions to discover better rewards or choosing known actions which have previously led to positive outcomes.

An example of work that used this method is the work [15]. In this paper on autonomous ship navigation, the key components of Reinforcement Learning (RL) are effectively applied to maritime infrastructure. The ship acts as the agent, interacting with its environment, which includes other ships and navigational factors like location, speed, and heading. The agent continuously adjusts its rudder angles to avoid collisions, embodying the RL principle of trial and error learning. Through this process, the agent evaluates different avoidance maneuvers, gradually learning optimal actions based on the outcomes. The system uses a reward-based feedback mechanism, where positive rewards are given when the ship avoids collisions and follows the designated path, while penalties are assigned for unsafe maneuvers or violations of maritime rules like the COLREGs. This forms the basis for the agent's sequential decision-making, where each rudder adjustment influences future states, allowing the system to plan long-term navigation strategies. Additionally, the method uses policy learning to develop an optimal strategy that guides the ship's actions based on its current state (location, speed, heading, and collision risk). This policy ensures the ship can autonomously generate avoidance paths that comply with international regulations.

(d) **Metaheuristics**

Metaheuristics are high-level optimization strategies designed to solve complex problems that are often difficult to handle using traditional methods. Mostly, the problems that are solved by using metaheuristics typically involve finding optimal or near-optimal solutions within large search spaces. Even though metaheuristics do not guarantee the best solution, the result of this method aims to find a good enough solution in a reasonable amount of time to adjust to real-world applications where exact solutions are computationally impractical.

Natural processes or phenomena inspire metaheuristics. For instance, some methods mimic the process of evolution (Genetic Algorithms), the behavior of animals searching for food (Ant Colony Optimization), or the movement of particles (Particle Swarm Optimization). These algorithms generally balance two key strategies: exploration and exploitation. Exploration refers to searching through new regions of the solution space to avoid getting stuck in local optima, while exploitation focuses on refining existing solutions to improve their quality. The key components of a metaheuristic algorithm often include an initialization phase, where initial solutions are generated, and an iterative process, where solutions are evaluated, modified, and improved over time. These modifications can be influenced by factors like randomness, neighborhood searches, or memory usage. Additionally, metaheuristics can employ mechanisms to escape local optima, such as simulated annealing's temperature parameter or Genetic Algorithms' crossover and mutation processes.

Metaheuristics are flexible and can be applied to a wide range of optimization problems across various fields, including marine infrastructure. For instance, in this example from the liner shipping business [16], the "Port Scheduling Problem" involving feeder vessels at multi-terminal ports is tackled using a metaheuristic approach. The problem is first formulated with Constraint Programming, providing an initial structure to generate feasible solutions, representing the initialization phase. The exploration of the solution space is carried out through an Adaptive Large Neighborhood Search heuristic, which broadens the search across various potential solutions. As new best-known solutions are found, exploitation occurs by applying Constraint Programming as an intensification technique to refine and improve these solutions. The iterative nature of the local search heuristic further enhances solution quality by continuously adjusting and optimizing the current solutions. To avoid getting stuck in local optima, the hybrid approach of combining local search with Constraint Programming serves as an escape mechanism, allowing the algorithm to escape suboptimal regions and find better global solutions. This combination results in significant improvements in solving large-scale instances of the Port Scheduling Problem.

3.1.3 Blockchain

Blockchain is a decentralized, distributed ledger technology that securely records transactions across a network of computers in a transparent and immutable way. It consists of blocks wherein each block contains a list of transactions, a timestamp, and a unique cryptographic hash. These blocks are linked together to form a chain for ensuring data integrity. Decentralization is a key feature, as the blockchain operates on a peer-to-peer network without a central authority, promoting transparency and trust. A consensus mechanism, such as Proof of Work (PoW) or Proof of Stake (PoS), is used to validate transactions and maintain agreement across the network. Cryptographic hashing secures each block by generating a unique identifier, making any tampering easily detectable. Some blockchains also support smart contracts, self-executing contracts that automate actions based on predefined conditions. One of blockchain's core strengths is its immutability once data is recorded which cannot be altered for ensuring a high level of security and trust in the system.

In the context of application, blockchain, a decentralized and distributed ledger technology, has significant potential in maritime infrastructure, as evidenced by various pilot projects at ports like Rotterdam, Veracruz, and Marseille [17]. The technology securely records transactions across a network of computers, ensuring data integrity through cryptographic hashing and decentralized operation. At the Port of Rotterdam, blockchain is being used in initiatives such as IBM and Maersk's TradeLens, as well as a pilot project to handle containers more efficiently by eliminating the need for a pin code which then enhances the safety and transparency of operations. These blockchain applications work by fostering decentralized collaboration between stakeholders, breaking down central control in port logistics, and offering low participation costs, making it especially beneficial for small and medium enterprises (SMEs).

3.1.4 Big Data Analytics

Big data analysis is the process of examining vast and complex datasets to extract meaningful insights, often too large or varied for traditional data processing systems to handle effectively. This involves leveraging advanced computational techniques, algorithms, and ML models to sift through large volumes of data, identify patterns, correlations, trends, and anomalies, and provide actionable insights that can inform decision-making and innovation. By integrating data from diverse sources and applying sophisticated analytical methods, big data analysis enables organizations to make data-driven decisions in real time to respond to dynamic market conditions, operational challenges, or environmental factors.

In the current status, big data is used to combine with IoT, AI, and other digitalization platforms and technology to become essential for data-driven decision-making across most industries. However, the maritime industry, one of the oldest and most traditional sectors, has historically relied more on intuition than data-driven approaches due to the complexity of its vast networks and planning processes. Recently, big data and AI have attracted significant attention, with numerous publications highlighting their potential. Still, some experts view "big data" as overly hyped. Big data generally refers to the handling of large volumes of information. Still, with the exponential growth in data generation, there has been a continuous push to develop novel techniques for analyzing this data. Many of these methods have become integral to AI for enabling advanced analysis and smarter decision-making as already mentioned in the previous section before.

One example of application to big data in marine infrastructure is vessel speed decision-making optimization [18] which benefits especially in addressing both economic and environmental challenges. Utilizing comprehensive weather and environmental data from sources like the Copernicus dataset, shipping companies can develop more accurate fuel consumption models that account for real-time conditions such as wind, sea currents, and temperature factors often overlooked in traditional theoretical models. By applying data mining techniques to analyze historical weather data and its impact on specific voyage routes, companies can better optimize speed to reduce fuel consumption while maintaining Service Level Agreements (SLA) with clients. Particle Swarm Optimization (PSO) is employed to find Pareto optimal solutions, balancing the conflicting goals of minimizing fuel use and maximizing schedule reliability. This approach enables shipping companies to consider a wide range of factors in real time, leading to more precise, data-driven decisions. Furthermore, big data enhances managerial decision-making by enabling simulations of different speed strategies, allowing companies to predict the effects on both operational costs and Greenhouse Gas (GHG) emissions.

3.2 The Application of Digital Transformation to Marine Infrastructure

3.2.1 Automation in Ports and Shipping

The traditional methods of managing ports and shipping operations relied heavily on manual labor and paper-based processes making them too time-consuming, prone to error, and less efficient. In conventional port operations, tasks like cargo loading, unloading, and vessel scheduling were largely dependent on human coordination and oversight. Crane operators, stevedores, and dock workers manually handled the movement of containers, often leading to bottlenecks and delays due to human limitations and unpredictable factors like weather or labor shortages. Similarly, documentation in traditional shipping was predominantly paper-based which required the physical exchange of bills of lading, shipping manifests, and customs forms. Those all not only slowed down the process of clearing goods through customs but also increased the likelihood of errors, misplacement of documents, and fraudulent activities. Additionally, communication between various stakeholders such as shipping companies, port authorities, customs agents, and freight forwarders was fragmented, often relying on phone calls, faxes, or emails, which then added more delay and led to inefficiencies in communication due to busy calls and other human factors.

As global trade expands and the demand for faster, more reliable shipping grows, the limitations of traditional port operations have become increasingly apparent. As mentioned before currently, it still relies upon manual processes, fragmented communication, and paper-based systems that struggle to keep pace with the scale and complexity of modern logistics. This is where digitalization and automation come into play to offer a much-needed solution to these inefficiencies. By integrating them with advanced technologies, connected devices, and data-driven operations, ports can eliminate bottlenecks and delays of manual operations and shift toward transitioning to faster, more accurate, and adaptable systems to pave the way for the future of global shipping.

One example of the transformation for automation in ports can be found in automation in container handling [19]. Automation in container handling has made significant strides in improving operational efficiency, sustainability, and cost-effectiveness at ports. Numerous studies have examined the potential and challenges associated with implementing automation in terminal operations. Automated terminals are typically equipped with systems such as automated quay cranes (QCs), automated rubber-tired gantry cranes (ARMGs), and automated guided vehicles (AGVs) for container transportation, all supported by integrated IT systems for efficient operation. A key advantage of automation is its ability to reduce human error and improve coordination, leading to higher consistency and reliability in cargo handling, particularly in large terminals with high throughput. However, achieving full automation benefits requires addressing operational challenges, such as ensuring sufficient cargo volume to support automated processes and dealing with the loss of flexibility when handling non-standard containers.

Another application is blockchain which is efficient for securing and handling documentation. This blockchain technology will bring transparency and security to maritime logistics by enabling secure, tamper-proof digital records of transactions and cargo movements. This is especially beneficial for international shipping, where multiple stakeholders, such as shipping lines, customs authorities, and freight forwarders, need to access and verify shipping documents.

Another key application of IoT in port operations extends beyond logistics to include real-time tracking and monitoring of port machinery itself. By leveraging IoT connectivity, ports can create a smart, interconnected environment where everything, from cranes to cargo containers, communicates seamlessly in real time. Sensors embedded in port equipment and containers transmit live data on machinery performance, cargo location, temperature, and environmental conditions, making operators monitor critical systems continuously. This real-time visibility will allow operators for quicker identification and response to potential issues which will improve operational safety, reduce energy consumption, and reliability of port activities.

3.2.2 Real-Time Data Monitoring and Decision-Making

Still related to the previous discussion, real-time data monitoring and decision-making are foundational elements in the transformation of port operations, especially in the context of automation support systems. By integrating IoT sensors across port assets, such as cranes, automated guided vehicles (AGVs), and containers, real-time data provides continuous insights into the status, condition, and movement of goods. This stream of data is essential for port operators to monitor and optimize operations in real time. IoT devices collect data from various factors, including cargo location, machinery status, temperature, and even environmental conditions such as weather, which are critical for operational decision-making. This data is then processed through advanced AI and ML to enable predictive maintenance, performance monitoring, and optimization of logistics processes, including container handling and vehicle routing. In this context, AI-powered decision-making systems analyze data patterns to predict failures before they occur. Within this concept, the ports can quickly respond to issues such as equipment malfunctions or hazardous conditions by identifying anomalies and operational bottlenecks early to reduce the risk of accidents. Furthermore, these AI technologies can also be used to optimize resource allocation by reducing idle times of machinery and improving the coordination of operations, which helps mitigate energy consumption and emissions, aligning with the growing emphasis on sustainability in port management. For example, AI can optimize the scheduling of cranes and AGVs to reduce energy usage during non-peak hours, while minimizing delays and idle times.

3.2.3 Predictive Maintenance

Predictive maintenance in port automation goes beyond traditional maintenance by using advanced data analytics to ensure continuous equipment operation and efficiency. This topic is also related to the effect of real-time data monitoring but is more focused on its application to decision-making. With the integration of IoT and AI technologies, this approach transforms how ports handle equipment upkeep. Real-time data collected from sensors on cranes, AGVs, and other critical port machinery processed by ML algorithms that identify patterns and anomalies for predicting any potential equipment malfunctions before they happen. This system will direct the scheduling of maintenance only when necessary and also help ports to avoid any unexpected breakdowns, reducing repair costs, and optimizing the availability of equipment. This will enhance the situation when relying on fixed maintenance schedules, which can result in unnecessary downtime for equipment that may not need immediate attention. Therefore, predictive maintenance uses data-driven insights to guide maintenance decisions based on IoT and AI: For instance, in the case of predictive models that can monitor ship engines, hydraulic systems, and other vital components, detecting signs of wear or stress long before these issues cause operational failures. This helps prevent costly mid-operation breakdowns. In marine infrastructure, predictive maintenance also can be set to ensure the stability of docks, piers, and mooring systems by providing real-time monitoring which helps with timely intervention when structural issues arise.

3.2.4 Enhanced Navigation Systems

Enhanced navigation systems are essential for optimizing port operations, especially as automation and digitalization become more widespread. These systems typically integrate advanced technologies such as GPS, radar, ML, and real-time data analysis to provide precise and efficient guidance for both ships and port machinery. By using these technologies, ports can significantly enhance their ability to manage the complex and dynamic environment of maritime logistics.

Incorporating IoT technology and diverse sensor data, driven by AI algorithms, can further improve navigation systems, especially for vessel operations. For instance, AI-driven solutions can help reduce the risk of collisions by providing real-time adjustments to vessel routes. This can also enhance docking precision and optimize the scheduling of ships entering and leaving the port. A practical application of this technology is AI-based predictive route optimization, which utilizes weather and traffic data to adjust shipping routes. This not only helps ships avoid delays caused by unfavorable conditions but also improves fuel efficiency and overall throughput.

Moreover, real-time monitoring systems enable continuous tracking of both ships' positions and port machinery. This level of tracking enhances operational efficiency, particularly during peak times, by reducing waiting times for vessels and improving the overall coordination of port activities. In automated port environments, these enhanced navigation systems are also critical for guiding automated cranes, Automated Guided Vehicles (AGVs), and other machinery. Precise coordination of container movements ensures smoother operations and minimizes human error to make an improvement to productivity and reliability in port logistics.

3.2.5 Reducing Emissions in the Maritime Sector

The maritime sector, a crucial component of global trade and transportation, is increasingly focusing on energy management to enhance efficiency and reduce environmental impact [20]. This effort is driven by the dual imperatives of economic sustainability and compliance with international environmental regulations. This section explores current energy consumption patterns in the maritime industry, the advent of digital solutions for energy efficiency, and case studies of successful implementation of these technologies.

A. **Current Energy Consumption Patterns**

The maritime sector is a significant consumer of energy, primarily in the form of marine fuels [21]. The industry's energy consumption is influenced by various factors, including vessel type, size, age, and operational patterns. Understanding these consumption patterns is essential for identifying areas where efficiency improvements can be made. The current reliance on fossil fuels presents both a challenge and an opportunity for innovation in energy management.

B. **Digital Solutions for Energy Efficiency**

Digital technologies are revolutionizing energy management in the maritime sector by providing tools for monitoring, analyzing, and optimizing energy use as in [22].

- **Smart Routing and Fuel Management**: Smart routing involves the use of advanced algorithms and real-time data to determine the most efficient paths for vessels, minimizing fuel consumption and reducing voyage times. Fuel management systems, on the other hand, provide comprehensive insights into fuel usage, enabling operators to implement fuel-saving measures and monitor their effectiveness.
- **Energy-Efficient Vessels**: New vessel designs and retrofitting existing ships with advanced technologies are critical aspects of improving energy efficiency. Innovations such as hull modifications, energy-saving devices, and the integration of alternative fuels contribute to the development of energy-efficient vessels. These

measures can significantly reduce the carbon footprint of maritime operations while also delivering cost savings.

C. **Case Studies of Successful Implementation**

Examining real-world examples highlights the potential and practical benefits of implementing energy-efficient solutions in the maritime sector.

- **Case Study 1**: The MariEMS Project, an EU-funded initiative, focused on developing a comprehensive training strategy for maritime energy management. This project included online learning platforms designed to equip cadets and existing seafarers with the necessary skills to implement regulatory requirements like the Ship Energy Efficiency Management Plan (SEEMP) effectively. As a result, it produced a more skilled workforce capable of managing ship efficiency performance over time, which led to a reduction in emissions and operational costs. The successful implementation of this program underscored the importance of continuous education and training in achieving sustainable maritime operations [23].
- **Case Study 2**: The system utilized satellite data and weather forecasting to optimize routes, emphasizing the importance of advanced technological interventions in achieving significant fuel economy. This is aligned with [24], which explores the relationship between energy emission parameters, fuel consumption, and the overall environmental impact. The study also highlights how the adoption of technological advancements and strategic routing can yield substantial reductions in energy use and emissions. The application of such smart routing systems showcases the effectiveness of combining real-time data and predictive analytics to enhance operational efficiency and environmental sustainability in the transport sector.
- **Case Study 3**: [25] has made significant contributions to maritime energy management through its specialized programs, such as the Master of Science in Maritime Affairs program with a focus on Maritime Energy Management (MEM). Graduates from these programs have returned to their respective countries equipped with the skills necessary to implement sustainable practices in their maritime operations. This effort has contributed to global decarbonization initiatives, demonstrating the critical role of higher education institutions in fostering a sustainable future for the maritime industry.

Effective energy management in the maritime sector is critical for enhancing operational efficiency, reducing costs, and minimizing environmental impact. The exploration of current energy consumption patterns provides a foundation for identifying improvement areas. Digital solutions, such as smart routing and energy-efficient vessel designs, offer practical tools for achieving these efficiencies. The successful implementation of these technologies, as demonstrated by various case studies, underscores their potential to transform the industry. Moving forward, the maritime sector must continue to innovate and invest in energy management solutions to ensure a sustainable future.

3.3 *Challenges and Opportunities in Digital Transformation*

3.3.1 Barriers to Digital Transformation

The digital transformation of ports, while offering significant potential for efficiency, sustainability, and cost savings, also faces several barriers that can hinder its successful implementation. Those two primary challenges are investment and cost issues, along with cybersecurity concerns.

A. **Investment and Cost Issues**

The transition to digital and automated systems in ports requires substantial upfront investment in technologies such as IoT sensors, AI systems, and other infrastructure. The initial cost of acquiring and installing these technologies can be prohibitively high, especially for smaller or less financially secure port operators. Additionally, ports need to invest in robust IT infrastructure to support the integration of these technologies, which may involve upgrading legacy systems and ensuring compatibility across various platforms. This creates a financial strain that may deter many stakeholders, especially in developing economies from pursuing such digital initiatives. However, despite these financial hurdles, the long-term benefits of automation and digitalization, such as reduced operational costs, higher efficiency, and increased throughput, can eventually offset the initial investment. Nevertheless, securing funding, justifying return on investment (ROI), and managing the financial risks associated with such large-scale transformation remain significant challenges for many ports.

B. **Cybersecurity Concerns**

As ports become more digitized, they become more vulnerable to cyber threats. The integration of IoT devices, AI, and other digital tools creates more entry points for potential cyberattacks, which can disrupt operations, steal sensitive data, or compromise the integrity of port systems. This situation will worsen when ports serving as critical infrastructure for global trade then face the risk of sophisticated cyberattacks that could have severe consequences for the supply chain. For instance, a ransomware attack could halt port operations, causing delays in container handling, shipping, and logistics, with cascading effects on global trade. This could lead to significant economic losses, as businesses dependent on timely shipments experience production delays, stock shortages, or missed delivery deadlines. Moreover, the disruption could create vulnerabilities in the supply chain, making it harder to recover swiftly. As the digitalization of port systems increases, so too does the attack surface, allowing cybercriminals to exploit weak security protocols, compromise sensitive data, or even take control of operational technologies. In extreme cases, this might result in accidents or environmental hazards due to disruptions in the handling of hazardous materials. The global interconnectedness of trade means that even a local port incident could have far-reaching consequences, affecting industries and consumers worldwide.

To mitigate these risks, ports must invest in cybersecurity measures, including robust encryption, firewalls, and continuous monitoring systems. Encryption ensures that sensitive data is protected during transmission, while firewalls serve as barriers to unauthorized access. Continuous monitoring enables the detection of suspicious activity in real time for helping to identify potential security breaches before they cause significant damage. Additionally, integrating AI for threat detection and response can also provide an advanced layer of protection. AI systems here are set to analyze large volumes of data quickly, recognize patterns, and detect anomalies that might signal a cyberattack. For example, AI could identify unusual network traffic or unauthorized attempts to access critical systems, enabling quicker responses to threats. Training personnel in cybersecurity best practices is equally important, as human error often plays a role in breaches. Staff should be educated on recognizing phishing attacks, managing secure passwords, and following proper procedures when handling sensitive information. Regular training sessions and simulations can help ensure that employees stay aware of emerging threats. Despite these efforts, the evolving nature of cyber threats means that ports must remain vigilant and proactive in their approach to cybersecurity, which further adds to the cost and complexity of the digital transformation process.

3.3.2 Future Trends and Opportunities

As ports and the broader maritime industry continue to evolve, there are emerging trends that signal both challenges and opportunities. Two of the most significant drivers of change are collaboration across the maritime industry and the rapid development of emerging technologies.

A. **Collaboration Across the Maritime Industry**

The digital transformation of ports cannot occur in isolation. It requires extensive collaboration among various stakeholders within the maritime ecosystem. This includes port authorities, terminal operators, shipping lines, technology providers, and regulatory bodies. Such collaboration is crucial for establishing universal standards, ensuring the interoperability of systems, and optimizing the flow of data across the entire supply chain. For example, integrated digital platforms that allow for real-time tracking of container shipments are only effective when different operators agree to share data and adopt common technological frameworks. One of the key benefits of such collaboration is the potential to address bottlenecks more efficiently. Real-time data exchange between stakeholders ensures a smoother flow of information and goods, helping to minimize delays caused by weather, equipment malfunctions, or labor issues. Moreover, this collaboration will be crucial for meeting the sustainability goals of the maritime industry. For example, coordinated efforts between ports and shipping lines can promote the adoption of green technologies such as electric cranes and automated vehicles, which reduce both carbon emissions and operational costs. Additionally, partnerships between ports and technology companies can drive the development of advanced solutions. For instance, collaborating with companies that specialize in blockchain, cloud computing, IoT, and AI-driven systems, especially those offering solutions that are easily integrated across any platform, could reshape the landscape of port automation. Nowadays, it is required to have a service that can connect devices and systems effortlessly without the need for extensive coding or complicated integration to save valuable time and resources. This integration will ease industries to streamline operations, adapt quickly, connect various technologies, and also ensure that ports can handle the growing complexity of global shipping demands with greater agility in the future and accommodate future needs for expanded services.

B. **Emerging Technologies**

The maritime industry is on the brink of a significant transformation as it adopts emerging technologies that promise to greatly enhance port operations. These advancements are crucial for addressing the growing complexities of global trade, which demands faster, more efficient, simpler, and safer processes. By streamlining operations, these technologies provide critical tools to tackle key challenges such as security, sustainability, and scalability, especially as ports face increasing pressure to handle larger volumes and meet the rapidly evolving demands of global shipping. As such, technologies like 5G/6G, AI, blockchain, and quantum computing are essential for staying competitive and adapting to the future needs of the industry, especially in this century.

1. **5G/6G Networks**: The rollout of 5G technology is already transforming communication in the maritime industry by enabling low-latency, high-bandwidth connections between port systems, ships, and autonomous vehicles. This allows for real-time data exchange, improving coordination between automated guided vehicles (AGVs), cranes, and other port equipment, which reduces cargo handling times and boosts throughput, particularly in high-traffic ports. As 6G technology begins to roll out, it promises to take these benefits even further by providing even faster data transfer speeds and the capacity to handle massive volumes of data. Designed for big data analysis and real-time decision-making, 6G will support the seamless transfer of vast amounts of information, enhancing the integration of advanced technologies such as AI and autonomous vehicles, ultimately driving even more efficient and responsive port operations.
2. **Artificial Intelligence and Machine Learning**: The future of AI in the maritime industry will need to evolve to handle the increasing complexity and volume of data generated by automated ports. As ports rely more on real-time decision-making and interconnected systems, AI will be required to process vast amounts of data quickly and accurately. While advancements like 6G will provide faster data transmission, AI itself must continue to develop into more lightweight and efficient systems capable of handling big data without overwhelming computational resources. These AI systems must not only process enormous datasets from sensors, cameras, and autonomous vehicles but also extract meaningful insights in real time to optimize port operations. For example, imagine a scenario where AI is tasked with managing the logistics of thousands of containers, each equipped

with sensors providing continuous data on their location, condition, and contents. The volume of this data would require AI systems capable of performing rapid, parallel computations and delivering highly accurate predictions about the best ways to load, unload, and store containers, all while maintaining safety and minimizing delays. AI will need to ensure that its algorithms can process this massive flow of data efficiently, without sacrificing the quality of decision-making. Therefore, to achieve this, future AI models need to be optimized for both speed and precision, incorporating advanced techniques like deep learning networks that can handle complex, multi-dimensional datasets. These AI models will need to be both scalable and adaptable to be able to continuously learn from new data inputs and refine their predictions over time. Furthermore, accuracy will be critical, especially in high-risk situations such as detecting potential safety hazards or predicting equipment failures. In these cases, even a small error in data interpretation can lead to costly mistakes or operational disruptions.

3. **Blockchain**: In the future, blockchain technology will play a pivotal role in the digitalization of the maritime industry for addressing challenges related to transparency, security, and efficiency in port operations. With the growing complexity of global trade, blockchain offers a decentralized and secure way to manage and verify transactions, agreements, and documentation in a manner that can be seamlessly integrated with other systems. In the context of port operations, blockchain will enable real-time tracking of goods, vessels, and containers, providing a transparent and immutable record of every movement and transaction. For example, in the scenario of managing thousands of containers with various sensors and data inputs, blockchain can serve as a secure ledger for tracking the entire journey of each container, from its point of origin to its destination. Each container's movement and condition data, verified through IoT sensors, could be stored on a blockchain, providing stakeholders with real-time visibility into the supply chain. This eliminates the need for paper-based documentation and manual verification processes, reducing human error and fraud while speeding up the clearance and customs process. Looking ahead, the future development of blockchain in the maritime industry expects to likely focus on creating more scalable and efficient solutions for managing large volumes of transactions and data. As ports become increasingly automated and interconnected, blockchain can help synchronize all parties involved, including shipping companies, port operators, customs authorities, and cargo owners, by providing a single, trusted source of truth. With the added advantage of smart contracts, blockchain can also automate certain processes, such as payments and regulatory compliance, ensuring that all parties fulfill their obligations in real time without the need for intermediaries. Moreover, blockchain's ability to provide a tamper-proof record will be critical for ensuring the integrity of data and transactions in highly sensitive areas such as cargo handling, security, and insurance. So, in the future, blockchain could help mitigate the risks of data manipulation and ensure that all parties have access to an accurate and transparent view of operations.
4. **Quantum Computing and Edge Computing**: Quantum computing and edge computing are two transformative technologies that can significantly enhance

the efficiency and capabilities of modern port operations. Quantum computing leverages the principles of quantum mechanics to perform calculations that are exponentially faster than classical computers, especially for complex problems involving large datasets and optimization tasks. Unlike classical computers, which process information in bits (0 and 1 s), quantum computers use qubits, which can represent multiple states simultaneously. This allows quantum computers to solve problems that would be practically impossible for classical computers to handle, such as large-scale optimization and complex simulations, which are essential in industries like shipping and logistics. Edge computing, on the other hand, refers to the practice of processing data closer to the source of the data rather than relying on distant cloud servers. This is particularly important for real-time applications where low latency is critical, such as in autonomous vehicles, industrial machines, and IoT sensors used in port operations. By processing data locally, edge computing reduces the time it takes to make decisions supposedly improving responsiveness and reducing the strain on network bandwidth before it is further processed to the main processing such as the cloud or any computational devices as main final decisions. This is especially useful in environments like ports, where thousands of sensors and automated systems generate massive amounts of data that need to be analyzed instantly to ensure smooth operations. In the context of port operations, these technologies can be integrated to address both immediate, real-time decision-making and complex, long-term optimization challenges. For instance, in a busy port, a fleet of autonomous vehicles (AGVs) and cranes equipped with IoT sensors generate continuous streams of data that need to be processed in real time. Edge computing processes this data locally on each device for immediate actions like avoiding obstacles, adjusting speeds, and optimizing routes without waiting for central server instructions to handle local critical decisions and improve time-sensitivity. However, as the system scales, the optimization of these decisions across the entire port becomes far more complex which then forces the system to coordinate across thousands of devices and systems which then also requires a more coordinated strategy, device communication network, and huge computational device for handling this situation. This is where quantum computing becomes valuable. Quantum algorithms can be used to process the vast amount of data generated by these devices and run simulations for complex scenarios, such as optimizing the movement of all AGVs and cranes at once to maximize throughput and minimize delays. Quantum computing can analyze millions of variables simultaneously, such as traffic congestion, container locations, equipment status, and weather conditions, and provide an optimized schedule or set of actions for the entire port. These tasks would be too computationally intensive for edge devices or classical systems alone. By combining both technologies, the port can achieve a dynamic, seamless system. Edge computing handles the immediate, real-time tasks at the device level, ensuring that operations run smoothly without delay. Quantum computing, meanwhile, performs the more computationally expensive tasks at the system-wide level, improving overall efficiency through global optimization.

4 Conclusion

Maritime infrastructure driven by digital transformation, especially through the adoption of IoT and AI, will be a key to advancing faster, safer, and more sustainable global trade. Port digitalization and the use of automation technologies not only improve operational efficiency but also enrich customer experience, reduce logistics costs, and ensure timely delivery of goods. With the use of IoT sensors that enable real-time monitoring and AI applications that can forecast needs and optimize routes, ports can respond better to dynamic market needs. In addition, the application of this technology can also support operational sustainability through more efficient resource management and minimize environmental impacts. For ports that want to remain competitive in the global market, investing in digital transformation is no longer an option, but a necessity. Connected and intelligent maritime infrastructure will play a pivotal role in supporting growing global trade, meeting new challenges, and capitalizing on opportunities.

References

1. Alderson DL, Funk D, Gera R (2020) Analysis of the global maritime transportation system as a layered network. J Transp Secur 13:295–325. https://doi.org/10.1007/s12198-019-00204-z
2. Wang X, Fai K, Diew Y, Li KX (2020) How can the maritime industry meet Sustainable Development Goals? An analysis of sustainability reports from the social entrepreneurship perspective. Transp Res Part D 78:102173. https://doi.org/10.1016/j.trd.2019.11.002
3. Bueger C, Liebetrau T (2023) Critical maritime infrastructure protection: What's the trouble? Mar Policy 155:105772. https://doi.org/10.1016/j.marpol.2023.105772
4. Bursztyński A (2020) Safety of maritime critical infrastructure facilities in the aspect of contemporary threats 14(1) (2020)
5. Ligteringen H (1999) Ports and terminals. TUDelft
6. Nordås HK, Piermartini R (2004) World trade organization economic research and statistics division
7. Sibali A, Jainuddin J (2024) Economics and digital business review maritime infrastructure development and its impact on national economic growth 5(2):911–920
8. Ayhan EE (2023) From ports to prosperity : leveraging maritime sector for poverty reduction Limanlardan Refaha: Yoksulluğun Azaltılması İçin Denizcilik Sektöründen Yararlanmak 3(2):99–109
9. Hargono S, Sutomo S, Alisyahbana J (2013) The 3rd international conference on sustainable future for human security the influence of the port to the economical growth of the Batam Island. Procedia Environ Sci 17:795–804. https://doi.org/10.1016/j.proenv.2013.02.097
10. Batalha E, Chen S, Pateman H, Zhang W (2023) Defining a social role for ports: managers' perspectives on whats and whys. Sustainability 15(3). https://doi.org/10.3390/su15032646
11. Khalid N, Joni ZM (2009) The importance of the maritime sector in socioeconomic development: measuring the marine economies of Southeast Asia in times of economic and The Marine Economy in East Asia's Marine Economy: Our history, our future 16(1)
12. Karatuğ Ç, Arslanoğlu Y (2022) Development of condition-based maintenance strategy for fault diagnosis for ship engine systems. Ocean Eng 256:111515. https://doi.org/10.1016/j.oceaneng.2022.111515

13. Xu X et al (2020) Machine learning-based wear fault diagnosis for marine diesel engine by fusing multiple data-driven models. Knowl-Based Syst 190:105324. https://doi.org/10.1016/j.knosys.2019.105324
14. Papandreou C, Ziakopoulos A (2022) Predicting VLCC fuel consumption with machine learning using operationally available sensor data. Ocean Eng 243:110321. https://doi.org/10.1016/j.oceaneng.2021.110321
15. Chun D-H, Roh M-I, Lee H-W, Ha J, Yu D (2021) Deep reinforcement learning-based collision avoidance for an autonomous ship. Ocean Eng 234:109216. https://doi.org/10.1016/j.oceaneng.2021.109216
16. Sacramento D, Solnon C, Pisinger D (2020) Constraint programming and local search heuristic: a matheuristic approach for routing and scheduling feeder vessels in multi-terminal ports. SN Oper Res Forum 1(4):32. https://doi.org/10.1007/s43069-020-00036-x
17. Samsung, Rotterdam port, ABN AMRO partner for trade blockchain - Ledger Insights - blockchain for enterprise. https://www.ledgerinsights.com/samsung-rotterdam-port-abn-amro-blockchain/. Accessed 02 Nov 2024
18. Lee H, Aydin N, Choi Y, Lekhavat S, Irani Z (2018) A decision support system for vessel speed decision in maritime logistics using weather archive big data. Comput Oper Res 98:330–342. https://doi.org/10.1016/j.cor.2017.06.005
19. Papadakis MN, Kopanaki E (2022) Innovative maritime operations management using blockchain technology & standardization. JICTS. https://doi.org/10.13052/jicts2245-800X.1041
20. Wang Y, Chen P, Wu B, Wan C, Yang Z (2022) A trustable architecture over blockchain to facilitate maritime administration for MASS systems. Reliab Eng Syst Saf 219:108246. https://doi.org/10.1016/j.ress.2021.108246
21. Harahap F, Nurdiawati A, Conti D, Leduc S, Urban F (2023) Renewable marine fuel production for decarbonised maritime shipping: pathways, policy measures and transition dynamics. J Clean Prod 415:137906. https://doi.org/10.1016/j.jclepro.2023.137906
22. Kouzelis K, Frouws K, Van Hassel E (2022) Maritime fuels of the future: what is the impact of alternative fuels on the optimal economic speed of large container vessels. J Shipp Trade 7(1):23. https://doi.org/10.1186/s41072-022-00124-7
23. Ziarati R, Bhuiyan Z, de Melo G, Koivisto H, Maritime Energy Management System (MARIEMS) online delivery platform. In: Presented at the International association of maritime universities conference. Mari EMS, Bulgaria. https://upcommons.upc.edu/bitstream/handle/2117/111730/MariEMS%20IAMU%2017%20Final%202%20AUGUST%20VER.3.pdf?sequence=1&isAllowed=y
24. Pašagić Škrinjar J, Abramović B, Bukvić L, Marušić Ž (2020) Managing fuel consumption and emissions in the renewed fleet of a transport company. Sustainability 12(12):5047. https://doi.org/10.3390/su12125047
25. (NEWS) WMU maritime energy management: making an impact for a sustainable maritime future. https://www.wmu.se/news/wmu-maritime-energy-management-making-an-impact-for-a-sustainable-maritime-future. Accessed 02 Nov 2024

Chapter 2
Challenges in Energy Management and Emission Reduction

Syafiuddin, Niki Veranda Agil Permadi, Izzul Fikry, Indri Santiasih, and Ovi Prina Gastriani

Abstract This article examines energy consumption and greenhouse gas (GHG) emissions in the maritime sector, highlighting strategies for improving energy efficiency and reducing environmental impacts. Ships rely on various energy forms—chemical, mechanical, electrical, and thermal—to operate, with propulsion consuming the largest share. However, burning fossil fuels generates significant CO_2, NO_x, SO_x, and particulate emissions, contributing to global climate change. Energy efficiency measures include optimizing hull design, adopting hybrid propulsion systems, and utilizing alternative fuels such as hydrogen, ammonia, and liquefied natural gas (LNG). Emission control technologies like scrubbers and carbon capture and storage (CCS) are also explored as critical solutions. Global regulations, such as the International Maritime Organization's (IMO) 2023 GHG Strategy, aim to reduce emissions by 80% by 2040, but implementation faces barriers including immature technologies, high initial costs, and insufficient port infrastructure for alternative fuels. Emerging green technologies—air lubrication systems, renewable energy integration, and digitalized operations—are promising but require further research to achieve broader adoption and efficiency. The article underscores the importance of collaboration among regulators, industry stakeholders, and academia to address these challenges. By fostering innovation, investment, and infrastructure development, the

I. Fikry
Department of Shipbuilding Engineering, Shipbuilding Institute of Polytechnic Surabaya, Surabaya, Indonesia
e-mail: izzulfikry@ppns.ac.id

Syafiuddin (✉) · N. V. A. Permadi · I. Santiasih · O. P. Gastriani
Department of Marine Engineering, Shipbuilding Institute of Polytechnic Surabaya, Surabaya, Indonesia
e-mail: syafiuddin@ppns.ac.id

N. V. A. Permadi
e-mail: niki.veranda@ppns.ac.id

I. Santiasih
e-mail: indri.santiasih@ppns.ac.id

O. P. Gastriani
e-mail: ovigastriani@ppns.ac.id

M. Elsisi et al. (eds.), *Maritime Infrastructure for Energy Management and Emission Reduction Using Digital Transformation*, Studies in Infrastructure and Control,
https://doi.org/10.1007/978-981-96-4438-4_2

maritime sector can transition toward more sustainable operations, aligning with global decarbonization goals.

Keywords Maritime energy efficiency · Greenhouse gas emissions · Alternative fuels · Hybrid propulsion systems · Decarbonization strategies

1 Introduction

1.1 Ship Energy Consumption

Ship is a vehicle with a very complex system. In general, a ship is divided into three parts, namely machinery space, accommodation, and cargo hold or tanks [1], as seen in Fig. 1. In the machinery space there are propulsion systems, fuel systems, auxiliary engines, and general services. In the accommodation there are cooling and navigation systems, and in the cargo hold there are loading and unloading systems and tank cleaning [1, 2]. These systems certainly require energy to drive the main engine, auxiliary engine, pumps, compressors, navigation equipment, air conditioners, and the other equipment. In general, propulsion is the largest consumer of energy in the ship [2, 3].

To meet the energy needs on the ship, several equipment are needed, namely the main engine, auxiliary engine and auxiliary boiler [2]. There are four types of energy on a ship, namely chemical energy, mechanical energy, electric energy, and thermal energy, as seen in Fig. 2 [2].

The components that use chemical energy are the main engine, auxiliary engine, and boiler. These three components will then produce mechanical energy, electric energy, and thermal energy. The mechanical energy produced by the main engine and auxiliary engine is transmitted to the ship's propeller and generator (shaft and auxiliary generator), where the generator will produce electrical energy which will later be distributed to the equipment in the engine room, deck (deck machinery and hull equipment), and accommodation space, as seen in Fig. 3.

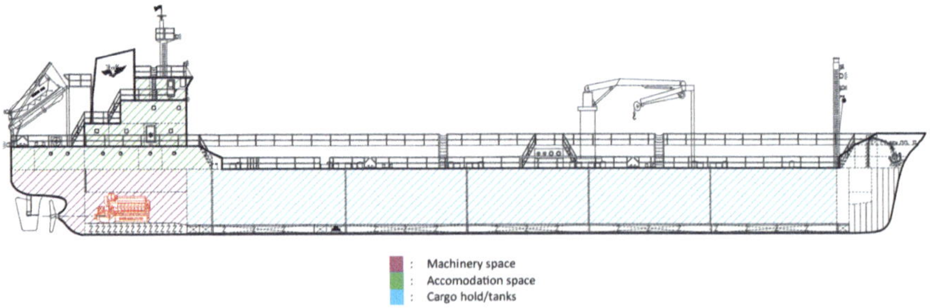

Fig. 1 Main parts of the ship

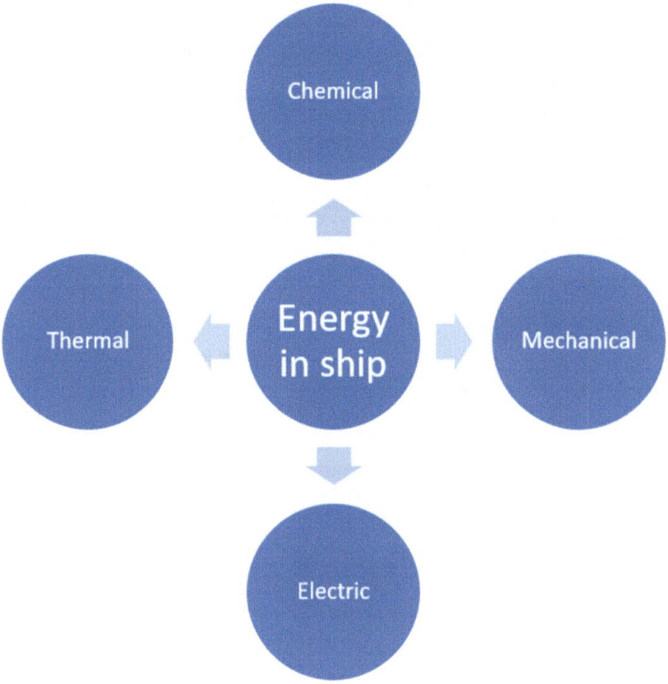

Fig. 2 Types of energy in ship

In addition to transmitting mechanical energy, the main engine and auxiliary engine also produce thermal energy with the boiler. The thermal energy is then used for tank cleaning (on tankers), heating the fuel, and entering the cooling system (high temperature and low temperature cooling). Heating the fuel is usually done to increase the viscosity of the fuel, so that the fuel is easier to flow [1].

The amount of energy required to operate a ship certainly cannot be separated from the negative effects it produces, where one of the CO_2 gas producers is the result of the combustion reaction from the ship's internal combustion engine [4]. The impact of global warming is now increasingly visible, so much needs to be done to minimize the causes of this warming, one of which is CO_2 gas. There are several ways to achieve energy efficiency on ships, namely improved hull design, propulsion optimization, hybrid propulsion systems, and alternative fuel sources.

Improved hull design is carried out by optimizing the shape of the bow and stern of the ship. One of the bow shapes that can be used is the X-Bow, as seen in Fig. 4. The use of X-bow can reduce the ship's resistance by around 4% [5]. In propulsion, the use of energy saving devices can affect the efficiency of the propeller, one of the energy saving devices is the propeller boss cap fins (PBCF). PBCF is said to be able to increase the efficiency of the propeller [6, 7].

Combining the main engine and electrical energy storage system, or what is usually called the hybrid propulsion system method, is also a form of energy

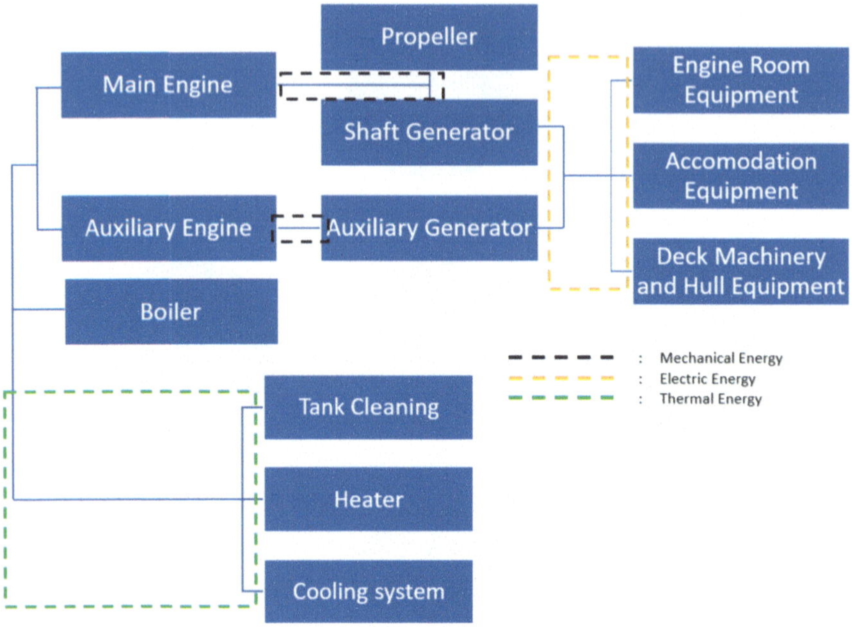

Fig. 3 Energy flow in the ship

Fig. 4 X-bow illustration

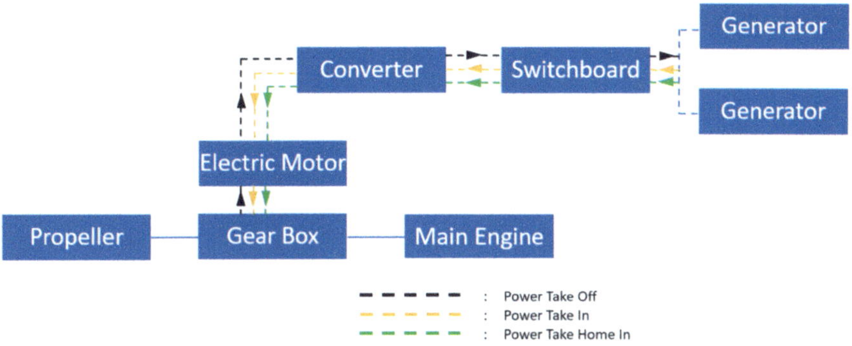

Fig. 5 Workflow PTI, PTO, and PTH

efficiency [4]. Generally there are three types of hybrid propulsion systems [8], namely:

A. **Power Take off (PTO)**

The propeller is driven by the main engine, where there is a generator connected to the gearbox which will then produce electric energy to be used as a source of electrical energy for the ship.

B. **Power Take in (PTI)**

It is a system where the electric motor will supply additional power to rotate the propeller. The energy needs of the electric motor are supplied by a diesel generator.

C. **Power Take Home In**

When the main engine is damaged, the propeller will be rotated by an electric motor whose energy is obtained from a diesel generator.

As shown in Fig. 5, alternative energy resources are also one way to make energy use more effective. The forms of alternative energy that have been used today are solar energy, wind energy, marine energy, hydro-power, biomass, and bio fuel. However, currently the alternative energy used as the main power on ships is wind and solar energy [9].

1.2 Ship Emission

The shipping industry faces an uncertain path forward as climate change challenges become increasingly urgent across all sectors. Applying a quantitative interpretation of the Paris Agreement's goals highlights significant obstacles for shipping, alongside other fossil fuel-dependent industries, to achieve successful decarbonization in the coming decades. The shipping sector faces significant challenges in achieving decarbonization, requiring robust policymaking to drive this transition. This presents an opportunity for academia to provide valuable insights and analyses to support the industry's efforts in addressing the global challenge of mitigating dangerous climate change, while ensuring shipping continues to play a vital role in global economic development [10].

The maritime industry is currently grappling with the urgent need for a rapid transition in energy and technology to reduce emissions of GHGs and air pollutants. This shift is primarily driven by international regulations, such as the International Maritime Organization's global fuel sulphur limit and its Greenhouse Gas Strategy. The long-term objective is to adopt carbon–neutral fuels; however, these fuels are not yet competitively priced. Additionally, the expansion of shipping ac-tivities and the development of new routes, particularly in the Arctic, are anticipated to have significant effects on these fragile environments [11] (Fig. 6).

Air pollutants emitted by ships (primary pollutants) mix with other atmospheric pollutants and undergo complex chemical and physical processes, resulting in the formation of secondary pollutants such as particulate matter. These pollutants interact with radiation, causing light absorption and scattering, which influence the climate. Particulate matter can also interact with clouds, indirectly impacting the climate. When black carbon settles on snow, it accelerates melting, while particles deposited in the surface ocean release nutrients and hazardous substances that affect ocean biogeochemistry, further influencing the climate indirectly [11].

Ship emissions consist of both complete combustion products, such as CO_2 and SO_x, and emissions from the oxidation of nitrogen in the air intaken (NO_x). Incomplete combustion leads to the release of hydrocarbons (HCs, including CH_4), carbon monoxide (CO), formaldehyde, particulate matter (PM) and its components

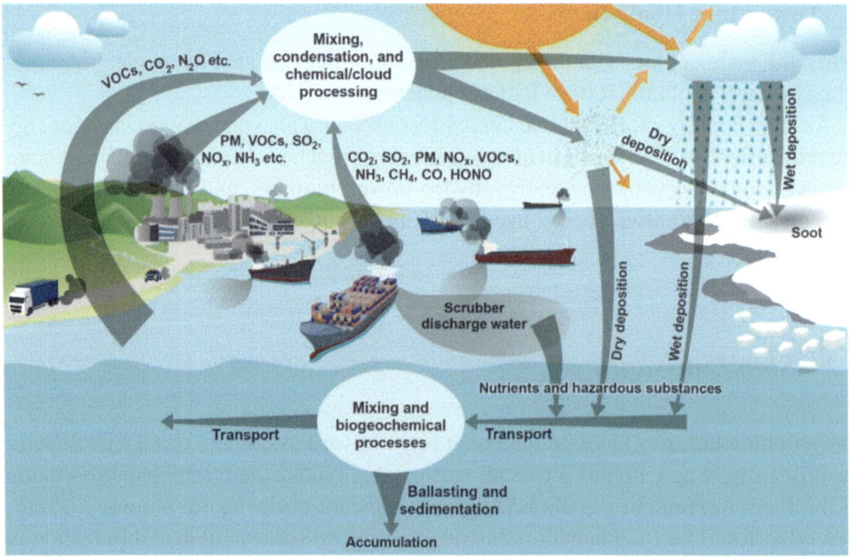

Fig. 6 An overview of the emissions, processes, and impacts of shipping [11]

(black carbon (BC), polyaromatic hydrocarbons (PAHs)), and particle number (PN) emissions. Some pollutants are formed in exhaust after-treatment systems, such as nitrogen dioxide (NO_2), ammonia (NH_2), and N_2O. Emissions of heavy metals may also arise from fuel, lubricating oil, and engine wear.

SO_x and NO_x emissions can harm terrestrial and aquatic ecosystems, leading to acid rain and the eutrophication of waters due to nitrogen-rich nutrients. NO_x in the atmosphere contributes to the formation of ground-level ozone when combined with organic compounds, heat, and sunlight. Engine-out NO eventually oxidizes into NO_2, which has harmful respiratory effects and reacts to form nitric acid (HNO_2) and other compounds. Although the NO_2 share in engine-out emissions is low, its formation increases significantly with the use of oxidative after-treatment devices, such as diesel particulate filters (DPFs) and catalysts, which must be considered when these technologies are implemented on ships [12].

According to Fourth IMO GHG Study [13], ship emission globally increases to 2.89% in 2018 compared to 2012 at 2.76%. The greenhouse gas (GHG) emissions–including carbon dioxide (CO_2), methane (CH_4) and nitrous oxide (N_2O)—expressed in CO_2e- of total shipping (international, domestics and fishing) have increased from 977 million tonnes in 2012 to 1,076 million tonnes in 2018 (9.6% increase). In 2012, 962 million tonnes were CO_2 emission, while in 2018 this amount grew 9.3% to 1,056 million tonnes of CO_2 emissions. Emissions, which were approximately 90% of 2008 levels in 2018, are expected to rise to between 90 and 130% of 2008 levels by 2050 under various plausible long-term economic and energy scenarios. Emissions may exceed or fall below the projections if economic growth rates are higher or lower than assumed, respectively, or if GHG reductions in land-based sectors are less or more substantial than required to limit global temperature rise to well below 2 °C (Fig. 7).

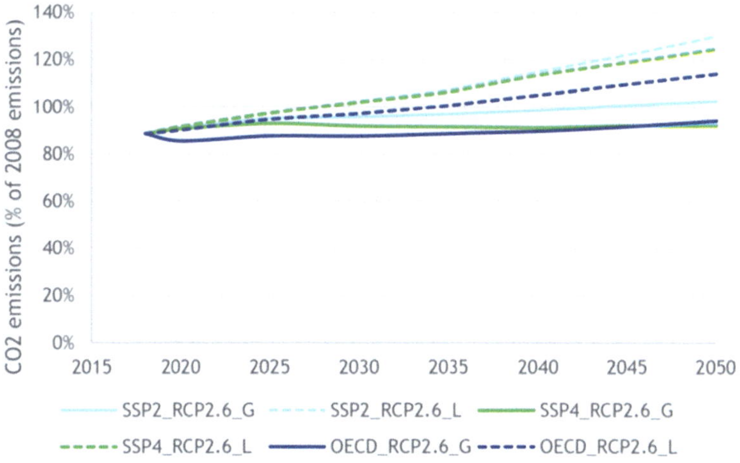

Fig. 7 Projections of maritime ship emissions as a percentage of 2008 emissions [13]

But, with the years going on, the ship emission would rise as transport capacity expands. To contribute its fair share to limiting global warming under the Paris Agreement, the shipping industry must prioritize energy conservation and emission reduction. The rapid growth of the global oceangoing fleet and the associated rise in GHG emissions present a compelling new question mark. According to [14] CO_2 and CH_4 emissions from shipping exhibit a negative allometric relationship with global fleet size, whereas N_2O emissions show a positive allometric relationship. After 2008, many shipping companies reduced navigation speeds or placed ships in idle status to lower costs. This indirectly decreased CO_2 and N_2O emissions from shipping but had minimal effect on CH_4 emissions.

The main source of air pollutant from ship is from the ship fuel consumption (SFC). Therefore, optimal fuel consumption is crucial for cost reduction, efficiency improvement of operational ships, and realising energy efficiency management included with reducing ship emissions. The factors influencing ship fuel consumption (SFC) are primarily internal factors such as main engine, specific fuel oil consumption (SFOC) and type of main engine. Generators, boilers, and steering gears also the main consumers of fuel and emits GHG and other pollutants [15]. Ship fuel consumption (SFC) has a lot of influencing factors that are summarised in Fig. 8 (Table 1).

Back to International Maritime Organization (IMO) that has adopted the "2023 IMO GHG Strategy on Reduction of GHG Emissions from Ships" (2023 IMO GHG Strategy) includes a "vision", "Levels of ambition", and indicative checkpoint" [16]. And for indicatives checkpoints to achieve net-zero GHG emissions target is 20–30%

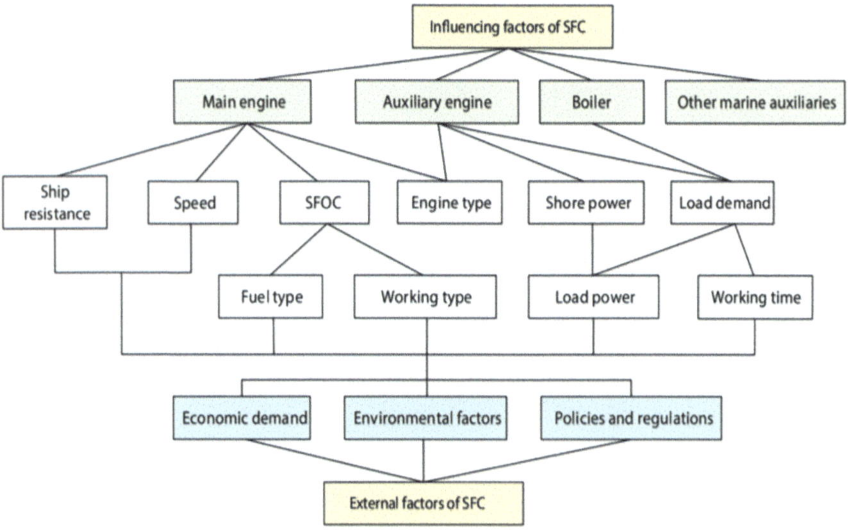

Fig. 8 Influencing factors of SFC [15]

Table 1 2023 IMO GHG strategy [16]

	Initial strategy (2018) (tank-to-wake)	**Revised strategy (2023)** (well-to-wake: lifr cycle)
Vision	**Phase out GHG emission** as soon as possible **in this century**	**Phase out GHG emission** as soon as possible
Levels of ambition	**Total annual GHG emission (compared to 2008)**	
	At least **50% reduction** by 2050	**At least 20% (striving for 30%)** reduction by 2030 (indicative checkpoint)
		At least 70% (striving for 80%) reduction by 2040 (indicative checkpoint)
		Reach **net-zero GHG emissions** by or around, i.e. close to 2050
	Total annual GHG emission (compared to 2008)	
		At least 5% (striving to 10%) by 2030
	Carbon intensity improvement (CO_2 emissions per transport work) (compared to 2008)	
	At least **40% reduction** by 2030	At least **40% reduction** by 2030
	At least **70% reduction** by 2050	

reduction in GHG emissions by 2030 (compared to 2008) and 70–80% reduction in GHG by 2040 (compared to 2008).

The 2023 IMO GHG Strategy is set to drive further progress toward decarbonizing the shipping industry. To align with this strategy, it's crucial to clearly understand and define the steps needed to meet the ambitious emission reduction targets right from the beginning. The 2023 IMO GHG Strategy highlights the need to focus on innovative solutions and regulations to meet its goals. Key areas of attention include life cycle assessments (LCAs) for zero or near-zero GHG emission technologies and fuels, the use of onshore power, and the implementation of carbon intensity indicator (CII) ratings, among other measures [17].

1.3 Regulatory Framework

Today, one major cause of environmental risk is greenhouse gas (GHG) emissions, particularly CO_2, largely from internal combustion engines is used in transportation, including the maritime sector. International shipping, which handles 78% of global trade in metric tonnes, contributes about 2.4% of global GHG emissions. These emissions are projected to increase significantly by 2050 if no measures are taken [18]

It is crucial to emphasize that there are no regulations specifically targeting energy efficiency on ships as their primary focus. Existing regulations on ship efficiency are not independent but are instead tied to concerns about environmental pollution, particularly greenhouse gas emissions. For instance, the main metric for energy efficiency, the Energy Efficiency Design Index (EEDI), is not suitable for certain types of ships or those employing specific power systems, as its method of measuring efficiency is directly linked to CO_2 emissions [19].

Understanding the origin and primary concerns of current regulations is essential to grasp the research and investigations conducted in the field of energy efficiency on ships. Consequently, various regulations will be outlined, detailing their scope and the main enforcement measures. Regulations and policies continue to be the primary forces driving the decarbonization of the shipping industry by imposing direct requirements and providing incentives for ships and shipping companies. Even regulations that do not directly address energy efficiency can restrict the development of more energy-efficient ship systems (Fig. 9).

This subchapter presents the regulations on Greenhouse Gas from the International Maritime Organization (IMO) and the European Commission of European Union (EU). International Maritime Organization (IMO) is United Nations specialized agency responsible for ensuring the safety and security of shipping, as well as preventing marine and atmospheric pollution caused by ships. Meanwhile the European Commission of European Union (EU) that invented FuelEU Maritime Regulation is one of European strategy aims to gradually incorporate maritime emissions into the Greenhouse Gases policy through systematic monitoring of ships' fuel consumption.

Fig. 9 Main regulator of ships

1.3.1 International Maritime Organization (IMO)

The International Maritime Organization (IMO) stands as the leading authority in overseeing the efficiency of the maritime transport sector. Its mission is to establish a level playing field, enabling shipowners to address financial challenges without resorting to budget cuts that compromise the safety of personnel, ship structures, or the environment. Additionally, the IMO actively promotes innovation and efficiency within the industry. Many other systems focusing on energy efficiency or the environmental impact of ships often adopt IMO criteria as the basis for their own measures [20].

A crucial aspect of the IMO's work in energy efficiency is its technical committees, particularly the Marine Environment Protection Committee (MEPC). The MEPC oversees matters related to the protection of the marine environment and addresses energy efficiency issues, primarily through the regulation of CO_2 emissions. [21] The IMO has also created several conventions that are very significant and important for navigation, such as International Convention for the Safety of Life at Sea (SOLAS), International Convention for the Prevention of Pollution from Ships (MARPOL), and International Convention on Standards of Training, Certification and Watchkeeping for Seafarers (STCW). The most important convention in the objective of energy efficiency is by far the MARPOL. It is the primary international convention aimed at preventing pollution of the marine environment by ships, whether caused by operational activities or accidental incidents. The most crucial aspect of MARPOL concerning energy efficiency is Annex VI: Prevention of Air Pollution from Ships. This was the last annex to be enforced, it includes mandatory technical and operational measures to enhance energy efficiency and reduce greenhouse gas emissions from ships (Fig. 10).

The IMO is currently focused on implementing the GHG strategy to ensure that the shipping industry meets the established targets. These targets include reducing total GHG emissions by 20% (with an aim for 30%) by 2030, and 70% (with an aim for 80%) by 2040, all relative to 2008 levels. The ultimate goal is to achieve net-zero GHG emissions by or around 2050. This goal entails zero well-to-wake GHG emissions, though discussions are ongoing about its precise definition and implementation. As of MEPC 81 in March 2024, the IMO has developed an overarching framework called the 'IMO net-zero framework' for the necessary regulatory changes to MARPOL Annex VI. These regulations are expected to be finalized by 2025 and come into effect around mid-2027.

1.3.2 European Commission of the European Union

In 2024, ships trading in the EU are subject to the EU Emissions Trading System (ETS) for the first time, necessitating a revision of regulatory responsibilities and contracts to ensure that the costs of allowances are passed through the supply chain to the responsible company. Shipping companies must purchase and surrender emission allowances for tank-to-wake (TtW) CO_2 emissions within EU and European

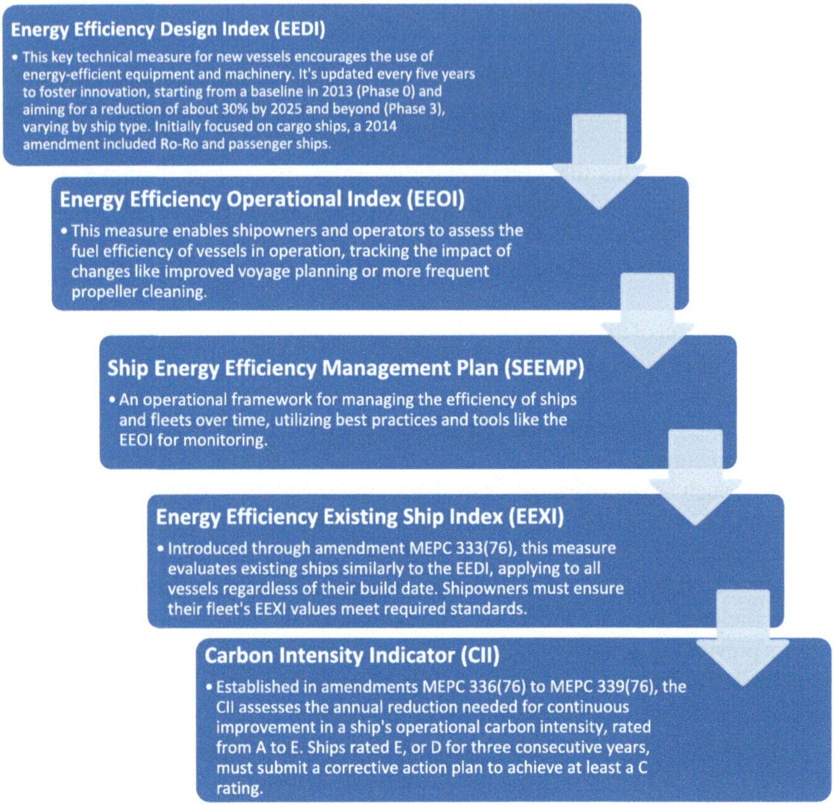

Fig. 10 Mandatory technical and operational measures

Economic Area (EEA) ports, emissions on voyages between these ports, and 50% of emissions on voyages into or out of these regions. The first deadline for surrendering allowances is in September 2025, based on 2024 emissions.

Starting in 2025, general cargo vessels ranging from 400 to 5,000 gross tonnage (GT) and offshore vessels from 400 GT must report their greenhouse gas (GHG) emissions, although they are not yet subject to the ETS. Offshore ships above 5,000 GT will be subject to the ETS from 2027, while the inclusion of general cargo and offshore ships between 400 and 5,000 GT will be considered after a review in 2026. As of July 2024, the EU has not defined "offshore ships."

Methane and nitrous oxide emissions must be reported starting in 2024 and will be included in the scope of the EU ETS from 2026. The ETS also considers the permanent storage of carbon.

Additionally, the upcoming FuelEU Maritime regulation in 2025 will impose a well-to-wake GHG intensity requirement on the energy used throughout the year, mandating the use of low GHG fuels. Starting January 1, 2025, ships over 5,000 gross tonnage (GT) carrying cargo or passengers for commercial purposes must comply with annual well-to-wake greenhouse gas (GHG) emissions intensity requirements. This intensity can be reduced through several methods, including the use of fossil LNG or LPG, qualified low GHG intensity fuels (with an additional reward for Renewable Fuels of Non-Biological Origin), shore power, or wind-assisted propulsion. Wind-assisted propulsion can decrease GHG intensity by up to 5%, though this reduction is based on design criteria and does not factor in the actual use of the system [22].

2 Technological Barriers in Green Ships and Green Shipping

The shipping industry plays a vital role in global trade, accounting for approximately 90% of international goods transportation [23]. However, its contribution to environmental pollution, particularly in terms of carbon dioxide (CO_2) has become a significant concern. The emission emitted from ship combustion engines is approximated to be one billion tons of carbon dioxide equivalents (CO_2e) annu-ally [13, 24–26]. The transition toward green shipping aims to reduce these emissions through the adoption of cleaner fuels, energy-efficient designs, and sustainable operational practices that drive shipping sector complete the International Maritime Organization's (IMO) deep decarbonization targets by 2050 [27]. Although green shipping offers vast potential, the industry faces several technological barriers that hinder its progress. This chapter explores the key technological challenges in green ship design, fuel and energy systems, propulsion technologies, emission control, digitalization, and supporting infrastructure. In general, the de-carbonizing ways can be categorized into various categories and sub-categories as illustrated in Fig. 11 [28]. In the subsequent discussion, each operational and technological barrier will be covered.

3 Operational and Technical Challenges to Decarbonize Shipping

Green ship design focuses on optimizing energy efficiency by adopting greener technology e.g. exhaust gas cleaning system (scrubber), alternative fuel systems (Ammonia, LNG, Hydrogen), and energy efficiency enhancement such as advance hull coating [29]. Traditional ship designs and older propulsion systems contribute significantly to air pollution, generating sulfur oxides (SO_x), nitrogen oxides (NO_x), carbon dioxide (CO_2), and particulate matter [30]. Retrofitting old ships with new technology can greatly decrease their environmental effect and ensure compliance with international environmental standards such as the Inter-national Maritime Organization's (IMO) 2020 sulfur cap and the Carbon Intensity Indicator (CII) criteria

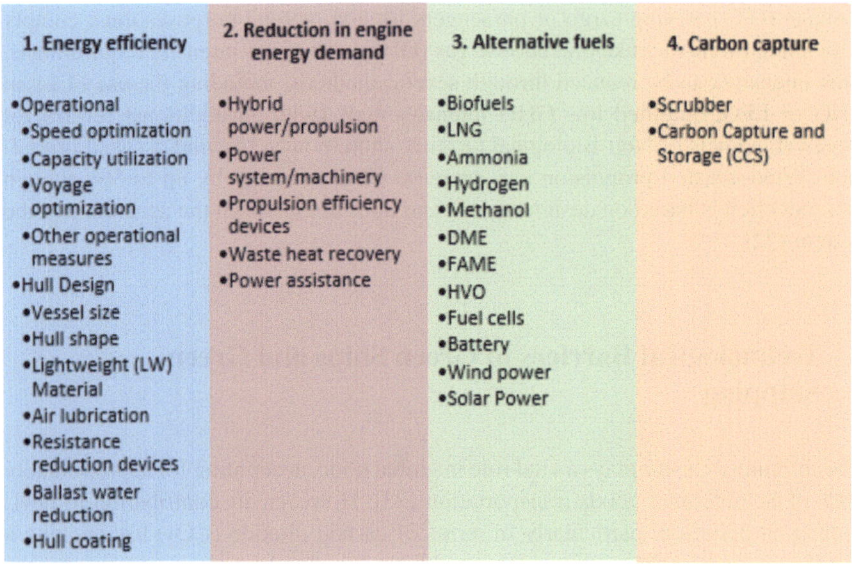

Fig. 11 Key retrofitting to decarbonize shipping

[31]. Retrofitting on ships requires lightweight materials, advanced hull designs, and hybrid systems. Nevertheless, these technologies significantly raise the cost of shipbuilding. Compared to traditional vessels, green ships demand higher initial investments, creating a major financial barrier for shipowners.

The majority of global shipping fleet consists of older vessels that were not designed with green technologies for consideration. Retrofitting existing ships which is aimed to meet new environmental regulations is expensive and technically challenging. The physical space required to install advanced systems like LNG tanks, new propulsion devices, or scrubbers is often limited, particularly on smaller or older vessels. Additionally, retrofitting involves significant downtime, causing operational disruptions that can be financially troublesome for shipping companies. Furthermore, there is no universally accepted standard for green ship technologies adopting alternative fuels and propulsion systems. For example, some studies indicate that all LNG utilization is advantageous to both GHG emissions and local air quality. Other studies, on the other hand, advocate against LNG because of its significant methane slip along the whole Well-to-Tank (WTT) chain. Consequently, shipowners are faced with a range of options, including LNG, hydrogen, and hybrid-electric propulsion systems, each with its own benefits and limitations. Companies find it challenging to commit to a particular solution due to the ambiguity around which technology will become the industry standard. The transition to cleaner shipping is delayed and investment in green shipbuilding is hindered by this lack of standards.

The historical trend of energy demand in the shipping sector shows that an increase trade requires an increase in energy use as presented in Fig. 12. To address the energy demand intensity issue, the IMO established four mandates aimed at

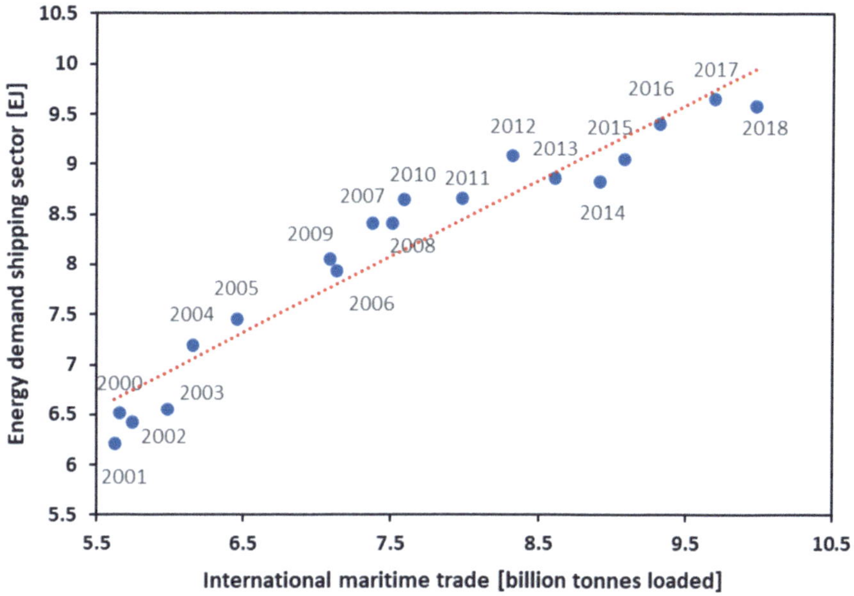

Fig. 12 Historical trend of energy demand in shipping sector [32–34]

improving energy efficiency (EE) through vessels: EEDI (Energy Efficiency Design Index), SEEMP (Ship Energy Efficiency Management Plan), EEXI (Energy Efficiency Existing Ship Index), EEOI (Energy Efficiency Operational Indicator), and CII (Carbon Intensity Indicator). As technology advances, the IMO's efforts to decarbonize the shipping sector evolve. The EEDI directly addresses the vessel design phase by providing measures that affect the ship's hull and superstructure, propulsion and power systems, mechanical technology integration, and the use of alternative energy sources. This contrasts with operational EE metrics, which are measured using EEOI and SEEMP.

4 Energy Efficiency

From operational point of view, reducing the service speed of a vessel can reduce the fuel consumption [35, 36]. Nevertheless, it makes transportation time longer, consequently more ships are needed to maintain the supply chain. Thus, it is still inefficient and a speed optimization is required to optimize the transportation time while reducing fuel consumption. Low speed or slow steaming operation: 20–22 knots are considerably an optimum operation to reduce emission up to 19% in average. However, on the other hand, slow steaming may cause serious damage to the engine since it operates for longer periods at load condition [37].

To achieve better energy efficiency hull design become unneglected consideration. A design of green ship is expected to experience low resistance when it is operating. Some design parameters and technologies can be enhanced to achieve a green ship such as hull shape optimization, lightweight material usage, air bubble hull lubrication, resistance reduction devices, hull coating, and no-ballast ship design. Some of these enhancement ways are shown in Fig. 13 Hull design optimization (a) air cavity; (b) hull form optimization; (c) air bubble lubrication. The design optimization and new technologies can only be applied on new shipbuilding. These technologies significantly raise the cost of shipbuilding. Compared to traditional vessels, green ships demand higher initial investments, creating a major financial barrier for shipowners. Therefore, for traditional vessels which are currently operating a retrofitting approach can be completed by changing some key component to reduce ship resistance which leads to emission reduction. Some resistance reduction devices can be a possible solution to change traditional ship into a green ship.

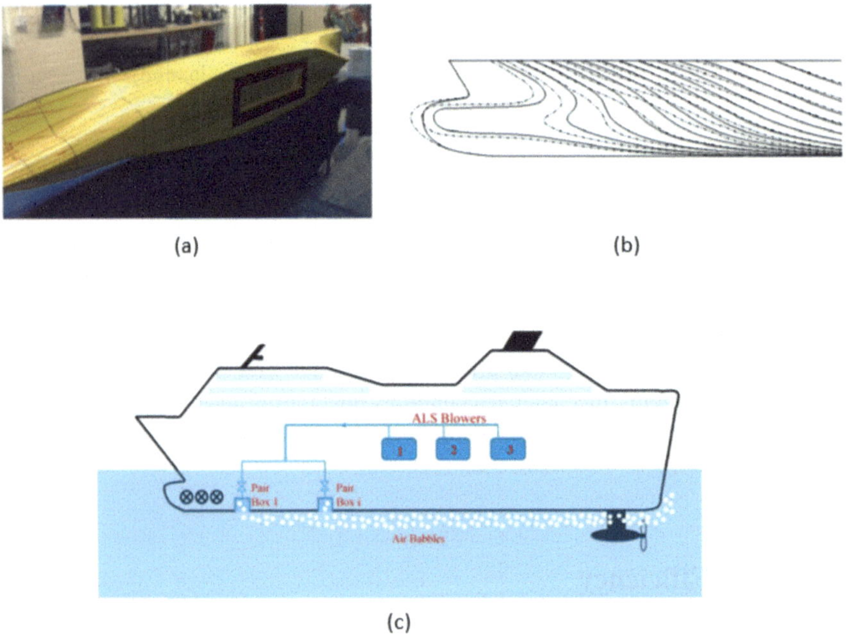

Fig. 13 Hull design optimization **a** air cavity; **b** hull form optimization; **c** air bubble lubrication [38–40]

5 Propulsion and Power Generation Systems

Combining an optimum hull design and new technology in the propulsion system is capable of building a new green ship. To achieve IMOs deep decarbonization targets by 2050, the use of electric and hybrid propulsion systems offers great promise for reducing emissions in shipping. Electric propulsion systems, powered by batteries or fuel cells, can drastically cut down on fuel consumption and emissions. However, current battery technology is limited in terms of energy density, which restricts its use to short-range vessels. Long-range cargo ships require more energy than current batteries can provide, making it necessary to develop more efficient and compact energy storage systems. Hybrid propulsion systems, which combine conventional fuel engines with electric power, present a middle ground. These systems can reduce fuel consumption and emissions, but they still rely on fossil fuels to some amount. Additionally, hybrid systems require sophisticated control technologies to switch between power sources, adding to the complexity and cost of implementation.

Propulsion efficiency devices or energy saving devices (ESD) may also be used to improve the performance of a ship propulsion system. The devices are primarily designed to optimize the flow around the propeller or eliminate losses arising on the propeller. Various ESDs, their working principal, as well as their potential on saving energy are summarized in Table 2. These type of devices may be used on an old vessel or a new ship, decreasing energy loss from the propulsion system. As a result, it becomes a potentially useful technology for reducing ship emissions. Furthermore, when compared to other cutting-edge technologies, it is very affordable and simple to install aboard ships. Figure 14 illustrates the shape of various ESDs.

Renewable energy resources are nowadays also integrated into the ship propulsion system. However, integrating renewable energy sources such as wind and solar power into shipping operations presents several challenges. For instance, solar panels, while capable of generating supplemental power, are limited by the available deck space on a ship and the amount of sunlight available at sea. The wind-assisted propulsion technologies, such as rotor sails and kites, are also weather-dependent and cannot provide

Table 2 ESD's type and their potential [41]

Working principle, reducing	Devices	Energy saving potential (%)
Rotational losses	Pre-swirl fins	3–5
	Twisted rudder	1–2
	Contra rotating propeller	4–7
Rotational losses and separation in the aft body	Wake equalizing duct	3–8
Hub vortex losses	Efficiency rudders	2–6
	Rudder bulb	2–5
	Hub cap fins	2–5
Tip vortex losses	Kappel propeller	3–6

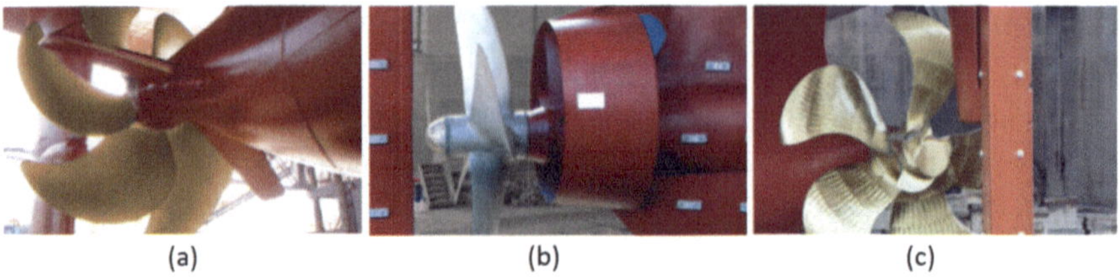

Fig. 14 ESD's type **a** pre-swirl fin; **b** wake equalizing duct; **c** hub cap fins [41]

consistent power. Although these technologies can improve fuel efficiency, they are not yet capable of fully powering large cargo ships on long voyages. However, those devices still have the potential to reduce energy consumption if in-depth research is carried out until they are ready for optimal use with low installation and maintenance costs.

6 Fuel and Energy Barriers

The introduction of alternative fuels like Liquefied Natural Gas (LNG), hydrogen, and ammonia has been hailed as a critical step toward greener shipping [42]. LNG comprises mostly of methane and is recommended because the emissions of air pollutants SO_x, PM, and NO_x from LNG engines are substantially lower than standard marine diesel engines using HFO or MGO/MDO. Due to this potential, the use of LNG on the new ship keeps increasing as presented in Fig. 15. Even though LNG is cleaner than traditional marine fuels, it still poses environmental concerns due to methane slip, a phenomenon where unburned methane escapes into the atmosphere.

Hydrogen and ammonia also offer zero-emission potential but they are not yet commercially viable on a large scale. Hydrogen requires advanced cryogenic storage systems, while ammonia presents challenges related to toxicity and energy density. Those alternative green fuels require specialized storage systems which pose challenges both in terms of ship design and global infrastructure. For instance, LNG and hydrogen must be stored at extremely low temperatures, necessitating the use

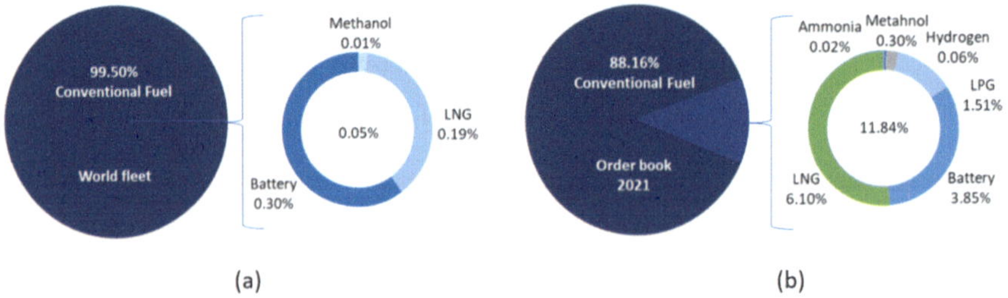

Fig. 15 Alternative fuels for the world fleet (June 2021) [33]

of cryogenic tanks. These tanks are bulky, reducing the available space for cargo. Furthermore, the global bunkering infrastructure for alternative fuels is underdeveloped, making it difficult for ships to refuel in all parts of the world. Until this infrastructure is built out, ships using alternative fuels will be restricted to specific routes limiting their operational flexibility.

7 Emission Control Technologies

Emission control technologies or also known as the exhaust gas treatment system (EGTS) may reduce the emission emitted by a ship which will comply with the IMO's SO_x and PM emission regulations [43]. One of these technologies is the exhaust gas scrubber. It has been widely adopted as a solution for reducing sulphur emissions from ships, particularly in regions with strict environmental regulations. However, scrubbers come with their own set of challenges. They are expensive to install and maintain, and their operation requires significant energy which reduces overall fuel efficiency. Additionally, scrubbers produce waste products that must be properly disposed of, creating additional environmental concerns. Open-loop scrubbers, in particular, have been criticized for discharging acidic wash water into the ocean, which can harm marine ecosystems.

Other new concept for EGTS is the so-called carbon capture and storage (CCS). It is a relatively new concept in the maritime industry. The idea is to capture CO_2 emissions from ship exhaust and store them onboard for later disposal. According to Global CCS Institute, there are 14 active CCS industrial projects on shore by the end of 2012. The existence of CCS can be potentially adopted on ship board. Three working principles of CCS system are frequently used i.e. pre-combustion, oxy-fuel, and post-combustion as illustrated in Fig. 16.

The CSS is still in its early stages regardless of its potential to significantly reduce greenhouse gas emissions. Current systems are bulky and energy-intensive making them impractical for widespread use particularly for a ship. Further research and development are needed to create compact and efficient CCS systems that can be integrated into ships without compromising cargo capacity or fuel efficiency.

8 Digitalization and Smart Shipping

Digitalization has the potential to revolutionize green shipping by improving operational efficiency and reducing fuel consumption. Smart shipping systems use sensors and data analytics to monitor fuel usage, engine performance, and weather conditions, allowing ships to optimize their routes and reduce emissions. However, integrating data from multiple systems can be technically challenging, especially on older vessels. Likewise, as ships become more connected, they are increasingly vulnerable to cyberattacks. Cybersecurity is a major concern in the digitalization

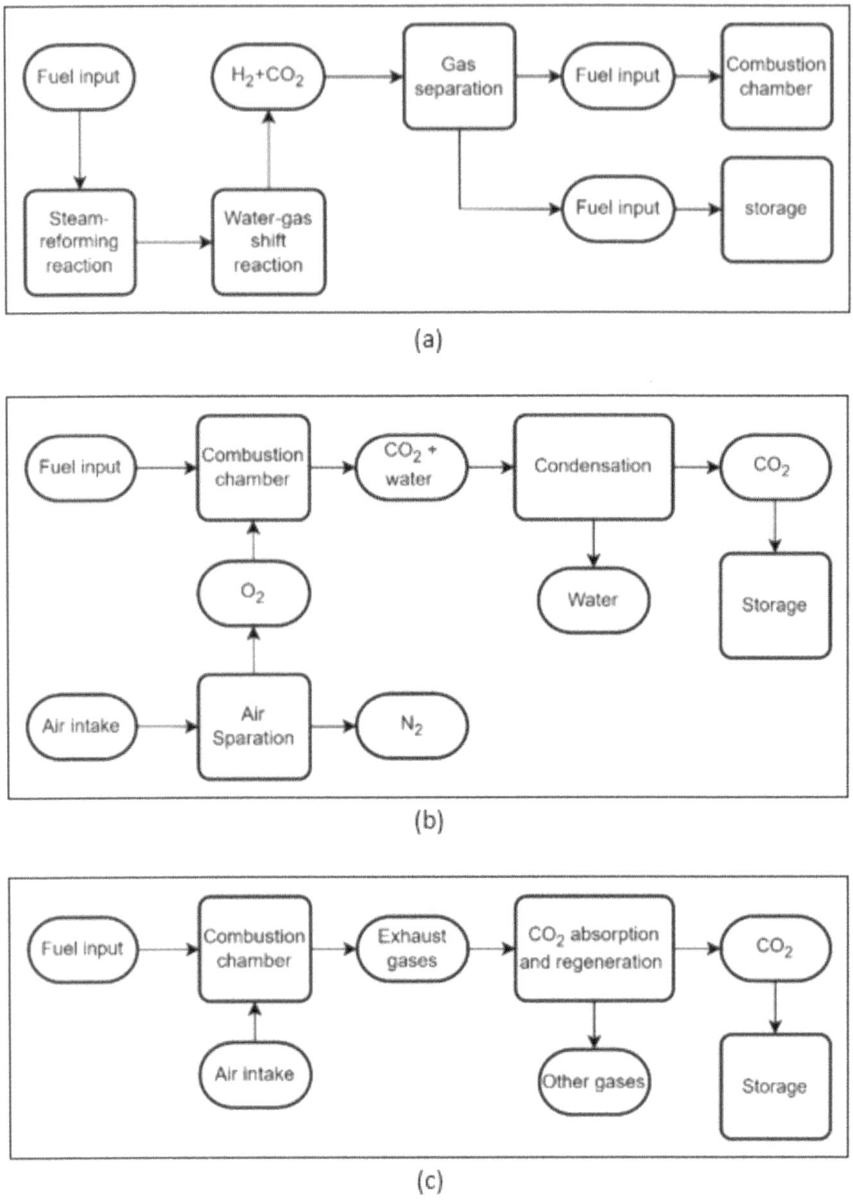

Fig. 16 Working principle of CCS systems **a** pre-combustion; **b** oxy-fuel; **c** post-combustion [44]

of shipping, particularly if more critical systems become automated. A successful cyberattack could disrupt operations, compromise safety, or lead to environmental damage. Ensuring robust cybersecurity measures is essential for the safe and efficient operation of smart ships yet the cost and the complexity of implementing these measures can be a barrier for many operators.

9 Infrastructure and Policy Barriers

The global shipping industry lacks the infrastructure needed to support the widespread adoption of green technologies. Ports are not yet equipped with the facilities required to refuel ships with alternative fuels like LNG, hydrogen, or ammonia. Additionally, electric charging stations for battery-powered ships are still rare. Until ports are upgraded to accommodate these technologies, green ships will face operational constraints that limit their adoption. In addition to infrastructure barrier, conflicting regulations imposed by regulators like IMO may occur [45]. For instance, ballast water treatment system reduces sea pollution but at the same time it increases fuel consumption which lead to high emission. Furthermore, environmental regulations vary significantly between regions, creating operational challenges for international shipping companies. Ships that comply with regulations in one region may face additional requirements in another which complicate global operations. This inconsistent enforcement of environmental regulations can create an uneven playing field where some companies avoid investing in green technologies to reduce costs.

In summary, the transition to green shipping is essential for reducing the environmental impact of the maritime industry. However, the technological barriers discussed in this chapter including challenges in ship design, fuel systems, propulsion technologies, emission control, digitalization, and infrastructure must be addressed to enable the widespread adoption of green shipping practices. Dealing with these constraints requires coordination among shipbuilders, fuel suppliers, port authorities, and regulators, as well as major investment in research and development.

10 Environmental Impact

10.1 The Emission Source from Marine Transportation

In 2007, the shipping industry accounted for 3.3% of worldwide greenhouse gas emissions, and over the next several decades, its CO_2 emissions are expected to rise sharply [24]. The effects of greenhouse gases released by ships on air quality and climate change have been the subject of numerous research. These studies used PM2.5, nitrogen oxide, and sulphur oxide to measure greenhouse gases and found a strong positive correlation between these three substances and environmental emissions or climate change (as indicated by carbon dioxide) in both Asian and European nations [46].

In terms of actual research on shipping emissions and their environmental impacts on Africa, Onwuegbuchunam et al. [47] used wastewater the physicochemical and microbiological analyses to assess the level of marine pollution at Nigerian seaports. They came to the conclusion that chemical emissions from ships were the primary cause of the pollution. Additionally, Olukanni and Esu [48] calculated the quantity of greenhouse gases released by port vessel activities in Nigeria's Tin Can Port and Lagos. From the time of their arrival, the emission estimates were based on the kind of vessels and their movements. The results showed that the biggest emitters of emissions are premium motor spirit carriers, followed by general cargo and container vessels. Additionally, a more recent study by Azubuike [49] evaluated the cost-effectiveness of marine emission control measures for crude oil tanker vessels shipping in Nigeria and demonstrated that in order to control excessive environmental pollution activities by crude oil tankers, a cost-effective method for shipping in Nigeria is required.

Deng et al. [50] investigated the 20% variation in CO_2 emissions reported by the reviewed inventories as a result of differences in the estimation methodology and study scope. Top-down approaches (like IEA) improve the timeliness of emission estimation, while bottom-up approaches (like CAMS-GLOB-SHIP and EDGAR) facilitate the availability of geospatial information. The urgency of decoupling the growth of seaborne trade from carbon emissions is underscored by the re-bound in CO_2 emissions by 2021. Source and process control measures will pro-vide the majority of the abatement potential, with carbon capture and out-of-industry transfers bearing the remaining abatement burden by 2050. However, present low-carbon development is challenged by secondary emissions, navigational safety, crew welfare, international cooperation, and technical and economic viability. There is still more work to be done in order to achieve the net-zero aim, which calls for the collaboration and coordination of all parties involved in the shipping industry's whole value chain. The Estimation of Green House Emission Released from Marine Transportation Source.

The CO_2 amounts emitted from the shipping sector follows the formula to determine how much CO_2 is released per ton nautical mile (ε) [24].

$$\varepsilon = \left(\frac{F}{DM}\right)K_e \tag{1}$$

F is the total amount of fuel consumed during the round-trip as defined by Eq. (2); where DM is the freight service in distance-weight units (ton nautical mile), where D is the distance and M is the average weight (tons) of the cargo on a particular round-trip; and K_e is the CO_2 emitted per unit of fuel burnt.

The fuel expended in formula (1) includes the fuel used at port and slow zones ($F_p\&_s$) and while sailing (F_s) for a round-trip, as stated by Linstad et al. [51]

$$F = F_{s+}F_{p\&s} = K_f\left(\left(\frac{PD}{v}\right) + \left(P_{p\&s}T_{p\&s}\right)\right) \qquad (2)$$

The desired power in formula (2) is denoted by P, the round-trip distance by D, the actual vessel speed by v, and the fuel quantity (in grams) per kWh produced by K_f. Additionally, $T_{p\&s}$ is the amount of time spent in port and slow zones on a round-trip basis, whereas $P_{p\&s}$ is the power need in these areas. This distinction between Linstad et al. [51] and the Second IMO GHG study (2009) is the formula computes the fuel as a function of speed, taking into account both the calm water resistance and the additional resistance from waves and propulsion efficiency.

It implies that the amount of fuel consumed is inversely proportional to the speed over the round-trip distance, while the desired power is directly proportional to the fuel spent. Meanwhile, the amount of fuel expended will affect the concentration of CO_2 emitted according to the formula (1). Consequently, the increase or decrease in speed during the ship's round-trip distance will greatly affect the concentration of CO_2 emitted into the environment.

10.2 Reduction in Greenhouse Gases Emission

Reducing CO_2 emissions is the main goal of energy efficiency initiatives in shipping. In 2009, the transportation industry accounted for 23% of global CO_2 emissions [52], and by 2035, the sector's fuel demand is expected to increase by around 40% [53]. In order to lower CO_2 emissions by increasing fuel economy, several emission control measures have been proposed in other transportation sectors in recent years, including Hickman and Banister [54], Chen and Cheng [55], Ferreira [56], and Morris et al. [57]. However, Parera and Mo [58], Olukanni [48], Armstrong et al. [59], Armstrong and Victor [60], Meng et al. [61], Lindstad et al. [51], Calleya et al. [53] and other researchers investigated the emission in marine transportation Fig. 17. Through the reduction of fuel emissions in sever-al transportation sectors, "the carbon price" can play a significant impact.

Parera and Mo [58] provide examples of the energy efficiency concepts of Energy Efficiency Design Index (EEDI), Energy Efficiency Operational Indicator (EEOI), and Ship Energy Efficiency Management Plan (SEEMP) as well as the emission control concepts of Emission Control Areas (ECAs), CO_2 reductions, SO_x limits, and NO_x Technical Codes.

The second portion consists of identifying possible energy-saving scenarios throughout the ship's operation phase in connection to the proposed energy flow route. Preliminary data evaluations of performance and navigation characteristics are provided for a selected vessel to support the previously specified scenarios along the energy flow channel. An essential component of the SEEMP is the identification of high-power consumption areas and possible energy-saving scenarios in relation to the EEOI that is provided. Finally, these results suggest that using appropriate navigation strategies (such as trip planning, draft, and trim optimization), especially

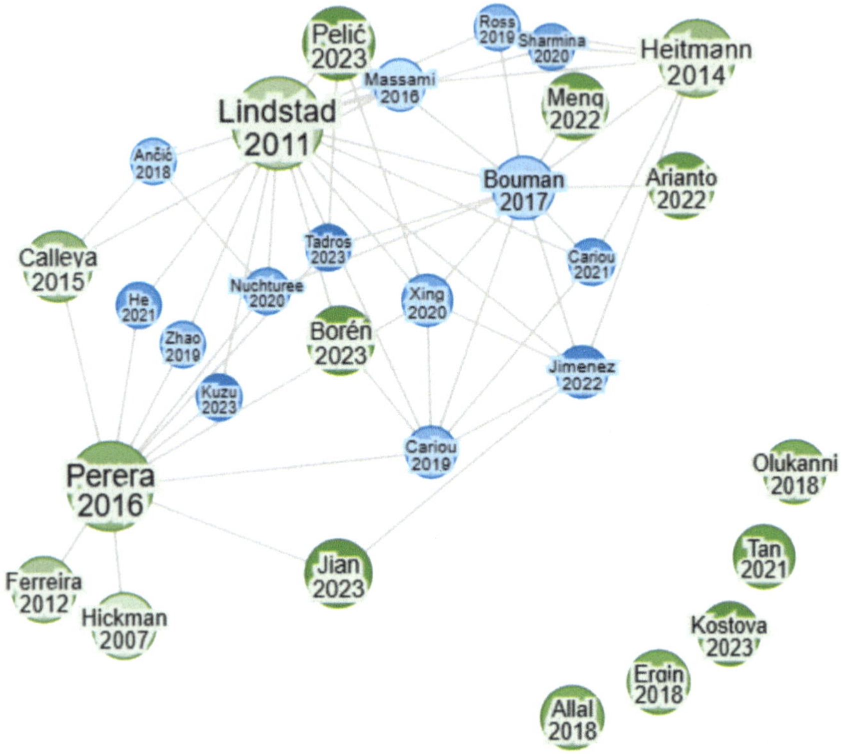

Fig. 17 The researchers who investigated the reduction of greenhouse gases emission in marine transportation

in the approved ECAs, could potentially reduce exhaust emissions (CO_2, SO_x, and NO_x).

Under MARPOL Annex VI, a number of required rules aimed at lowering CO_2 emissions from the maritime sector were adopted. This includes the Energy Efficiency Design Index (EEDI) for new ships and the Ship Energy Efficiency Management Plan (SEEMP) for all ships, which are two crucial components for increasing energy efficiency in shipping. All new ships that weigh 400 gross tons or more are subject to the EEDI; however, the relevant ship administration may suspend this requirement for new ships for a maximum of four years [58]. As part of the SEEMP, each ship shall, however, have an International Energy Efficiency Certificate (IEEC) for routine audits and inspections. Moreover, CO_2 emissions from the current shipping fleets are measured and compared using the Energy Efficiency Operational Indicator (EEOI).

Olukanni [48] estimated the quantity of greenhouse gases released by port vessel activities in Nigeria's Tin Can and Lagos ports. The type of vessel and its movement from the time of arrival (400 km from the coast) were taken into consideration while estimating emissions. Using data on vessels processed by both ports in the first and second quarters of 2017, the bottom-up approach was used to estimate emissions based on individual vessel characteristics. The largest emitters among the several kinds of vessels are Premium Motor Spirit (PMS) carriers, which are followed by general cargo and container vessels. According to the results for the first and second quarters of 2017, roughly 16,335 t and 773 t.

The shipping sector also contributes to air pollution through nitrous oxides (NO_x) and sulfur oxides (SO_x), which are related to engine operating conditions and bunker fuel types. Marine diesel engines' operating parameters, such as operating temperature, are restricted by the NO_x emission limits, while the SO_x emission restrictions limit the fuel's sulphur content. Tighter SO_x and NO_x emission limits are implemented in the designated Emission Control Areas (ECAs) despite the introduction of general SO_x and NO_x emission limits worldwide. To reduce emission-related pollution, the IMO and other environmental agencies in the designated ECAs are expected to progressively tighten these emission limitations [62].

11 Conclusion

The maritime industry faces significant challenges in reducing greenhouse gas (GHG) emissions due to its reliance on fossil fuels and increasing operational demands. Strategies such as optimizing hull designs, adopting hybrid propulsion systems, utilizing alternative fuels (e.g., LNG, hydrogen, ammonia), and integrating renewable energy show promise for enhancing energy efficiency and reducing emissions. However, these advancements are hindered by high costs, technological immaturity, and inadequate global infrastructure. Regulatory frameworks, such as the IMO's GHG Strategy, provide essential guidance, but achieving decarbonization targets requires greater innovation, collaboration, and investment. Addressing these challenges is crucial to transitioning toward sustainable maritime operations.

References

1. Taylor DA (1990) Introduction to marine engineering. Elsevier Butterworth-Heinemann, Oxford
2. Baldi F, Hannes J, Gabrielii C, Anderson K (2014) Energy analysis of ship energy systems – the case of a chemical tanker. Energy Procedia 61:1732–1735
3. Aijjou A, Bahatti L, Raihani A (2019) Study on container ship energy consumption. Energy Sustain VIII:25–36
4. Barreiro J, Zaragoza S, Casas VD (2022) Review of ship energy efficiency. Ocean Eng 257:111594
5. Samuel S, Simanjuntak F, Hadi ES, Iqbal M, Kurniawan WA, Dinda NK, Chaniago MA (2023) Analyzing the influence of water depth on total resistance in ulstein x-bow configuration: a numerical study. E3S Web Conf
6. Arief IS, Bidowi A, Pradana AN (2021) Analysis of the effects from adding propeller boss cap fins to skewed propeller performance with CFD methods. Int J Mar Eng Innov Res 6(2):140–151
7. Arief IS, Jadmiko E, Fikry I (2019) The effect of span to chord ratio PBCF on C440 Wageningen series propeller with CFD method. Int J Mar Eng Innov Res 4(3):183–190

8. Xiao N, Zhou R, Lin X (2016) Type selection and design of hybrid propulsion system of ship. IOP Conf Ser: Mater Sci Eng 157:012034
9. Arief IS, Fathalah AZ (2021) Review of alternative energy resource for the future ship power. IOP Conf Ser: Earth Environ Sci 972:012073
10. Larkin A, Smith T, Wrobel P (2017) Shipping in changing climates. Marit Policy 75:188–190
11. Endres S, Shi Z, Rutgersson A, Al-Hajjaji S, Brynolf S, Booge D, Hassellov I-M, Kontovas C, Kumar R, Liu H, Marandino C, Matthias V, Moldanova J, Salo K, Sebe M, Yi W, Yang M, Zhang C (2023) Perspective on shipping emissions and their impacts on the surface ocean an dlower atmosphere: an environmental-social-economic dimension. Elemnta 11(1):00052
12. Aakko-Saksa PT, Lehtoranta K, Kuittinen N, Järvinen A, Jalkanen JP, Johnson K, Jung H, Ntziachristos L, Gagné S, Takahashi C, Karjalainen P (2023) Reduction in greenhouse gas and other emissions from ship engines: Current trends and future options. Prog Energy Combust Sci 94:101055
13. Faber DJ (2021) Fourth IMO greenhouse gas study. International Maritime Organization (IMO), London
14. Chen J, Fei Y, Wan Z (2019) The relationship between the development of global maritime fleets and GHG emission from shipping. J Environ Manag 242:31–39
15. Fan A, Yang J, Yang L, Wu D, Vladimir N (2022) A review of ship fuel consumption models. Ocean Eng 264:112405
16. Class NK (2023) Pathway to zero-emission in international shipping - understanding the 2023 IMO GHG strategy
17. Zhang C, Zhu J, Guo H, Xue S, Wang X, Wang Z, Chen T, Yang L, Zeng X, Su P (2024) Technical requirements for 2023 IMO GHG STRATEGY. Sustainability 16(7):2766
18. Hüffmeier J, Johanson M (2021) State-of-the-Art methods to improve energy efficiency of ships. J Mar Sci Eng 9(4):447
19. Ančić I, Vladimir N, Luttenberger LR (2018) Energy efficiency of ro-ro passenger ships with integrated power systems. Ocean Eng 166:350–357
20. International Maritime Organization (2024) Technical and operational measures. https://www.imo.org/en/Ourwork/Environment/Pages/Technical-and-Operational-Measures.aspx. Accessed October 2024
21. Barreiro J, Zaragoza S, Diaz-Casas V (2022) Review of ship energy efficiency. Ocean Eng 257:111594
22. European Commission of the European Union (2023) Decarbonising maritime transport – FuelEU Maritime. https://transport.ec.europa.eu/transport-modes/maritime/decarbonising-maritime-transport-fueleu-maritime_en. Accessed November 2024
23. Xiao G, Wang Y, Wu R, Li J, Cai Z (2024) Sustainable maritime transport: a review of intelligent shipping technology and green port construction applications. J Mar Sci Eng 12(10):1728
24. Buhaug Ø, Corbett J, Endresen Ø, Eyring V, Faber J (2009) Second IMO GHG study 2009. International Maritime Organization, London
25. Lindstad E, Eskeland GS, Rialland A, Valland A (2020) Decarbonizing maritime transport: the importance of engine technology and regulations for LNG to serve as a transition fuel. Sustainability 12(21):8793
26. Lindstad E, Lagemann B, Rialland A, Gamlem GM, Valland A (2021) Reduction of maritime GHG emissions and the potential role of E-fuels. Transp Res Part Transp Environ 101:103075
27. Mallouppas G, Yfantis EA (2021) Decarbonization in shipping industry: a review of research, technology development, and innovation proposals. J Mar Sci Eng 9(4):415
28. Tadros M, Ventura M, Guedes Soares C (2023) Review of the IMO initiatives for ship energy efficiency and their implications. J Mar Sci Appl 22(4):662–680
29. Kolios A (2024) Retrofitting technologies for eco-friendly ship structures: a risk analysis perspective. J Mar Sci Eng 12(4):679
30. Suner M (2024) Analysis of air pollution from three main transportation vehicles: a case study. Energy Sources Part Recov Utiliz Environ Eff 46(1):1890–1906
31. Murugan K, Md Arof A (2023) Compliance to IMO sulphur cap regulations for vessels of 10 years of age and below. In: Materials and technologies for future advancement. Springer, pp 147–153

32. IRENA (2021) A pathway to decarbonise the shipping sector by 2050. International Renewable Energy Agency
33. Veritas DN, Lloyd G (2020) Maritime forecast to 2050
34. UNCTAD (2020) Information and data center. In: United Nations Conference on Trade and Development. UNCTAD, Geneva
35. Lee CY, Lee HL, Zhang J (2015) The impact of slow ocean steaming on delivery reliability and fuel consumption. Transp Res Part E Logist Transp Rev 76:176–190
36. Balcombe P, Brierley J, Lewis C, Skatvedt L, Speirs J, Hawkes A, Staffell I (2019) How to decarbonise international shipping: options for fuels, technologies and policies. Energy Convers Manag 182:72–88
37. Yfantis E, Pariotis E, Zannis T, Katsanis J, Roumeliotis I (2016) CO2 Emissions from ships: reduction methods and technologies. In: Proceedings of the international conference energy in transportation
38. Giernalczyk M, Kaminski P (2021) Assessment of the propulsion system operation of the ships equipped with the air lubrication system. Sensors 21(4):1357
39. Kim H, Yang C (2010) A new surface modification approach for CFD-based hull form optimization". J Hydrodyn 22(S1):503–508
40. Butterworth J, Atlar M, Shi W (2015) Experimental analysis of an air cavity concept applied on a ship hull to improve the hull resistance. Ocean Eng 110:2–10
41. MAN (2023) Basic principles of ship propulsion. MAN Energy Solutions
42. N. Government (2019) The government's action plan for green shipping. Norwegian Ministry of Climate and Environment
43. Notteboom T, Delhaye E, Vanherle K (2010) Analysis of the consequences of low sulphur fuel requirements. ITMMA–Universiteit Antwerp Transp
44. Wang H, Zhou P, Wang Z (2017) Reviews on current carbon emission reduction technologies and projects and their feasibilities on ships. J Mar Sci Appl 16(2):129–136
45. Jafarzadeh S, Utne IB (2014) A framework to bridge the energy efficiency gap in shipping. Energy 69:603–612
46. Ayesu EK (2023) Does shipping cause environmental emissions? Evidence from African countries. Transp Res Interdiscip Perspect 21:100873
47. Onwuegbuchunam DE, Ebe TE, Okoroji LI, Essien AE (2017) An analysis of ship-source marine pollution in Nigeria seaports
48. Olukanni DO, Esu CO (2018) Estimating greenhouse gas emissions from port vessel operations at the Lagos and Tin Can ports of Nigeria. Cogent Eng. 5(1):1507267
49. Azubuike R (2019) Cost effective assessment of marine emissions control measures for crude oil tanker shipping in Nigeria Port. https://www.academia.edu/76383979/Cost_Effective_Assessment_of_Marine_Emissions_Control_Measures_for_Crude_Oil_Tanker_Shipping_in_Nigeria_Port [Diakses November 2024]
50. Deng S, Mi Z (2023) A review on carbon emissions of global shipping. Mar Dev 1:4
51. Lindstad H, Asbjørnslett BE, Strømman AH (2011) Reductions in greenhouse gas emissions and cost by shipping at lower speeds. Energy Policy 39:3456–3464
52. Trodden DG, Murphy AJ, Pazouki K, Sargeant J (2015) Fuel usage data analysis for efficient shipping operations. Ocean Eng 110:75–84
53. Calleya J, Pawling R, Greig A (2015) Ship impact model for technical assessment and selection of Carbon dioxide Reducing Technologies (CRTs). Ocean Eng 97:82–89
54. Hickman R, Banister D (2007) Looking over the horizon: transport and reduced CO2 emissions in the UK by 2030. Transp Policy 14(5):377–387
55. Chen B, Cheng HH (2010) Review of the applications of agent technology in traffic and transportation systems. IEEE Trans Intell Transp Syst 11(2):485–497
56. Ferreira M, d'Orey PM (2012) On the impact of virtual traffic lights on carbon emissions mitigation. IEEE Trans Intell Transp Syst 13(1):284–295
57. Morris BT, Tran C, Scora G, Trivedi MM, Barth MJ (2012) Real-time video-based traffic measurement and visualization system for energy/emissions. IEEE Trans Intell Transp Syst 13(4):1667–1678

58. Perera LP, Mo B (2016) Emission control based energy efficiency measures in ship operations. Appl Ocean Res 60:29–46
59. Armstrong VN, Banks C (2015) Integrated approach to vessel energy efficiency. Ocean Eng 110:39–48
60. Armstrong VN (2013) Vessel optimisation for low carbon shipping. Ocean Eng 73:195–207
61. Meng L, Liu K, He J, Han C, Liu P (2022) Carbon emission reduction behavior strategies in the shipping industry under government regulation: a tripartite evolutionary game analysis. J Clean Prod 378:134556
62. IMO (2024) Sustainable development knowledge platform. International Maritime Organization. https://sustainabledevelopment.un.org/index.php?page=view&type=6&nr=57&menu=1359&template=375. [Diakses November 2024]

Chapter 3
Digital Technologies for Energy Monitoring and Optimization

Imam Sutrisno, Dimas Pristovani Riananda, Gusma Hamdana Putra, Wahyu Mulyo Utomo, and Dinda Pramanta

Abstract This article explores the critical role of energy monitoring and management within the maritime sector, emphasizing the need for sustainable and efficient energy practices to mitigate environmental impacts. The document outlines the significant contribution of maritime operations to global carbon emissions and highlights international regulatory frameworks like the IMO's MARPOL and SEEMP to address these challenges. Advanced energy monitoring systems, such as nanogenerators and IoT-enabled devices, are discussed as transformative tools to optimize energy consumption and emissions. The integration of digital technologies, including machine learning and AI-driven optimization models, demonstrates potential for predictive maintenance, real-time energy usage analytics, and enhanced operational efficiency. By aligning innovative technologies with sustainability goals, the maritime industry can achieve cost reductions, regulatory compliance, and

I. Sutrisno (✉) · D. P. Riananda
Department of Marine Electrical Engineering, Shipbuilding Institute of Polytechnic Surabaya, Surabaya, Indonesia
e-mail: imam_sutrisno@ppns.ac.id

D. P. Riananda
e-mail: dimaspristovani@ppns.ac.id

G. H. Putra
Department of Marine Engineering, Shipbuilding Institute of Polytechnic Surabaya, Surabaya, Indonesia
e-mail: hamdana.putra@ppns.ac.id

W. M. Utomo
Department of Electrical and Electronics Engineering, Tun Hussein Onn University, Parit Raja, Malaysia
e-mail: wahyu@uthm.edu.my

D. Pramanta
Department of Information and Network Sciences, Kyushu Institute of Information Sciences, Fukuoka, Japan
e-mail: pramanta@kiis.ac.jp

M. Elsisi et al. (eds.), *Maritime Infrastructure for Energy Management and Emission Reduction Using Digital Transformation*, Studies in Infrastructure and Control,
https://doi.org/10.1007/978-981-96-4438-4_3

environmental stewardship. This research provides actionable insights for implementing energy-efficient systems and future-oriented strategies in maritime energy management.

Keywords Energy monitoring · Maritime industry · Carbon emissions · IoT Technology · Energy optimization

1 Fundamentals of Energy Monitoring and Management

Energy users have intensively conducted studies on transition energy in the last decade due to its relationship with climate problems and the increase in the earth's temperature. This phenomenon occurs due to the use of energy derived from fossils, resulting in an increase in the greenhouse effect due to carbon gas emissions (CO2). Although the world's energy needs decrease by 0.5% annually, carbon gas emissions (CO2) production continues to increase. The contribution of fossil fuels to carbon emissions is still relatively high, around 2% every year, which is far different compared to the contribution of land management such as deforestation, agriculture, livestock, etc.

One of the contributors to carbon gas emissions is the maritime industry sector. It is recorded that the operational contribution of ships is 3% of the total contribution of carbon gas emissions (CO2), where the operational effectiveness level reaches 90% [12]. In addition, the operation of the maritime industry until now still relies on fossil fuels. Therefore, several strategies have been established by both the government and international organizations to control carbon emissions in the maritime industry sector and limit the use of energy derived from fossils.

IMO, as an international organization in the maritime sector, has set several rules to control carbon gas emissions, such as The International Convention for the Prevention of Pollution from Ships (MARPOL) in 1973, the Ship Energy Efficiency Management Plan (SEEMP) and Energy Efficiency Design Index (EEDI) in 2011, and adopted The Initial GHG Strategy in 2018 [9]. In addition, several organizations have implemented energy management optimization and energy monitoring. The sub-chapter below explains the vital role of energy monitoring, the basic concept of energy monitoring, and energy management systems (EMS).

2 The Importance of Energy Monitoring in Modern Systems

Energy monitoring involves systematically recording and analyzing energy usage. This activity is essential for companies because it can provide many benefits. Real-time monitoring capabilities, optimization of use to be more efficient, reduction of operational costs, and minimization of environmental impacts can be applied in several sectors [11]. In addition, other positive impacts include ease of decision-making due to data availability, easy identification of performance inefficiencies, and facilitation of development and maintenance according to existing conditions.

2.1 Overview of Energy Monitoring Significance Across Various Sectors

The energy hierarchy in the maritime industry has several structured approaches to its management and maintenance, especially in shipping operations and infrastructure. This hierarchy includes individual ship energy management up to port and fleet operations, focusing on efficiency, desirability, and compliance.

First, ship-level management begins on each ship, where owners and operators are responsible for implementing energy efficiency and saving technologies and practices. For example, the Energy Management System (EMS) integrates and optimizes energy consumption in real-time to guarantee ship operational efficiency. In addition, fuel efficiency, such as optimizing the ship's hull, propeller, and routine maintenance schedules, has a significant impact. Second, fleet management involves coordinating multiple vessels. Utilizing data from all ships will make it easier to eliminate energy consumption patterns. Data can be analyzed to optimize energy management by making routes, schedules, and maintenance decisions. Third, port operations provide the infrastructure required to manage energy efficiently. For example, shore connection, which allows ships to connect to the electricity network while anchored, will reduce carbon emissions. In addition, cargo operations such as systematic loading and unloading will impact energy use efficiency and operational time. Fourth, maritime sector regulations, one related to carbon emissions set by the IMO, ensure that all ships in operation meet clean environmental standards. In addition, Incentives for Cleaner Technologies is an incentive institution for operators who adopt low-emission technology, and many investors donate to this program.

Several types of energy monitoring applied in the maritime sector, both on ships and in ports, include electrical energy monitoring systems, fuel monitoring systems, renewable energy integration systems, condition monitoring systems, and development management systems. Electrical energy monitoring systems are designed to collect and analyze data by measuring energy consumption. This technology can be applied to any system. This monitoring system provides real-time data based on several parameters such as electrical power, power factor, total harmonic distortion, and many more that can be developed (Fig. 1).

One of the electrical energy monitoring systems is nanogenerator technology for power systems, which has a broad scope. The system can be applied to external insulation monitoring (such as transmission lines monitoring, electrical equipment monitoring), internal insulation monitoring (such as insulating gas monitoring, transformer oil monitoring), and environmental energy harvesters (vibration energy,

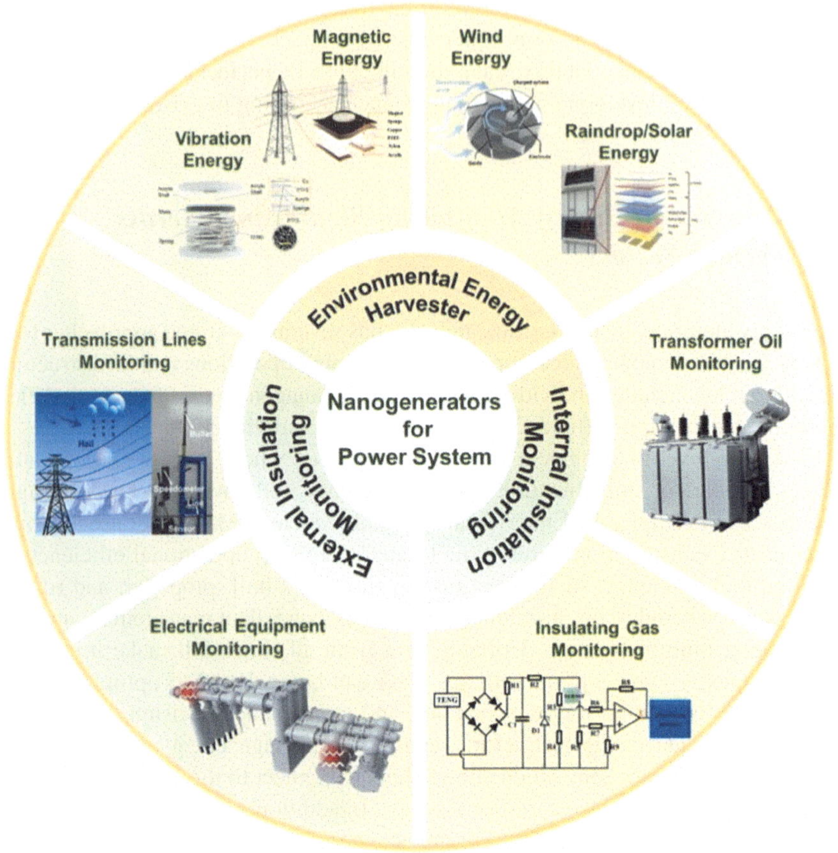

Fig. 1 Nanogenerator for power system

magnetic energy, wind energy, raindrop/solar energy, etc.) as shown in the image upper.

Based on energy harvesting, this technology can replace conventional electromagnetic induction and photovoltaic power generation, which are considered less effective because they are easily affected by disturbances (blackouts, heavy rain, lightning, etc.). In addition, based on the networked sensing aspect, this technology can increase the accuracy and depth of analysis, the efficiency of operations and maintenance, and the safety and availability of the power grid. Finally, based on monitoring intelligence, this technology can provide detailed information regarding the unit's working conditions, making it easier for operators to improve their working environment.

Fuel monitoring is a system that measures the amount of fuel consumed in a generating unit or vehicle. Components include a flow meter that reads the flow rate for later consumption calculation. Some add level sensors, temperature sensors,

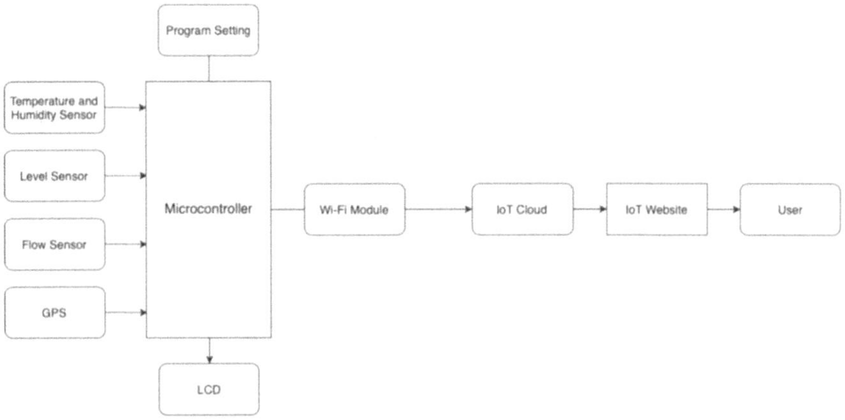

Fig. 2 Fuel monitoring system [13]

humidity sensors, and GPS to be processed to obtain more informative data, as in the research conducted by Hapsari et al. in the Fig. 2.

Some energy monitoring applications in the maritime sector include Variable Frequency Drive (VFD), which is used to operate seawater cooling pumps. The phenomenon that occurs when the pump operates is the varying water flow due to dependence on the water level. This causes back pressure and impacts other systems, such as condensers to turbines and main/auxiliary engines. This technology can adjust the pump speed to reduce power consumption [5].

2.2 Key Goals in Energy Management: Efficiency, Sustainability, and Cost Savings

Efficiency is one of the most critical aspects of energy management in the maritime sector. It will impact optimal energy consumption, reduce operational costs, and reduce environmental impacts. In addition, operational efficiency will be more economical. Operators can substantially reduce fuel, maintenance, and repair expenditures by optimizing fuel consumption and minimizing energy waste. For instance, integrating Variable Frequency Drives (VFDs) into pumping systems can lead to a 92.12% reduction in energy consumption, as seen in a commercial sea vessel case study.

Efficient energy management will have an increasingly optimal operational impact. For example, an advanced monitoring system with real-time energy consumption time will impact the quality of the operator's recommendations. This proactive condition could increase the equipment's life function. In addition, proper electrical usage monitoring helps identify faults early, preventing failures that could

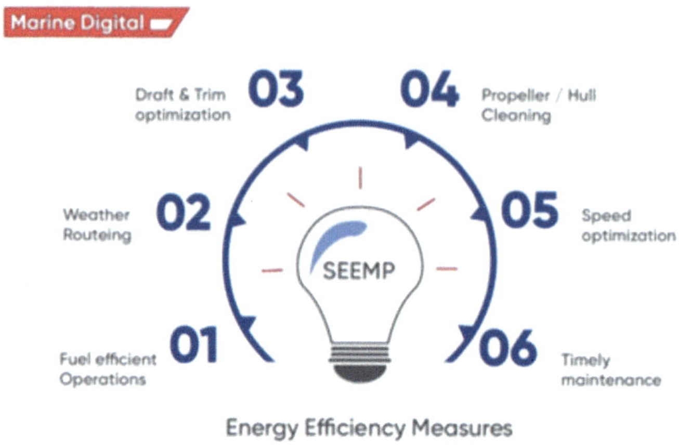

Fig. 3 Ship energy efficiency management plan. (*Source:* Marine digital)

endanger crew and passengers. Furthermore, optimal energy use can minimize the likelihood of accidents caused by equipment overheating.

Modern maritime regulation emphasizes sustainability and environmental stewardship. An Efficient Energy Management System (EEMS) helps ensure compliance with stringent regulations such as the Ship Energy Efficiency Management Plan (SEEMP). By regularly monitoring and improving energy efficiency, operators can demonstrate adherence to international guidelines, avoiding penalties and reputational damage [10] (Fig. 3).

Energy management in the maritime sector prioritizes efficiency, and the principal basis of efficiency is the responsibility for environmental safety and health. Consumption of fossil fuels can reduce greenhouse gas emissions, contributing to sustainability goals. Integration between renewable energy sources such as solar panels and wind turbines minimizes dependence on fossil fuels, which aligns with the commitment to reduce reliance on fossil fuels.

Several strategic approaches can be applied to achieve energy management efficiency in the maritime sector. First, comprehensive mechanisms for data collection and analysis capabilities increase the chances of efficient energy management. The results of the analysis can also be the basis for policymakers' decisions. Second, if appropriately implemented with well-established planning, advanced technologies such as VFD, intelligent lighting systems, and renewable energy integrations can significantly increase efficiency. In addition, this method will provide the opportunity for a longer equipment life. Third, human resource development includes providing training and certification and cultivating disciplined work to increase opportunities for efficient energy management. With data play skills, crews/workers will efficiently and quickly provide information and recommendations on occurring conditions. Efficiency can be a fundamental principle in today's maritime energy management because it has a comprehensive scope. Energy management can be optimal with implementation based on integrity, discipline, and total responsibility.

Attached are several aspects of sustainability as objectives in energy management in the maritime sector, including reducing greenhouse gas emissions, collaboration of renewable energy sources, increasing environmental compliance, protecting marine ecosystems, promoting sustainability, and involving policymakers to be able to provide correct recommendations that prioritize the community. First, emissions produced by the maritime industry contribute around 2–3% of global emissions, so efforts to reduce emissions in this sector are critical. Second, collaboration between renewable energy and the maritime industry can potentially reduce dependence on fossil fuels. Collaboration is also in line with the IMO's objectives regarding lowering emissions. Third, all crews must comply with international regulations related to emissions both on board and at the port. One example is implementing the ISO 50001 standard to support energy management by building a framework for sustainability. Fourth, implementing environmental concerns such as waste management systems, considerations in ballast water operations, and the use of environmentally friendly materials will improve the quality of the marine environment because waste and pollution are reduced. Fifth, companies that are conscientious about sustainability will be more competitive because consumers like companies that care about the environment. In addition, the company is committed to the sustainability of the surrounding environment, making it easy to develop. Sixth, the involvement of stakeholders such as government, society, regulations, and environmental organizations in innovating comprehensive and practical sustainability will potentially achieve goals quickly due to high awareness and willingness.

3 Core Concepts in Energy Monitoring

Energy monitoring is essential for tracking, measuring, and comprehensively analyzing energy management in the maritime sector. Traditional energy monitoring relies on manual readings or basic sensors, which produce limited information on energy consumption patterns, overload conditions, or overheating. Handling them takes time and effort.

Here are some optimization principles of energy monitoring and management, including real-time data collection, data acquisition system, benchmarking and target setting, energy monitoring software, integration with building management systems (BMS), and sustainability goals.

3.1 Basic Principles of Energy Monitoring and Measurement Techniques

First, real-time data collection is one of the energy monitoring systems that uses sensors and metering to collect data. Its application is in electricity, gas, air, and other energy source metering. This data provides information related to usage patterns, which impacts the rapid identification of inefficiencies and decisions when anomalies occur or even energy use. Second, a data acquisition system combines data from several measuring devices, which ensures that energy usage information is collected accurately and on time. Its applications include data loggers, gateways, and communication networks. The purpose of this system is as a data transmission medium that can be analyzed. Third, benchmarking and target setting are performance benchmarks that allow measuring work efficiency against industry and national/international standards. By developing information on energy use, companies/agencies/organizations can set specific targets for operational efficiency. In addition, benchmarks can help track anomalies and identify areas requiring special attention. Fourth, energy monitoring software is one of the software that can perform data analysis to results and visualize energy consumption patterns. Further levels of this software incorporate algorithms to identify patterns, anomalies, and optimization opportunities that support decision-making based on information on energy management strategies. Fifth, integration with Building Management Systems (BMS), a system capable of providing holistic energy usage information related to industrial operations. Integrating energy resources can increase efficiency and optimize operational costs. Last, sustainability goals are the ultimate basis for implementing various companies/agencies/organizations where energy use can be optimal and sustainable, thus contributing to environmental sustainability.

Utilizing the above concept can optimize energy use effectively, thus impacting minimal operational costs and positively contributing to the goal of a green environment. Therefore, this energy monitoring serves as the basis for energy use management strategies in various sectors.

3.2 Types of Data and Metrics in Energy Management

Effective energy management relies on various data types and metrics that provide insights into energy consumption patterns, operational efficiency, and financial implications. Several types of data and metrics are needed for energy management, such as energy consumption data, cost analysis metrics, performance metrics, environmental impact metrics, operational data, behavioral analysis, and benchmarking data. First, energy consumption data includes total energy consumption during a specific period, measured in kilowatt-hours (kWh) for electricity or therms for gas. Data indicators consist of total energy consumption and peak demand, which describe overall energy usage within a defined timeframe and the highest energy demand observed during a specific period. This information is crucial for understanding load requirements and potential cost implications. Second, cost analysis metrics, which are financial metrics related to energy expenditures, help organizations understand the economic impact of their energy use. Data indicators consist of energy expenditure and cost per energy unit, which describe total costs associated with energy consumption and the average cost incurred for each energy consumed unit. This information is helpful for budgeting and forecasting. Third, performance metrics evaluate energy usage efficiency and effectiveness across systems and processes. Data indicators consist of energy intensity and equipment efficiency, which describe energy used per unit of output or square foot and indicate how efficiently energy is utilized relative to production or space, and performance metrics for individual pieces of equipment, such as the efficiency rating of HVAC systems or motors. Fourth, the environmental impact metrics track the environmental implications of energy consumption, particularly regarding greenhouse gas emissions. Those indicators consist of carbon footprint and the percentage of renewable energy used. Those are also described as total greenhouse gas emissions produced from energy consumption, typically measured in CO2 equivalents, and the proportion of total energy sourced from renewable resources, reflecting a commitment to sustainability. Fifth, operational data related to facilities' operational aspects, including schedules and production levels. Operational hours are used to calculate energy consumption and monitor production results to analyze efficiency levels.

4 Digital Technologies for Energy Monitoring

Energy monitoring in the maritime industry is a critical aspect of modern operations, focusing on tracking, measuring, and managing energy consumption across various systems. It serves as a cornerstone for reducing operational costs and mitigating environmental impacts, aligning the industry with global sustainability goals. By accurately assessing energy usage, maritime operators can identify inefficiencies, optimize processes, and transition toward cleaner energy sources. This process is particularly vital as the maritime sector faces increasing scrutiny regarding its carbon footprint and energy-intensive activities.

The implementation of energy monitoring systems in maritime operations has evolved significantly with the advent of digital technologies. These technologies facilitate real-time data collection from engines, auxiliary systems, and other energy-intensive equipment onboard ships or in port facilities. The data collected is not only vast but also granular, enabling operators to gain detailed insights into energy consumption patterns. By leveraging digital tools such as IoT sensors, big data analytics, and artificial intelligence, maritime stakeholders can transition from reactive to proactive energy management. For instance, real-time monitoring allows for immediate adjustments in operations, such as optimizing engine performance or switching to more energy-efficient modes during idle periods. Moreover, digital dashboards and visualizations provide decision-makers with actionable intelligence, ensuring that energy management strategies are data-driven and aligned with broader sustainability targets.

The integration of digital technologies in energy monitoring represents a transformative shift in the maritime industry. By enabling real-time data acquisition and analysis, these technologies empower stakeholders to make informed decisions that enhance energy efficiency and reduce environmental impact. As the industry moves toward greener operations, digital energy monitoring systems will play an indispensable role in achieving cost savings and compliance with environmental regulations. This underscores the importance of continued innovation and investment in digital solutions to support sustainable maritime development.

4.1 The Role of the Internet of Things (IoT)

Digital technologies, particularly the Internet of Things (IoT), have revolutionized energy monitoring by enabling more efficient and real-time management of energy resources. The IoT system in energy management typically consists of three main components that work together seamlessly to collect, process, and display data for effective decision-making. These components include the Monitoring System, the Telecommunication System, and the Data Processing and Display System.

4.1.1 Monitoring System (Data Collection)

The first critical component of an IoT-based energy monitoring system is the Monitoring System, which is directly responsible for gathering energy-related data from various sources. This system is composed of a range of hardware devices, sensors, and actuators, all strategically placed on energy-consuming objects or within specific locations (such as engines, HVAC systems, or lighting equipment). The sensors are designed to collect data on variables like temperature, pressure, humidity, energy usage, and even fuel consumption. The actuators, in turn, allow the system to make adjustments, such as regulating the speed of engines or controlling the operation of lighting systems. Additionally, a data management unit processes the raw data from sensors and transmits it for further analysis. The monitoring system serves as the foundation of the IoT network, providing the data that fuels the entire energy management process.

4.1.2 Telecommunication System (Internet or ETC)

The second essential component is the Telecommunication System, which enables the seamless transmission of data from the monitoring system to the central data processing units. The telecommunications infrastructure includes internet connections, communication networks, and telecommunication systems that ensure data is transmitted reliably from the sensors and actuators to remote servers or cloud platforms. This system allows for real-time monitoring and ensures that data collected from different locations (e.g., ships, factories, or warehouses) can be sent and accessed from anywhere in the world. The communication layer is crucial for maintaining the flow of information between the different components of the IoT system, ensuring that data is always available for immediate analysis and action.

4.1.3 Processing and Visualization

Finally, the Data Processing and Display System is where the collected data is analyzed, processed, and presented in an understandable format for decision-makers. This system typically includes servers, mobile applications, web applications, and databases that store the vast amounts of energy data collected by the monitoring system. The backend of the system processes the data, applying various algorithms and models, often including machine learning techniques, to detect patterns and generate insights. These insights can help identify energy inefficiencies or predict future consumption trends. The frontend (such as dashboards) displays this processed data in a visual format, making it easy for users to interpret and take action. Additionally, middleware integrates various software components, ensuring smooth communication between the front end, back end, and external systems. This complete data processing chain enables operators, facility managers, and stakeholders to make data-driven decisions that optimize energy use, reduce costs, and minimize environmental impact.

4.2 Fundamental of Sensors in Energy and Emission Monitoring System

In the maritime industry, energy efficiency and emission control have become crucial focal points for reducing operational costs, improving sustainability, and complying with stringent environmental regulations. As ships operate in highly dynamic environments, the continuous monitoring of energy consumption and emissions is vital for optimizing performance and reducing environmental impact. Sensors are at the heart of this monitoring process, providing real-time data that allows operators to make informed decisions. These sensors not only track how much energy a vessel consumes but also measure the emissions produced by its engines, ensuring that the ship meets international standards set by organizations such as the International Maritime Organization (IMO).

4.2.1 Key Roles of Sensors in Energy and Emission Monitoring

Sensors play a pivotal role in the effective monitoring of energy consumption and emissions aboard ships. With the growing demand for sustainability and regulatory compliance in the maritime industry, the integration of advanced sensor technology into energy and emission monitoring systems is essential. These sensors provide real-time, accurate data that is crucial for optimizing fuel usage, reducing energy waste, and minimizing harmful emissions. The role of sensors in this context goes beyond simple measurements; they are integral in improving overall operational efficiency, ensuring environmental protection, and helping vessels comply with international emission standards.

In energy monitoring, sensors are used to measure various parameters such as fuel consumption, electrical power usage, temperature, pressure, and battery health. By continuously monitoring these factors, sensors help operators assess the performance of the ship's energy systems. For example, fuel flow sensors provide insights into how efficiently fuel is being used by the engines and auxiliary systems, while power meters measure electrical consumption across various onboard systems. These sensors enable operators to identify inefficiencies, adjust operational settings, and optimize fuel distribution, ultimately leading to better energy management and reduced operational costs.

In emission monitoring, sensors play a critical role in tracking pollutants such as CO2, NOx, and particulate matter that are produced during combustion. CO2 sensors, for instance, measure the concentration of carbon dioxide in exhaust gases, which helps assess the combustion efficiency of the engines and identify areas where fuel is being wasted. Similarly, NOx sensors monitor nitrogen oxide emissions, which are a major contributor to air pollution and smog. By providing real-time data on these emissions, sensors enable ship operators to take corrective measures, such as optimizing engine performance or adjusting fuel usage, to ensure that the ship complies with environmental regulations like IMO MARPOL Annex VI.

Moreover, sensors are crucial for ensuring the health of the ship's systems and preventing potential issues that could lead to inefficient energy use or excessive emissions. Temperature and pressure sensors monitor critical components like engines and fuel systems, ensuring that they operate within optimal parameters. By detecting any anomalies, such as overheating or pressure irregularities, these sensors help prevent system failures that could result in higher fuel consumption or increased emissions. Battery monitoring sensors are another example, ensuring that hybrid energy systems are functioning efficiently by preventing overcharging or deep discharges, which could affect energy storage and consumption.

The data collected by these sensors is often integrated into the ship's energy management systems, allowing operators to monitor and analyze energy usage and emissions in real-time. This integration not only aids in making immediate operational adjustments but also contributes to long-term performance analysis, predictive maintenance, and optimization. By continuously tracking key energy and emission parameters, sensors help operators identify trends and areas for improvement, enabling more informed decisions that enhance the ship's energy efficiency and environmental performance.

The role of sensors in energy and emission monitoring also intersects with a number of international regulations designed to reduce the maritime sector's environmental footprint. Several key regulations are worth noting:

1. IMO MARPOL Annex VI (Air Pollution): The International Maritime Organization's (IMO) MARPOL Annex VI sets limits on the emissions of air pollutants from ships, including sulfur oxides (SOx), nitrogen oxides (NOx), and particulate matter. The regulations require ships to continuously monitor their emissions and ensure compliance with global and regional air quality standards. NOx and CO2 sensors are particularly important in this context, as they help ensure that emissions are within prescribed limits.
2. IMO Energy Efficiency Existing Ship Index (EEXI): The EEXI is a regulatory standard that requires ships to meet certain energy efficiency criteria, which are based on their design and operational parameters. Sensors used in energy monitoring systems, such as fuel flow meters, power meters, and temperature sensors, provide the necessary data to help operators calculate and improve energy efficiency in compliance with EEXI standards.
3. IMO Carbon Intensity (CII): The CII regulation, part of the IMO's decarbonization strategy, sets annual targets for reducing carbon intensity in shipping. This regulation requires ships to track their carbon emissions per distance traveled, making real-time monitoring of CO2 emissions vital. CO2 sensors integrated into emission monitoring systems provide critical data to help ships meet these stringent carbon intensity goals.
4. EU Emissions Trading System (ETS): The EU ETS imposes a carbon price on the maritime sector, requiring ships operating in European waters to monitor and report their CO2 emissions. The data collected by CO2 sensors plays a crucial role in ensuring compliance with this trading scheme, as ships are penalized for exceeding emission thresholds.
5. US EPA Emission Standards: The Environmental Protection Agency (EPA) in the United States sets strict emission standards for ships operating in U.S. waters. This includes regulations on nitrogen oxides (NOx) and sulfur oxides (SOx), both of which contribute to air pollution and health risks. NOx sensors and other emission-related sensors play an essential role in helping vessels comply with these standards by providing real-time data on pollutant levels.
6. EU Ship Recycling Regulation: In addition to emissions, the European Union has established regulations for the safe and environmentally sound recycling of ships. These regulations ensure that ships are dismantled in a manner that minimizes environmental damage, and onboard systems like sensors that monitor energy consumption and emissions can help in verifying the sustainability of operations throughout the vessel's life cycle.

By integrating sensors into energy and emission monitoring systems, ships can ensure that they meet these and other regulatory requirements. The continuous monitoring and data collection capabilities offered by sensors not only assist in operational decision-making but also contribute to global efforts to reduce the environmental impact of the maritime industry.

4.2.2 Integration of Sensors with Data Processing Systems

The integration of sensors with data processing systems is a crucial aspect of modern energy and emission monitoring aboard ships. Sensors collect real-time data on a variety of parameters, such as fuel consumption, power usage, engine performance, and pollutant emissions. However, the raw data collected by sensors only provides limited insights unless it is processed, analyzed, and visualized through advanced data processing systems. By integrating sensors with robust data processing platforms, operators can gain a comprehensive understanding of ship performance, optimize energy use, reduce emissions, and ensure compliance with international environmental regulations.

Data processing systems, such as Energy Management Systems (EMS), Automated Monitoring Systems (AMS), and Emission Control Systems (ECS), are designed to aggregate the data collected from sensors and provide actionable insights. These systems process large volumes of sensor data to generate reports, alerts, and dashboards that give operators a clear view of the ship's energy and emission performance. For example, a ship's EMS might combine data from fuel flow meters, power meters, and temperature sensors to analyze fuel efficiency and identify areas where energy consumption can be optimized. Similarly, emission control systems utilize data from CO2, NOx, and SOx sensors to monitor and control pollutant levels in real-time, ensuring that emissions remain within regulatory limits.

One of the most significant benefits of integrating sensors with data processing systems is the ability to make real-time decisions. By continuously analyzing the data provided by the sensors, these systems can trigger alerts or adjustments to optimize operations. For instance, if a sensor detects that fuel consumption is higher than expected, the system can alert the operator and suggest adjustments to engine settings or fuel load distribution. Similarly, if emissions exceed predefined thresholds, the system can initiate corrective actions, such as reducing engine output or adjusting combustion settings, to bring emissions back within acceptable limits.

Furthermore, the integration of Machine Learning (ML) and Artificial Intelligence (AI) into these systems enhances their predictive capabilities. By analyzing historical data, ML algorithms can forecast trends in energy consumption and emissions, allowing operators to make proactive decisions. For example, AI can predict when a specific system or sensor may fail based on previous performance data, enabling predictive maintenance and reducing the risk of costly breakdowns. Additionally, AI can identify patterns in fuel consumption and emissions, suggesting long-term operational adjustments to further improve efficiency and sustainability.

With sensor data being continuously processed and analyzed, operators can gain real-time insights into energy efficiency and emissions performance. This level of integration allows ships to make necessary operational adjustments quickly and efficiently. Furthermore, the integration of sensors into data systems enables automatic reporting of energy and emissions data to regulatory bodies, ensuring compliance without the need for manual reporting. For example, sensor data can be used to generate compliance reports for the IMO's Carbon Intensity Index (CII), ensuring that vessels meet the required reduction targets for carbon emissions. The integration of sensors with data processing systems not only aids in real-time decision-making but also facilitates long-term operational improvements and strategic planning. By continuously analyzing trends and identifying inefficiencies, vessels can take proactive measures to reduce their carbon footprint, optimize fuel consumption, and extend the life of their equipment.

4.3 Telecommunication System in Maritime and Port Operations

Telecommunication systems play a vital role in ensuring safe, efficient, and reliable communication in the maritime industry. Effective communication between ships, ports, and other stakeholders is essential for navigation, cargo handling, security, and emergency response. Over the years, several technologies have been developed to meet the unique demands of maritime communication, ranging from traditional radio systems to satellite-based solutions. This chapter will explore the key telecommunication systems used in maritime and port operations, including Cellular Telecommunication, Radio Frequency (RF) Communication (VHF, UHF, MF), and Satellite Communication, including emerging technologies like Starlink.

4.3.1 Cellular Telecommunication

Cellular telecommunication has become a crucial method for ship-to-shore and ship-to-ship communication, especially when vessels are within range of coastal areas. Cellular networks, like 4G and 5G, offer fast, reliable communication for both voice and data transmission, enabling ships to stay connected while at port or within coastal regions. Application in Maritime:

- Voice and Data Communication: Cellular networks allow for real-time voice communication for crew members and operational staff. Data applications for tracking, monitoring, and remote management of onboard systems, including energy monitoring, navigation, and logistics, are also possible via cellular networks.
- Internet Access: Cellular networks provide internet connectivity, enabling ships to access weather updates, communication with shipping companies, or even offer crew member internet access for leisure.
- Port Operations: Cellular networks facilitate communication between ships and port authorities for scheduling arrivals and departures, managing cargo, and optimizing vessel traffic.

One of the key limitations of cellular network technology in maritime communication is its **limited range**. Cellular networks rely on land-based towers to provide coverage, which means that ships can only access the network when they are within range of these towers, typically near coastal areas and ports. Once a vessel moves beyond the reach of these towers, it loses its connection to the cellular network. This limitation makes cellular communication unsuitable for long-distance navigation in open oceans, where ships require alternative communication systems, such as satellite communication or radio frequency (RF) systems, to maintain connectivity. As a result, while cellular networks are ideal for coastal and port operations, they cannot provide continuous, global coverage, requiring vessels to rely on more specialized systems for extended voyages.

4.3.2 Radio Frequency (RF) Communication

Radio frequency (RF) communication is vital for ensuring reliable, real-time communication across maritime operations. The RF spectrum used in maritime communications encompasses a wide range of frequencies, each serving distinct purposes based on the operational needs of the maritime sector. This includes Very High Frequency (VHF), Ultra High Frequency (UHF), Medium Frequency (MF), High Frequency (HF), and Low Frequency (LF) bands.

a. **Low Frequency (LF) Radio**

LF radio with Frequency Range: 30–300 kHz, operates at the lowest end of the radio spectrum and is primarily used for very long-range communication, especially in navigational systems. LF signals can travel vast distances, even over thousands of kilometers, making them valuable for communication in remote regions. Application:

- Navigational Aids: LF radio is used in LORAN (Long Range Navigation) systems to help ships determine their position in remote areas, especially where GPS signals may be weak or unavailable. LORAN stations transmit LF signals that ships can use for dead reckoning navigation.
- Emergency and Distress Signals: In extreme circumstances, LF radio can be used for distress signals or communication in cases where other communication technologies (like VHF or HF) are unavailable or inoperable.
- Limited Data Transmission: LF is not ideal for high-data-rate communication, as it is primarily used for simple navigational signals or emergency alerts.
- Limited Bandwidth: The bandwidth available at these frequencies is narrow, which limits the amount of data that can be transmitted.

b. **Medium Frequency (MF) Radio**

MF radio operates between 300 kHz and 3 MHz and is typically used for **long-range communication**, especially when ships are far from the coast and beyond the range of VHF or UHF systems. MF radios are primarily used for long-range communication, particularly when ships are far from the coast and beyond the range of VHF or UHF systems. MF radio can transmit signals over vast distances, making it valuable for communication in remote maritime areas. Application:

- Ship-to-Coastal Station Communication: MF radio is typically used for communication between ships and coastal stations, especially when vessels are beyond the VHF range. This is especially important in areas with little to no infrastructure, such as the open ocean.
- Backup Communication: In cases where VHF or UHF is unavailable (e.g., when ships are far offshore or during poor weather conditions), MF radio can serve as a backup communication method, ensuring reliable contact with shore stations.
- Emergency Communication: MF radio remains a reliable means of communication for emergencies, especially for distress alerts when other radio systems are out of range or not functioning.
- Limitations Atmospheric Interference: MF signals are susceptible to atmospheric conditions, such as ionospheric disturbances and sunspots, which can cause signal degradation or interference.
- Limitations Power Requirements: MF radios require more power than VHF or UHF radios to transmit signals over long distances, which can be a limitation on ships with limited power sources.

c. **High Frequency (HF) Radio**

High Frequency (HF) radio, with a frequency range of 3–30 MHz, is widely used in maritime communication, particularly for long-range communication. HF radio provides critical communication capabilities for vessels operating beyond the range of VHF and UHF systems, especially in remote areas or during deep-sea voyages. Here's a more detailed explanation of its use in the maritime industry:

- HF communication is used for very long-distance communication, enabling ships to stay in contact with coastal stations or other vessels when they are out of range of VHF and UHF systems.
- It is essential for operational communication over vast distances, particularly in remote oceanic areas.
- HF is still widely used in emergency situations or when satellite communication is not available.
- Limitations Atmospheric Variability: HF signals can be impacted by atmospheric conditions, such as solar activity or weather.
- Limitations Quality: While HF offers long-range capabilities, its signal quality can be variable, leading to communication challenges during certain weather conditions.

d. **Very High Frequency (VHF) Radio**

VHF radios are the most commonly used form of communication in maritime operations. Operating between 30 and 300 MHz, this frequency band provides short-range communication and is essential for ship-to-ship, ship-to-shore, and ship-to-port communication. VHF is the preferred medium for operational safety, distress signals, and basic voice communications. Application:

- Ship-to-Ship Communication: VHF radios are used for navigation safety, collision avoidance, and general operational coordination between vessels that are within sight of one another. These communications help ships safely navigate congested waters, such as in busy port areas or along shipping lanes.
- Ship-to-Shore Communication: VHF radios facilitate communication with shore stations for navigational updates, weather forecasts, and operational coordination. They also support communication with shore-based rescue and emergency services.
- Emergency Communication: VHF radios play a key role in maritime distress systems, particularly for the Global Maritime Distress and Safety System (GMDSS). In emergency situations, ships use VHF radios to broadcast distress signals and coordinate rescue efforts.
- Regulation GMDSS Compliance: VHF radios are required on all ships as part of mandatory equipment under SOLAS (Safety of Life at Sea) regulations. These radios are used for distress communication and safety alerts in emergencies.
- Regulation International Telecommunication Union (ITU): The ITU allocates frequencies within the VHF range to avoid interference and to ensure that communication is effective across international waters.
- Limitations Range: VHF communication is limited to line-of-sight range, typically around 20–30 miles, depending on the height of the antenna and terrain. This makes VHF unsuitable for long-distance communication, especially when ships are beyond visual range.
- Limitations Susceptibility to Interference: VHF signals can be subject to interference from nearby buildings, terrain, or atmospheric conditions, which can affect signal quality in port environments.

e. **Ultra High Frequency (UHF) Radio**

UHF radios, operating between 300 MHz and 3 GHz, are often used for communication in more confined environments, such as ports, where ships may not be within the range of VHF. UHF radios are used for communication in more confined environments such as ports, offshore platforms, and between ships and port facilities. UHF communication offers clear signals even in areas with physical obstructions such as tall buildings, docks, and cargo structures. Application:

- Port Operations: UHF radios are used in ports for coordinating activities such as cargo loading/unloading, vessel arrivals, and departures. Port authorities, tugboats, and container ships often rely on UHF for communication, especially in environments with a lot of physical interference.
- Ship-to-Port Communication: UHF radios are used for communication between ships and port authorities in specific areas like bulk terminals, oil rigs, and offshore platforms where VHF communication might not reach due to obstacles.
- Communication in Confined Spaces: UHF signals are better able to penetrate obstacles like buildings, containers, and metal structures in port environments, making them ideal for local communications in complex environments.

- Benefits Penetration of Obstacles: UHF signals can pass through obstacles like buildings and docks, making it suitable for communication in environments where VHF might fail.
- Benefits Clear Communication in Short Range: UHF radios provide clear and reliable communication within short distances, making them ideal for localized, high-density areas like ports.
- Limited Range: UHF radios are typically used for short-range communication and do not have the long-distance capabilities of VHF or MF radios.
- Limitations Possible Interference: UHF signals are more likely to suffer interference from nearby electronic devices, industrial equipment, and other systems that may operate in similar frequency ranges.

4.3.3 Satellite Communication (Satcom)

Satellite communication is a vital part of maritime communication, especially for vessels that operate in international waters or regions far from coastal networks. With advancements in satellite technology, communication capabilities for ships have dramatically improved, enabling **global coverage** and reliable data transmission, even in the most remote parts of the ocean.

a. **Traditional Satellite Communication**

Traditionally, ships use **geostationary satellites (GEO)** to enable communication via satellite-based systems such as **Inmarsat** and **Iridium**. These systems provide both voice and data services, including email, internet access, and emergency alerts.

- Application: These systems are used for voice communication, internet access, weather data, fleet management, and tracking systems.
- Benefits: Satellite communication offers global coverage and is essential for ensuring continuous communication while navigating in areas without any terrestrial coverage, including deep oceans and Polar Regions.
- Limitations: Traditional satellite systems often require large antennas and are expensive in terms of operational costs, including high data transfer fees.

b. **Low Earth Orbit (LEO) Satellite Communication**

Low Earth Orbit (LEO) satellite systems, such as OneWeb and Starlink, are emerging as more cost-effective alternatives to traditional GEO satellite systems. Unlike GEO satellites, LEO satellites orbit closer to the Earth (approximately 1,200 km above the surface), allowing for lower latency and faster data speeds.

- Starlink: Starlink, developed by SpaceX, is a LEO satellite system designed to provide high-speed internet and low-latency communication to ships and remote locations worldwide. As Starlink expands, it is poised to offer global connectivity for ships at sea, enabling seamless voice communication, real-time data sharing, video conferencing, and internet browsing.

- Benefits: Starlink offers reliable, high-speed internet and a lower cost of ownership for vessels compared to traditional satellite systems. It also provides high-quality video and data transmission in remote locations, improving both operational efficiency and crew welfare.
- Challenges: While Starlink promises low-latency internet, coverage in high-latitude regions near the poles can still be a challenge as the network evolves.

4.3.4 Role of Telecommunication in Port Operations

Telecommunication systems are critical not only for ship-to-ship and ship-to-shore communication but also for managing port operations. Ports require robust telecommunication systems to coordinate various activities, including cargo handling, vessel scheduling, traffic management, and safety procedures.

- Communication between ships and port control: Vessels must coordinate their arrivals and departures with port authorities to ensure the safe and efficient use of berth space and avoid congestion. VHF and satellite communication systems are commonly used for this purpose.
- Cargo and logistics: Ports use telecommunication networks to track cargo movements, monitor the status of containers, and manage logistics operations. Wireless communication systems, including RFID tags and barcode scanners, work in tandem with telecommunication infrastructure to provide real-time updates.
- Security and emergency response: Communication systems are also used in port security to monitor the movement of goods and people, ensuring compliance with safety standards and preventing illegal activities.

4.4 Data Collection, Processing, and Visualization in Maritime Energy and Emission Monitoring

In the context of energy and emission monitoring systems for maritime operations, the process of data collection, processing, and visualization plays a crucial role in enabling efficient and informed decision-making. This section explains the various components involved in managing, processing, and displaying data that allow operators and decision-makers to monitor and evaluate energy consumption and emissions on ships in real time.

4.5 Server and Database Management

At the heart of the system is the **server** that manages all the data flow. The server stores all the information gathered from different sensors (such as fuel flow meters, temperature sensors, CO2 sensors, etc.) deployed throughout the ship. It is responsible for managing large volumes of time-series data generated continuously by these sensors. The data can include measurements such as energy consumption, fuel usage, emission levels, system performance metrics, and operational parameters. Key Considerations for Server Management:

- Data Storage: Servers must handle the storage of large amounts of data efficiently. This is typically done through databases such as SQL (for structured data) or NoSQL (for unstructured or semi-structured data). The database ensures data integrity and reliability for long-term storage.
- Real-time Data Processing: Servers must also support real-time data processing. As energy consumption and emission data is collected continuously, it needs to be processed without delay to ensure accurate insights and actions. This can be achieved using streaming data platforms or data processing frameworks like Apache Kafka or Apache Spark.

Backend Processing: The backend is where most of the data processing and analytical functions happen. Using advanced algorithms, it processes the collected data, identifies patterns, and generates insights. It can involve:

- Data aggregation: Summarizing raw data into meaningful metrics.
- Predictive Analytics: Using machine learning (ML) or AI models to predict future energy usage patterns, identify energy inefficiencies, and forecast emissions levels.
- Anomaly Detection: Identifying unusual patterns or outliers in energy use or emissions, which may indicate operational issues or inefficiencies.

4.5.1 Backend and Application Logic

Once the data is collected and stored, the backend system processes it and applies business logic to transform raw data into actionable insights. The backend typically involves several key components:

- APIs (Application Programming Interfaces): These are used for communication between the backend server and external systems, enabling data exchange and integration. APIs allow the system to pull external data sources (e.g., weather data or ship operational schedules) or push data to other systems (e.g., enterprise management software).
- Algorithms and Machine Learning Models: To enhance decision-making, machine learning models are often deployed in the backend to predict energy consumption trends, detect inefficiencies, and recommend optimization actions. For example, models can learn from historical fuel consumption data and suggest adjustments to reduce fuel use or minimize emissions.

- Integration with External Systems: The backend integrates with other onboard systems like the SCADA system (Supervisory Control and Data Acquisition) or fleet management platforms. This ensures seamless data sharing and synchronization across all systems for comprehensive monitoring.

4.5.2 Frontend: Dashboards and Visualization Tools

The frontend is the user interface that presents processed data to the ship's operators, facility managers, or decision-makers. The goal is to provide a user-friendly and intuitive interface that makes it easy to monitor, interpret, and act upon energy and emission data.

- Dashboards: Dashboards serve as the primary tool for visualizing real-time data. These dashboards can display energy consumption metrics, emission levels, fuel usage patterns, and more, in the form of graphs, charts, heat maps, and tables. The design of the dashboard should prioritize clarity and ease of use.
- For example:
  - A fuel consumption graph could display current consumption rates compared to historical averages.
  - Emission levels might be shown in a real-time bar chart, with green, yellow, or red indicators to alert the user if emissions exceed acceptable thresholds.
- Interactive Features: The dashboard should allow users to interact with the data, enabling drill-down into specific parameters (e.g., viewing energy usage for individual systems like lighting, HVAC, or propulsion). It should also allow the user to customize views and set thresholds for alerts.
- User Profiles and Access Control: Depending on the role of the user, the frontend system can offer different levels of access to various data. For example, operational staff may have access to real-time performance data, while management may only have access to high-level reports and insights. This helps ensure that users only see data that is relevant to their responsibilities.
- Mobile and Web Applications: In addition to desktop dashboards, having mobile applications and web applications can provide greater flexibility and accessibility. Operators can monitor energy consumption and emissions on-the-go, receive alerts on their phones or tablets, and make quick decisions without being on-site.

4.5.3 Middleware: Ensuring System Integration

Middleware refers to the software layer that enables smooth communication between different system components. In the case of energy and emission monitoring, middleware facilitates communication between:

- Sensors collecting data from onboard systems
- Backend servers that process and analyze the data
- Frontend applications that visualize the data

Middleware is crucial for ensuring that all systems—whether onboard or external—work together in a coordinated manner, without data loss or delay. It often includes data synchronization protocols, message queues, and data format converters to ensure that data can flow smoothly between systems.

4.5.4 User Empowerment and Decision Making

The ultimate goal of data collection, processing, and visualization is to empower users with the tools and insights they need to make informed decisions regarding energy consumption and emissions management. With the help of these systems:

- Operators can continuously monitor energy performance and emissions levels, allowing them to quickly identify inefficiencies or performance deviations.
- Facility managers can access historical trends to plan for improvements, such as adjusting ship operation schedules or optimizing fuel consumption practices.
- Decision-makers can make data-driven decisions to optimize energy usage, reduce operational costs, and meet environmental goals, ensuring compliance with regulations and improving sustainability.

5 Energy Optimization Using Advanced Technologies

5.1 Machine Learning for Energy Consumption Forecasting

5.1.1 Introduction to Machine Learning in Energy Optimization

Energy optimization, the process of minimizing energy consumption while maximizing efficiency, has become a critical concern in today's world. As energy demand continues to rise, the need for innovative solutions to address this challenge is more pressing than ever. Machine learning (ML), a subset of artificial intelligence, offers a powerful toolkit for tackling complex energy optimization problems. By leveraging large datasets and advanced algorithms, ML models can identify patterns, make predictions, and optimize energy usage in various applications, from smart grids to building automation.

Machine learning is a field of study that gives computers the ability to learn without being explicitly programmed. It involves the development of algorithms that allow computers to learn from data and make intelligent decisions. ML can be broadly categorized into three types:

1. **Supervised Learning**: In supervised learning, algorithms are trained on labeled data, where the input data is paired with the desired output. The goal is to learn a mapping function that can predict the output for new, unseen input data. Common

supervised learning techniques include linear regression, logistic regression, decision trees, and neural networks.

2. **Unsupervised Learning**: Unsupervised learning involves training algorithms on unlabeled data, where the goal is to discover hidden patterns and structures within the data. Common unsupervised learning techniques include clustering and dimensionality reduction.
3. **Reinforcement Learning**: Reinforcement learning involves training agents to make decisions in an environment by interacting with it and receiving rewards or penalties. The agent learns to make decisions that maximize the cumulative reward over time.

Machine learning has the potential to revolutionize energy optimization by enabling more accurate predictions, efficient control strategies, and innovative solutions. Some of the key applications of ML in energy optimization include:

- **Demand Forecasting**: ML models can accurately predict energy demand patterns, allowing utilities to optimize generation and distribution.
- **Renewable Energy Forecasting**: ML models can forecast the output of renewable energy sources like solar and wind power, enabling better integration into the grid.
- **Building Energy Management**: ML models can optimize energy consumption in buildings by analyzing sensor data and adjusting HVAC systems, lighting, and other equipment.
- **Smart Grid Optimization**: ML models can improve the efficiency of smart grids by optimizing power flow, voltage control, and fault detection.
- **Energy Efficiency in Industries**: ML models can identify energy-saving opportunities in industrial processes, such as optimizing production schedules and equipment maintenance.

While ML holds great promise for energy optimization, there are several challenges that need to be addressed:

- **Data Quality and Availability**: High-quality and sufficient data is essential for training ML models.
- **Model Interpretability**: Understanding the decision-making process of ML models is crucial for building trust and ensuring transparency.
- **Computational Cost**: Training and deploying complex ML models can be computationally expensive.

Despite these challenges, the future of ML in energy optimization is bright. As ML techniques continue to advance and computational resources become more affordable, we can expect to see even more innovative applications that will help us achieve a more sustainable and efficient energy future.

5.1.2 Common Algorithms for Predicting Energy Demand

Predicting energy demand is crucial for efficient energy management. Several algorithms have proven effective in this task. Here are some of the most common ones:

Traditional Statistical Methods

1. **Time Series Analysis**:
   - **ARIMA (AutoRegressive Integrated Moving Average)**: This model captures the temporal dependencies in the data and is widely used for short-term forecasting.
   - **Exponential Smoothing**: This method assigns exponentially decreasing weights to past observations, making it suitable for smoothing out noise and trend identification.
2. **Regression Models**:
   - **Linear Regression**: This model assumes a linear relationship between the dependent variable (energy demand) and independent variables (e.g., temperature, humidity, economic indicators).
   - **Polynomial Regression**: This model captures non-linear relationships between variables, often useful for complex patterns.

Machine Learning Algorithms

1. **Artificial Neural Networks (ANNs)**:
   - **Multi-Layer Perceptron (MLP)**: This model can learn complex patterns and is effective for both short-term and long-term forecasting.
   - **Recurrent Neural Networks (RNNs)**: These models are well-suited for time series data, as they can capture temporal dependencies.
   - **Long Short-Term Memory (LSTM)**: A special type of RNN that can learn long-term dependencies, making it ideal for complex time series patterns.
2. **Support Vector Regression (SVR)**:
   - This model finds the optimal hyperplane to separate data points, making it suitable for both linear and non-linear regression tasks.
3. **Random Forest**:
   - This ensemble method combines multiple decision trees to improve accuracy and reduce overfitting.
4. **Gradient Boosting Machines (GBM)**:
   - This ensemble method builds models sequentially, with each model correcting the errors of the previous one.

Hybrid Models

Combining multiple algorithms can often lead to improved performance. For example, a hybrid model might use a statistical model for long-term trends and a machine learning model for short-term fluctuations.

Factors Affecting Energy Demand Prediction

- Weather conditions: Temperature, humidity, and wind speed can significantly impact energy demand, especially for heating and cooling systems.
- Economic indicators: Economic growth, GDP, and industrial production can influence energy consumption patterns.
- Social factors: Population growth, urbanization, and lifestyle changes can affect energy demand.
- Policy and regulatory factors: Government policies, energy pricing, and regulations can impact energy consumption behavior.

By carefully considering these factors and selecting appropriate algorithms, accurate energy demand predictions can be achieved, leading to more efficient energy management and reduced costs.

5.1.3 Case Studies in Predictive Maintenance

Case Study 1: Oil and Gas Industry

- **Challenge**: Oil and gas companies face significant downtime and maintenance costs due to unexpected equipment failures.
- **Solution**: Implementing predictive maintenance solutions that leverage sensor data and machine learning algorithms to predict potential failures before they occur.
- **Benefits**:
  - Reduced unplanned downtime
  - Optimized maintenance schedules
  - Increased equipment lifespan
  - Improved safety and environmental compliance

Case Study 2: Manufacturing Industry

- **Challenge**: Manufacturing plants rely on complex machinery that requires frequent maintenance to ensure optimal performance.
- **Solution**: Utilizing predictive maintenance to monitor the health of equipment and proactively address potential issues.
- **Benefits**:
  - Increased production efficiency
  - Reduced maintenance costs
  - Improved product quality

Case Study 3: Automotive Industry

- **Challenge**: Automotive manufacturers need to ensure the reliability of their vehicles to maintain customer satisfaction.
- **Solution**: Employing predictive maintenance to predict potential failures in vehicles and proactively schedule repairs.
- **Benefits**:
  - Enhanced vehicle reliability
  - Improved customer satisfaction
  - Reduced warranty claims

5.1.4 Case Studies in Usage Forecasting

Case Study 1: Energy Industry

- **Challenge**: Energy providers need to accurately forecast energy demand to optimize power generation and distribution.
- **Solution**: Using machine learning algorithms to analyze historical data, weather patterns, and other relevant factors to predict future energy consumption.
- **Benefits**:
  - Improved grid stability
  - Reduced operational costs
  - Enhanced renewable energy integration

Case Study 2: Retail Industry

- **Challenge**: Retailers need to accurately forecast product demand to optimize inventory levels and avoid stockouts or overstock.
- **Solution**: Employing machine learning algorithms to analyze historical sales data, customer behavior, and external factors to predict future demand.
- **Benefits**:
  - Reduced inventory costs
  - Improved customer satisfaction
  - Increased sales revenue

Case Study 3: Transportation Industry

- **Challenge**: Transportation companies need to accurately forecast passenger or freight demand to optimize resource allocation.
- **Solution**: Using machine learning algorithms to analyze historical data, economic indicators, and other factors to predict future demand.
- **Benefits**:
  - Improved operational efficiency
  - Reduced costs
  - Enhanced customer service

By leveraging advanced analytics and machine learning techniques, organizations can gain valuable insights into the performance of their assets and the behavior of their customers, leading to significant cost savings, improved efficiency, and enhanced customer satisfaction.

5.2 *AI-Driven Energy Optimization Models*

5.2.1 Optimization Techniques Using AI in Smart Energy Systems

Artificial intelligence (AI) has emerged as a powerful tool for optimizing various aspects of smart energy systems. By leveraging AI algorithms, we can enhance energy efficiency, reduce costs, and improve the overall sustainability of our energy infrastructure.

Here are some key optimization techniques using AI in smart energy systems:

1. **Predictive Analytics for Energy Demand Forecasting**
   - **Machine Learning Models**: Employing machine learning algorithms like time series analysis, regression, and neural networks to analyze historical data and predict future energy demand.
   - **Benefits**:
     - Optimized energy generation and distribution
     - Reduced peak demand and load shedding
     - Improved grid stability
2. **Real-Time Optimization of Energy Grids**
   - **Reinforcement Learning**: Training AI agents to make real-time decisions about energy flow and distribution in response to changing conditions.
   - **Benefits**:
     - Enhanced grid reliability and resilience
     - Reduced energy losses
     - Improved power quality
3. **Energy Efficiency in Buildings**
   - **AI-Powered Building Automation Systems**: Using AI to optimize heating, ventilation, and air conditioning (HVAC) systems, lighting, and other energy-consuming devices.
   - **Benefits**:
     - Reduced energy consumption
     - Improved indoor comfort
     - Lower operating costs

4. **Integration of Renewable Energy Sources**
   - **AI-Driven Renewable Energy Forecasting**: Predicting the output of renewable energy sources like solar and wind power to optimize grid operations.
   - **Benefits**:
     - Enhanced grid stability
     - Increased renewable energy penetration
     - Reduced reliance on fossil fuels
5. **Electric Vehicle Charging Optimization**
   - **AI-Based Charging Scheduling**: Optimizing the charging schedules of electric vehicles to minimize peak demand and maximize grid efficiency.
   - **Benefits**:
     - Reduced strain on the power grid
     - Improved vehicle charging experience
     - Enhanced grid flexibility
6. **Demand Response and Price Optimization**
   - **AI-Enabled Demand Response Programs**: Encouraging consumers to adjust their energy consumption patterns in response to real-time pricing signals.
   - **Benefits**:
     - Balanced energy supply and demand
     - Reduced peak demand
     - Lower energy costs for consumers

5.2.2 Applications of AI in Reducing Energy Waste and Enhancing Efficiency

Artificial Intelligence (AI) is revolutionizing the energy sector, offering innovative solutions to reduce energy waste and enhance efficiency. Here are some key applications:

1. **Predictive Maintenance**
   - **Real-time monitoring**: AI-powered systems can continuously monitor the health of energy infrastructure, such as power plants, wind turbines, and solar panels.
   - **Early detection of anomalies**: By analyzing sensor data, AI algorithms can identify potential issues before they escalate into major breakdowns.
   - **Optimized maintenance schedules**: Predictive maintenance reduces unplanned downtime and minimizes the need for excessive maintenance.

2. **Energy Demand Forecasting**
    - **Accurate predictions**: AI models can analyze historical data, weather patterns, and other relevant factors to forecast future energy demand.
    - **Optimized energy generation**: Utilities can adjust their power generation to match predicted demand, reducing energy waste.
    - **Improved grid stability**: Accurate forecasting helps balance supply and demand, ensuring a stable and reliable power grid.
3. **Smart Grid Optimization**
    - **Real-time optimization**: AI algorithms can optimize power flow, voltage control, and load balancing in real-time.
    - **Enhanced grid resilience**: AI can help identify and mitigate potential disruptions, such as power outages and cyberattacks.
    - **Integration of renewable energy**: AI can facilitate the integration of renewable energy sources like solar and wind power by optimizing their output and storage.
4. **Energy-Efficient Buildings**
    - **Intelligent building automation**: AI-powered systems can optimize heating, ventilation, and air conditioning (HVAC) systems, lighting, and other energy-consuming devices.
    - **Occupancy-based control**: AI can adjust energy usage based on occupancy levels, reducing energy waste in unoccupied spaces.
    - **Predictive maintenance for building systems**: AI can identify and address potential issues in building systems, such as HVAC and plumbing, before they lead to breakdowns.
5. **Electric Vehicle Charging Optimization**
    - **Smart charging**: AI algorithms can optimize the charging schedules of electric vehicles to minimize peak demand and maximize grid efficiency.
    - **Vehicle-to-grid (V2G) technology**: AI can enable V2G technologies, where electric vehicles can feed energy back into the grid during periods of high demand.
6. **Industrial Energy Efficiency**
    - **Process optimization**: AI can analyze industrial processes to identify inefficiencies and optimize energy consumption.
    - **Predictive maintenance for industrial equipment**: AI can help prevent equipment failures and reduce downtime.
    - **Energy-efficient manufacturing processes**: AI can optimize production schedules, material usage, and energy consumption.

By leveraging AI, we can unlock significant potential for energy savings, reduce carbon emissions, and build a more sustainable future.

5.2.3 Case Studies in Smart Buildings

Case Study 1: The Intelligent Office Building

- **Challenge**: High energy consumption, inefficient space utilization, and lack of occupant comfort.
- **Solution**: Implementing an AI-powered building automation system that optimizes HVAC, lighting, and security systems.
- **Benefits**:
  - Reduced energy consumption by up to 30%
  - Improved indoor air quality and occupant comfort
  - Enhanced security through real-time monitoring and access control
  - Increased productivity and employee satisfaction

Case Study 2: The Sustainable Data Center

- **Challenge**: High energy consumption, heat dissipation, and potential downtime.
- **Solution**: Utilizing AI-driven cooling systems, predictive maintenance, and energy-efficient server configurations.
- **Benefits**:
  - Reduced energy consumption and operational costs
  - Improved server reliability and uptime
  - Enhanced data center sustainability

5.2.4 Case Studies in Industrial Automation

Case Study 1: The Smart Factory

- **Challenge**: Inefficient production processes, quality control issues, and high labor costs.
- **Solution**: Deploying AI-powered robotic automation, predictive maintenance, and quality control systems.
- **Benefits**:
  - Increased production efficiency and output
  - Improved product quality and consistency
  - Reduced labor costs and operational expenses
  - Enhanced workplace safety

Case Study 2: The Autonomous Warehouse

- **Challenge**: Inefficient inventory management, labor-intensive operations, and slow order fulfillment.
- **Solution**: Implementing AI-powered autonomous robots for inventory management, order picking, and material handling.
- **Benefits**:

- Increased warehouse efficiency and productivity
- Reduced labor costs and errors
- Improved order accuracy and faster fulfillment times

By leveraging AI, smart buildings and industrial facilities can achieve significant improvements in energy efficiency, operational efficiency, and overall performance.

5.3 The Future of Digital Optimization in Energy Management

5.3.1 Emerging Trends in Digital Energy Optimization

The energy sector is undergoing a significant digital transformation, driven by a growing emphasis on sustainability, efficiency, and resilience. Several key trends are shaping the future of digital energy optimization:

1. **Artificial Intelligence and Machine Learning**

   - **Predictive Maintenance**: AI algorithms can analyze sensor data to predict equipment failures, allowing for proactive maintenance and reducing downtime.
   - **Demand Forecasting**: Accurate forecasting of energy demand helps utilities optimize generation and distribution, minimizing waste.
   - **Optimized Energy Management**: AI can optimize energy consumption in buildings and industrial facilities by analyzing real-time data and adjusting settings accordingly.

2. **Internet of Things (IoT)**

   - **Real-time Monitoring**: IoT devices enable continuous monitoring of energy systems, providing granular data for analysis.
   - **Remote Control and Automation**: Remotely controlling and automating energy systems can improve efficiency and reduce human error.
   - **Data-Driven Insights**: IoT data can be used to identify inefficiencies and optimize energy usage.

3. **Blockchain Technology**

   - **Secure and Transparent Energy Trading**: Blockchain can facilitate peer-to-peer energy trading, enabling consumers to buy and sell energy directly.
   - **Enhanced Grid Security**: Blockchain can improve the security of energy grids by providing a decentralized and tamper-proof record of transactions.

4. **Digital Twins**

   - **Virtual Representation of Physical Assets**: Digital twins can simulate the behavior of energy systems, allowing for testing and optimization without affecting physical assets.

- **Predictive Analytics**: Digital twins can be used to predict future performance and identify potential issues.

5. **Edge Computing**

- **Real-time Decision Making**: Edge computing enables data processing and analysis closer to the source, reducing latency and improving response times.
- **Enhanced Security**: By processing data locally, edge computing can mitigate security risks associated with data transmission.

6. **Cybersecurity**

- **Protecting Critical Infrastructure**: As energy systems become increasingly digital, cybersecurity becomes paramount.
- **Advanced Threat Detection**: AI-powered cybersecurity solutions can detect and respond to cyber threats in real-time.

By embracing these emerging trends, the energy sector can achieve significant improvements in efficiency, sustainability, and reliability.

5.3.2 Potential Impacts of New Technologies

New technologies are reshaping the energy landscape, offering significant potential for increased efficiency, sustainability, and resilience. Here are some of the potential impacts:

Blockchain Technology

- **Peer-to-Peer Energy Trading**: Blockchain can facilitate direct energy trading between consumers and producers, bypassing traditional intermediaries.
- **Enhanced Grid Security**: By providing a decentralized and transparent ledger, blockchain can improve the security of energy grids.
- **Transparent Energy Origin**: Blockchain can track the origin and journey of energy, ensuring transparency and accountability.

Edge Computing

- **Real-Time Decision Making**: Edge computing enables faster data processing and decision-making at the edge of the network, improving the responsiveness of energy systems.
- **Enhanced Grid Reliability**: By processing data locally, edge computing can help identify and mitigate potential issues before they escalate.
- **Improved Cybersecurity**: Edge computing can reduce the risk of cyberattacks by limiting the amount of sensitive data transmitted over the network.

Artificial Intelligence and Machine Learning

- **Predictive Maintenance**: AI can analyze sensor data to predict equipment failures, enabling proactive maintenance and reducing downtime.

- **Demand Forecasting**: Accurate demand forecasting can help utilities optimize generation and distribution, reducing energy waste.
- **Optimized Energy Management**: AI can optimize energy consumption in buildings and industrial facilities, leading to significant energy savings.

Internet of Things (IoT)

- **Real-time Monitoring**: IoT devices can collect data on energy consumption, production, and grid conditions, enabling real-time monitoring and control.
- **Remote Control and Automation**: IoT devices can be used to remotely control and automate energy systems, improving efficiency and reducing human error.
- **Data-Driven Insights**: By analyzing IoT data, energy providers can identify opportunities for optimization and innovation.

5.3.3 Potential Challenges and Considerations

While these technologies offer significant benefits, there are also challenges to consider:

- **Cybersecurity**: As energy systems become increasingly interconnected, the risk of cyberattacks also increases. Robust cybersecurity measures are essential to protect critical infrastructure.
- **Interoperability**: Ensuring seamless integration between different technologies and systems can be complex.
- **Regulatory Framework**: Developing appropriate regulatory frameworks to support the adoption of new technologies is crucial.

By addressing these challenges and leveraging the potential of these technologies, we can build a more sustainable, efficient, and resilient energy future.

5.3.4 Future Challenges and Opportunities in Sustainable Energy Management

The transition to a sustainable energy future presents both significant challenges and opportunities. As we move towards a low-carbon economy, several key challenges and opportunities are emerging:

Challenges

1. **Grid Integration of Renewable Energy**:
   - **Intermittency**: Renewable energy sources like solar and wind are intermittent, making it challenging to balance supply and demand.
   - **Grid Infrastructure**: Aging grid infrastructure may not be adequately equipped to handle the increased integration of renewable energy sources.

2. **Energy Storage**:
   - **Cost-Effective Storage Solutions**: Developing cost-effective and efficient energy storage technologies is crucial for storing excess renewable energy for later use.
3. **Policy and Regulatory Frameworks**:
   - **Consistent and Supportive Policies**: Implementing supportive policies and regulations can accelerate the transition to a sustainable energy future.
   - **Balancing Economic and Environmental Goals**: Striking a balance between economic growth and environmental protection is essential.
4. **Public Acceptance and Education**:
   - **Addressing Public Concerns**: Addressing public concerns about the potential impacts of renewable energy projects and energy efficiency measures is crucial.
   - **Promoting Energy Literacy**: Educating the public about the benefits of sustainable energy can foster support and engagement.

Opportunities

1. **Innovation and Technological Advancements**:
   - **Emerging Technologies**: Advancements in technologies like AI, machine learning, and IoT can revolutionize energy management.
   - **Energy Efficiency**: Innovative technologies can improve the efficiency of energy consumption in various sectors.
2. **Green Jobs and Economic Growth**:
   - **Job Creation**: The transition to a sustainable energy future can create new jobs in renewable energy, energy efficiency, and related industries.
   - **Economic Growth**: Investing in sustainable energy can stimulate economic growth and innovation.
3. **International Cooperation**:
   - **Global Partnerships**: International cooperation can facilitate the sharing of knowledge, technologies, and best practices.
   - **Climate Change Mitigation**: Collaborative efforts can help address global climate change challenges.
4. **Consumer Empowerment**:
   - **Energy Efficiency Programs**: Empowering consumers to make informed choices about energy consumption can drive demand for sustainable energy solutions.
   - **Community-Based Energy Initiatives**: Community-based initiatives can foster local ownership and engagement in energy projects.

By addressing these challenges and capitalizing on these opportunities, we can build a more sustainable and resilient energy future.

6 Conclusion

Based on the explanation above, we know the importance of digital technology in changing energy management in the maritime sector. By adopting advanced tools like IoT, AI, and machine learning, the industry can enhance energy efficiency, reduce operational costs, and meet environmental standards. This approach supports sustainability goals, minimizes carbon emissions, and ensures a more efficient and eco-friendly future for maritime operations.

References

1. Ahmad A et al (2020) Machine learning-based energy demand forecasting: techniques and applications. Renew Sustain Energy Rev 124:109786. https://doi.org/10.1016/j.rser.2020.109786
2. Ahmed M, Awan A (2020) Wireless communications in maritime systems: applications, challenges, and future trends. Telecommun Policy 44(7):101845. https://doi.org/10.1016/j.telpol.2020.101845
3. Albrecht T, Zander A (2020) Radio frequency (RF) technologies for maritime communication: a study of current practices and future innovations. Marit Technol 35(3):45–60. https://doi.org/10.1016/j.martech.2020.02.004
4. Goodfellow I, Bengio Y, Courville A (2016) Deep learning. MIT Press. https://www.deeplearningbook.org
5. Harish R, Subhramanyan EE, Madhavan R, Vidyanand S (2010) Theoretical model for evaluation of variable frequency drive for cooling water pumps in seawater based once through condenser cooling water systems. Appl Therm Eng 30(14–15):2051–2057. https://doi.org/10.1016/j.applthermaleng.2010.05.012
6. IMO (International Maritime Organization) (2020) International convention for the control and management of ships' ballast water and sediments (BWM Convention). International Maritime Organization. https://www.imo.org
7. Kharabsheh R, Nasser R (2020) Telecommunication and IoT in maritime communica-tion systems: a review of challenges and future directions. Ocean Eng 206:107–123. https://doi.org/10.1016/j.oceaneng.2020.107123
8. Pimm AJ et al (2015) Energy storage for low-carbon grids: dynamic analysis and optimization. Energy Policy 85:442–456. https://doi.org/10.1016/j.enpol.2015.06.005
9. Shi J et al (2024) Clean energy ship application to promote carbon neutrality in global maritime industry: state governance or international supervision? Ocean Coast Manag. https://doi.org/10.1016/j.ocecoaman.2024.107356
10. Team Amsa (2023) Ship Energy Efficiency Management Plan (SEEMP), Canberra, December 2023
11. ur Rehman A et al (2025) A solar-powered multi-functional portable charging device (SPMFPCD) with internet-of-things (IoT)-based real-time monitoring-An innovative scheme towards energy access and management. Green Technol Sustain 3. https://doi.org/10.1016/j.grets.2024.100134

12. Urban F, Nurdiawati A, Harahap F (2024) Sector coupling for decarbonization and sustainable energy transitions in maritime shipping in Sweden. Energy Res Soc Sci 107. https://doi.org/10.1016/j.erss.2023.103366
13. Winanda A, Prastowo H, Pitana T (2021) Real-time fuel consumption monitoring system integrated with internet of things (IOT), Kapal. Jurnal Ilmu Pengetahuan dan Teknologi Kelautan 18. https://doi.org/10.14710/kapal.v18i2.37180

Chapter 4
Advanced Data Analytics for Maritime Energy Efficiency

Ryan Yudha Adhitya, Noorman Rinanto, Syafiuddin, Fitroh Resmi, Nurvita Arumsari, and Sryang Tera Sarena

Abstract The maritime industry is transforming significantly to enhance energy efficiency and reduce environmental impacts, driven by global sustainability goals and stringent regulations. Advanced data analytics is pivotal in achieving these objectives by enabling data-driven decision-making and operational optimization. This study explores integrating advanced analytics techniques, including statistical methods, machine learning, and predictive modeling, to improve maritime energy efficiency. Key focus areas include fuel consumption optimization, emissions reduction, and route planning. By leveraging real-time data from ship sensors, weather forecasts, and operational records, advanced analytics provides actionable insights

R. Y. Adhitya (✉) · N. Rinanto
Department of Marine Electrical Engineering, Shipbuilding Institute of Polytechnic Surabaya, Surabaya, Indonesia
e-mail: ryanyudhaadhitya@ppns.ac.id

N. Rinanto
e-mail: noorman.rinanto@ppns.ac.id

F. Resmi
Department of Shipbuilding Engineering, Shipbuilding Institute of Polytechnic Surabaya, Surabaya, Indonesia
e-mail: fitroh.resmi@ppns.ac.id

Syafiuddin · N. Arumsari
Department of Marine Engineering, Shipbuilding Institute of Polytechnic Surabaya, Surabaya, Indonesia
e-mail: syafiuddin@ppns.ac.id

N. Arumsari
e-mail: arum@ppns.ac.id

S. T. Sarena
Power Electronics, Machines and Control (PEMC) Research Institute, Department of Electrical and Electronics, University of Nottingham, Nottingham, UK
e-mail: sryang.sarena@nottingham.ac.uk

M. Elsisi et al. (eds.), *Maritime Infrastructure for Energy Management and Emission Reduction Using Digital Transformation*, Studies in Infrastructure and Control,
https://doi.org/10.1007/978-981-96-4438-4_4

for stakeholders, leading to cost savings and enhanced compliance with environmental standards. The findings underscore the potential of data analytics to revolutionize maritime energy management, paving the way for a more sustainable and efficient future in the shipping industry.

Keywords Maritime energy efficiency · Data analytics · Machine Learning (ML) · Predictive modeling · Statistics analysis

1 Identification of Opportunities for Enhancing Energy Efficiency and Reducing Emissions

Enhancing energy efficiency and minimizing carbon dioxide (CO2) emissions in the international shipping industry are primary objectives for the maritime sector. Ship exhaust emissions impact the climate, air quality, health, and atmosphere composition. These emissions result from a combination of different engines. Large diesel engines are used for ship propulsion, while small diesel engines are utilized to generate electricity. Most of the fuel is entirely burned, resulting in the formation of carbon [1, 2]. This carbon dioxide primarily drives climate change, extreme weather, and future global warming due to increased greenhouse gas levels in the atmosphere. Emissions in international waters not only contaminate water resources but also threaten the survival of marine life.

Conversely, ships consume significant amounts of fuel oil, resulting in substantial greenhouse gas emissions that severely impact the environment, climate, and human health [3]. The International Maritime Organization (IMO) estimates that between 2007 and 2012, international shipping contributed around 3.1% of world emissions, or 1016 million tons of CO2 annually. Greenhouse gas (GHG) emissions from shipping are projected to rise by 90–130% compared to 2008 levels by 2050, driven by the expansion of global maritime trade. International naval trade has reached 11 million tonnes and is continuing to grow. Since marine transportation is the most cost-effective and energy-efficient form of transportation and the foundation of global supply chains and international trade, it has expanded by an average of 5% per year over the last few decades. Maritime transport offers the benefits of high transport volumes and low costs per volume. The United Nations Conference on Trade and Development (UNCTAD) reports that shipping accounts for over three-quarters of global trade volume [1, 2].

Since the shipping sector is one of the biggest emitters, there is a lot of pressure on it to lower societal emissions—for instance, rising fuel costs and stricter marine laws. Marine gasoline prices have more than doubled in the past ten years, with an average yearly increase of 16%, putting ship owners under tremendous financial strain [1, 2]. Fuel costs account for over 60% of a ship's operating expenses [4].

To cut expenses and adhere to new rules, the shipping sector must figure out how to make their ships more fuel-efficient. IMO has proposed two measures to address this issue, namely technical measures and operational measures. One technical solution to lower ship fuel usage and save energy on board is the Energy Efficiency Design Index (EEDI), which lowers the primary greenhouse gas emissions (CO2). At MEPC 62, in July 2011, the Ship Energy Efficiency Management Plan (SEEMP) and Energy Efficiency Operational Indicators (EEOI) were approved as operational measures and made mandatory. Efficient operation leads to a reduction of greenhouse gas (GHG) and a reduction of operating costs. Improving ship energy efficiency can effectively lower energy consumption and CO2 emissions. Ship energy efficiency can be optimized by adjusting ship speed, optimizing sailing routes, implementing economic fleet navigation schedules, and refining ship design (including dimensions, cargo layout, propulsion systems, engine specifications, propeller design, lightweight materials, and hull/steel structure). Additional strategies for improving energy efficiency include regular hull maintenance, optimizing cargo weight management, and fine-tuning trim, draft, and displacement. Other measures involve enhancing hull performance, scheduling drydocking effectively, upgrading propulsion systems, selecting efficient power systems and engines, and transitioning to alternative fuel types. Additional options include installing scrubbers, retrieving waste heat, utilizing renewable energy sources such as solar or wind power, and incorporating shore power. For instance, optimizing a ship's trim can improve fuel efficiency by 3–7% [3]. Optimization of cruise ship routes and power can minimize ship fuel consumption by about 3.1% [5]. Enhancing the propeller design, optimizing ship size, and refining the hull shape can help minimize ship resistance. Efficient fleet routing [6] and scheduling [7] can reduce fuel consumption and boost profits for cargo ships.

Ships can improve energy efficiency by reducing fuel consumption and CO2 emissions by about 4% by taking into account a combination of route, speed, and environmental factors, including wind speed and direction, wave height and direction, weather, currents, water and temperature of air, air pressure, and water level. For ship design, it only applies to new ships and requires significant investments. Optimizing operating modes such as route planning, speed adjustments, and trim optimization is an efficient method for existing vessels to conserve energy and lower emissions without modifying their structure [8]. Constantly adjusting engine settings to maintain a specific speed is both energy-inefficient and impractical for sailors. Over 90% of open-ocean navigation depends on autopilot systems, which automatically adjust the ship's course to follow a predetermined route [9]. As a result, engine settings are usually adjusted manually only once per day to modify the ship's speed [1, 2].

Ship fuel consumption prediction models can be divided into three categories based on modeling methods: data-based, simulation-based, and physical-based. From an ocean engineering perspective, physics-based models provide a more intricate representation of ship fuel consumption and air emissions. These models typically encompass the simulation of ship mechanical systems, propulsion systems, and thermodynamic processes. The physics-based ship fuel consumption model applies empirical formulas to estimate the ship's resistance, determines the necessary thrust by balancing opposing forces, and ultimately integrates the thrust with the specific fuel oil consumption (SFOC) to calculate the required fuel amount [1, 2]. A new interdisciplinary method that combines computer science and hydrodynamics to determine ship fluid resistance under varied sailing conditions is computational fluid dynamics (CFD), used in simulation-based ship fuel consumption modeling [3]. This model performs well when the ship is cruising in calm waters. However, in practice, ships hardly ever cruise in peaceful waters. As a result of inaccurate navigation, this approach could result in forecast performance that is not correct. The time-consuming nature of simulation trials, which typically take hours or days, makes it challenging to meet real-time prediction and optimization needs. To avoid complicated physical modeling and time-consuming simulations, data-driven models can be the best in modeling. Machine learning and regression techniques based on ship data are gaining traction in estimating fuel and emission issues. Machine learning has been employed as a powerful tool for classification, pattern recognition, and prediction in ship energy efficiency management. Numerous researchers have utilized artificial neural networks (ANNs) within deep learning to model ship fuel consumption data, leveraging their robust non-linear fitting capabilities to achieve high accuracy. In ensemble learning and ANN, support vector regression (SVR) was the next best predictor behind Extra Tree Regression (ETR) and Random Forest Regression (RFR). The estimation of a ship's Fuel Oil Consumption (FOC) target value is performed using the Least Absolute Shrinkage and Selection Operator (LASSO) regression model. In a case study using low-frequency data from a fleet of container ships, this model was found to perform best when compared to ANN, SVR, and Gaussian Processes (GP) models [1, 2].

Model fusion may effectively overcome the shortcomings of a single model and generate exceptionally precise and dependable forecasts by combining the benefits of multiple distinct models. Wind speed prediction [1, 2] and projecting the fuel consumption of cargo ships [3] are two examples of early research that show hybrid models, or model fusion, are more accurate and dependable than single models.

Ship operations' sustainability and profitability depend heavily on accurate fuel consumption forecasting, which is also critical for improving ship energy efficiency and lowering pollutants and greenhouse gas emissions. Minor inaccuracies in fuel calculations can result in operating costs that deviate significantly from projections, potentially leading to reduced expected revenue or increased wear and tear on equipment. For instance, a shipowner's primary expense in a chartering operation is fuel. Considering a 360-day annual run with a daily consumption of 60 tons and a fuel cost of 500 USD per ton, a 5% error in fuel calculation can quickly add up to 540,000 USD annually, or roughly 1,500 USD per day increase in charter rate.

To improve ship energy efficiency, we must be able to accurately predict and analyze ship performance during operation and during the design of new ships. Feature selection can decrease needless model complexity and increase accuracy [1, 2]. A popular practice before creating machine learning models is feature selection, which involves using correlation analysis to find and exclude features that have a high degree of association with one another. The main goal of existing models for predicting ship fuel usage and consumption is to provide more accurate estimates of fuel consumption rates under various conditions, including different trim settings, speeds, sea and weather conditions, as well as hull and propeller degradation.

This chapter begins with the identification of potentials for marine operational data analysis. The following section will elucidate the specific data. The study will employ statistical techniques to identify opportunities for improving energy efficiency and reducing emissions. Furthermore, the following subchapters will explore applying advanced data analytics techniques like machine learning and predictive modeling. The final subchapter will serve as the conclusion.

2 Maritime Operational Data

The shipping industry is responsible for about 75% of global trade. Recently, concerns about green practices in maritime transport have increased, prompting the proposal and implementation of various emission control regulations to reduce pollution and greenhouse gas (GHG) emissions from ships. These regulations are designed to steer the industry towards a low-carbon and low-pollution future by encouraging optimizing fuel consumption and minimizing energy use. Simultaneously, the global economic downturn and high ship fuel prices (bunker) have heightened the urgency of the shipping industry to enhance operational efficiency while continuing to meet international trade demands. Bunker fuel consumption Heavy fuel oil (HFO) and liquefied natural gas (LNG are the primary contributors to emissions and constitute a significant portion of operational costs. Consequently, shipping companies are striving to optimize energy efficiency [1, 2].

Fuel consumption modeling is critical in ship energy efficiency simulations, optimization, and carbon emission predictions. Existing predictive models for fuel consumption include the White Box Model (WBM) and the Black Box Model (BBM). The primary differences between these models are: (1) WBM operates more comprehensively, whereas BBM is more difficult to interpret; (2) the relationship between fuel consumption parameters in WBM is embedded within existing equations, while in BBM, it depends on the availability of data; and (3) WBM requires fewer data, whereas BBM demands a substantial amount [10]. Model testing requires adequate technical and operational data, as demonstrated by [11], who used two years of historical data collected via onboard sensors. Given the importance of this modeling and considering that fuel consumption monitoring serves as a key source of such data, a reliable monitoring system is essential.

Fuel consumption modeling can be categorized into three types: the White Box Model (WBM), rooted in mechanism analysis; the Black Box Model (BBM), dependent on data analysis; and the Grey Box Model (GBM), which integrates aspects of both approaches.

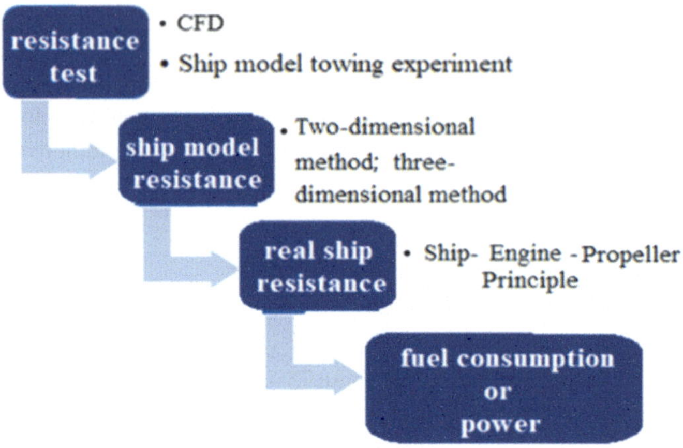

Fig. 1 White Box Model (WBM)

2.1 White Box Model (WBM)

The White Box Model (WBM) is based on the mechanism analysis of fuel consumption, encompassing the ship's static and dynamic characteristics. This Analysis can be further categorized into floating state and hydrodynamic assessments. Fuel usage data can be derived from ship specifications and sea trial results. Since the principles underlying the WBM are well understood, fuel consumption can be predicted quickly.

While sailing, a ship is influenced by numerous factors, and applying physical equations within the WBM framework is highly complex. The accuracy of predictions can be significantly reduced under extreme environmental conditions.

Fuel consumption in WBM is modeled using the ship-engine-propeller principle and ship resistance laws. The first step involves calculating ship resistance, which is crucial for ensuring WBM accuracy. The subsequent step is to account for the numerous influencing factors. However, despite these approaches, the increased workload on the engine makes the WBM more challenging to comprehend (Fig. 1).

2.2 Black Box Model (BBM)

The Black Box Model (BBM) can be categorized as a statistical and machine learning approach. The illustration of these models can be seen in Fig. 2. Various statistical and machine learning methods can be used to estimate ship fuel usage. The distinctions between BBM and WBM are as follows:

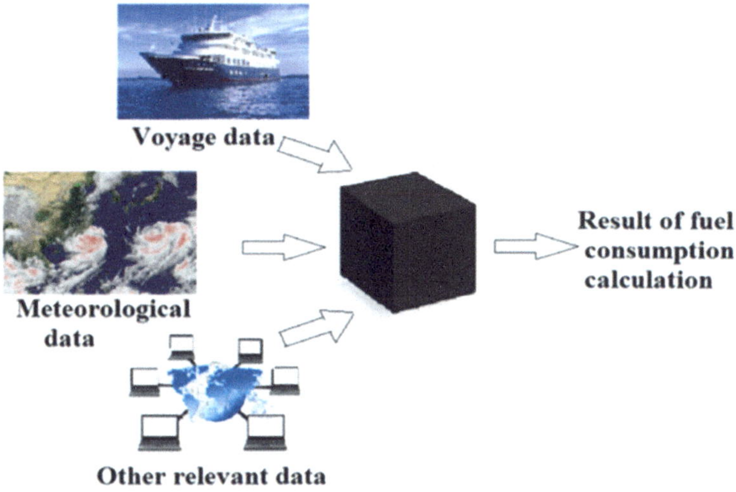

Fig. 2 Black Box Model (BBM)

1. WBM operates more comprehensively, whereas BBM is more difficult to interpret.
2. In WBM, the relationships between fuel consumption parameters are defined within established equations, whereas in BBM, these relationships depend on data availability.
3. WBM requires relatively little data, while BBM demands substantial data for accurate modeling.

2.3 Grey Box Model (GBM)

The prediction of fuel consumption using the Grey Box Model (GBM) is based on fluid dynamics combined with parameter adjustments from available actual data. Theoretically, GBM integrates the features of both WBM and BBM, offering better results than either method alone. By gathering data like ship speed, trim, wind, and wave conditions and integrating relevant mechanisms, the connection between energy-saving strategies, emission reduction, and fuel consumption reduction can be validated and optimized.

GBM can achieve similar outcomes to BBM but requires nearly half the data. Compared to BBM, GBM is more stable and demands less data. Typical data sources for GBM include navigation data, Noon reports, data logging systems, and Automatic Identification System (AIS) records (Fig. 3).

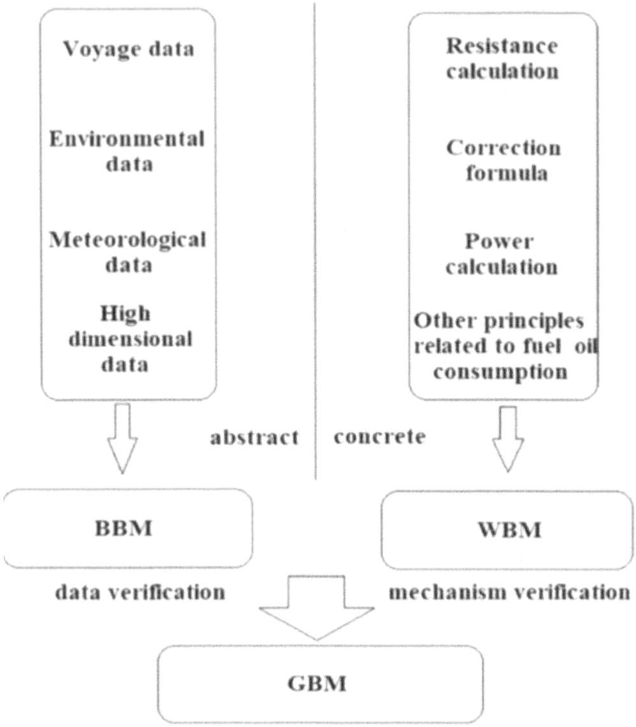

Fig. 3 Grey Box Model (GBM)

3 Data Analytics Technique Using Statistics

Analyzing fuel use on vessels is crucial for optimizing marine operations. Due to elevated fuel expenses and a rising requirement for energy efficiency, comprehending fuel use has been a focus for numerous transportation organizations. This study aids budget planning and enhances environmental initiatives by minimizing carbon emissions through improved fuel management. In this context, statistical techniques like linear and non-linear regression provide a means to discern and forecast fuel consumption [12] trends based on diverse operational aspects of ships [13].

Linear regression analysis is a fundamental statistical method [14] extensively employed to examine the relationship between variables [15]. The primary variable forecast is fuel consumption, influenced by ship speed, cargo load, weather conditions, and sea currents. Linear regression posits that the connection between these variables adheres to a linear pattern. This model is constructed by establishing an equation in which fuel consumption relies on one or more independent variables, each exerting a particular and direct impact on fuel consumption.

It is possible to use a conventional linear regression model when it is thought that a single variable, such as ship speed, influences fuel consumption. According to this theory, the amount of fuel consumed will be immediately and proportionally affected by each increase in speed. As a result of the fact that several factors influence fuel consumption in practice, multiple linear regression is employed more frequently because of this. It is clear from this model that fuel consumption is affected by several different elements, such as speed, the cargo's weight, and the sea's conditions. These aspects contribute to the model's complexity [16] while producing more comprehensive results [17].

On the other hand, the association between these components does not always follow a linear pattern in many different situations. It is possible that an increase in ship speed would initially result in a linear increase in fuel consumption. Still, after the ship reaches a certain velocity, the fuel consumption increases significantly, surpassing the predicted levels. As an illustration, this is an example of a situation in which non-linear regression is better appropriate for application. The correlation between gasoline consumption and other variables is not linear but a complex curve or equation in nonlinear regression. This is because the relationship between the two variables is not linear. Because of this, a more adaptive modeling is made possible, which precisely reflects the actual patterns of fuel consumption based on the data that has been seen.

When the relationship is curvilinear or incremental, non-linear models, such as polynomial or exponential regression, are employed. The model may demonstrate, by exponential regression [18], which fuel consumption increases exponentially at elevated speeds [19]. In scenarios where maximum consumption is constrained by engine capacity or efficiency, logistic or logarithmic models may be utilized. These models enhance the understanding of fuel use, enabling more precise estimations tailored to specific conditions.

Non-linear regression models necessitate more intricate optimization processes compared to linear regression. Non-linear regression typically necessitates iterations and optimal parameter search methodologies. These methods can be executed by sophisticated algorithms, such as the Newton–Raphson technique, or alternative strategies to ascertain parameter values that reduce prediction errors. Consequently, non-linear regression typically necessitates increased time and processing resources, yet it has the benefit of accurately forecasting fuel consumption dynamics that are more realistic.

Valuable insights can be derived by analyzing ship fuel consumption using linear and non-linear regression techniques. Linear regression is favored for its simplicity and ease of interpretation when the data pattern is relatively straightforward. The non-linear regression method is appropriate for intricate scenarios where the relationship between variables does not adhere to a linear pattern. The primary distinction between these two systems is in their versatility. Linear regression is more constrained; however, it is simpler to implement. Non-linear regression, conversely, offers more flexibility and is more appropriate for data displaying diverse consumption patterns.

Examining fuel consumption data through these two regression models can aid ships in adjusting their routes and speeds to ensure fuel efficiency. Transport firms can benefit from predictive models based on linear or non-linear regression.

3.1 Linear Regression in Fuel Consumption Analysis

Linear regression is a statistical technique designed to represent the relationship [20] between a dependent variable (e.g., fuel usage) and one or more independent variables (e.g., ship speed, cargo load, and weather conditions) [21]. Linear regression can be employed to develop straightforward and interpretable predictive models for fuel consumption, facilitating the identification of the primary elements that linearly affect fuel usage. The Simple Linear Regression model posits a single independent variable exerting a linear influence on fuel usage. For instance:

$$Y = \beta_0 + \beta_1 X + \epsilon,$$

In this context, Y represents fuel consumption, X denotes ship speed, β_0 signifies the intercept, and β_1 indicates the regression coefficient. The analytical procedures are delineated as follows:

- Variable Selection: Identifying pertinent independent factors that may influence fuel usage.
- Model Fitting: Employing the least squares method to determine the regression coefficients.
- Statistical Testing: Employing t and F statistics to evaluate the significance of the model and regression coefficients.
- Model Validation: Verifying regression assumptions, including linearity, homoscedasticity, and independence of residuals.

The linear regression method is simple to apply and offers high interpretability. Despite this, it cannot capture nonlinear correlations between variables, which is a significant limitation.

3.2 Non-linear Regression in Fuel Consumption Analysis

Non-linear regression is employed when the association between the dependent and independent variables is non-linear [22]. In examining ship fuel usage, consumption patterns are frequently intricate and necessitate models capable of fitting curves or exponentials [23]. The Polynomial Model employs elevated powers of the independent variable, such as:

$$Y = \beta_0 + \beta_1 X + \beta_2 X^2 + \epsilon.$$

The subsequent steps for analysis are as follows:

- Variable Transformation: Certain non-linear regression models necessitate variable transformation to facilitate their management.
- Parameter Estimation: Unlike linear regression, non-linear regression typically necessitates iterative techniques, such as the Newton–Raphson method or alternative optimization strategies, to ascertain optimal parameters.
- Model Fit Assessment: Alongside conventional statistical tests, residual analysis is essential for evaluating model fit.

The non-linear regression method is more adaptable when capturing intricate data patterns. This model, on the other hand, is difficult to interpret and calls for very complicated computations.

3.3 Case Study

A dataset comprising information on distance (nm), average wind speed (knot), vessel speed (knot), and fuel usage (Litre) is presented in Table 1.

Linear and non-linear regression analyses can be employed to ascertain the influence of each variable on fuel consumption and to forecast fuel requirements under certain situations. The subsequent results pertain to a Matlab pseudo-code that contrasts simple linear regression modeling with third-order polynomial regression:

Table 1 Fuel consumption dataset

ID	Date	Distance (NM)	Average wind speed (knot)	Vessel speed (knot)	Fuel consumption (liter)
1	19/02/2014	49.64	11	12.93	1610
2	27/02/2014	47.9	7	13	1550.512405
3	10/03/2014	103.7	12	13	3358.013012
4	19/03/2014	73.5	11	12.99	2376.886762
5	25/03/2014	49.2	13	8.5	4864.215772
…	…	…	…	…	…
46	27/04/2014	50.6	0	13	42

```
clear all; clc;
data = xlsread('dataset.xlsx','Sheet1','C2:F47');
dist = data(:,1);
avgwind = data(:,2);
vspeed = data(:,3);
fuel = data(:,4);
X1 = dist;
X2 = avgwind;
X3 = vspeed;
Y = fuel;
x = [X1 X2 X3];
y = Y;
b = polyfitn(x,y,3);
polyn2sympoly(b)
```

```
yp1 = 0.0043564.*X1.^3 - 0.042156.*X1.^2.*X2 + 0.23269.*X1.^2.*X3 -
3.6102.*X1.^2 - 0.18706.*X1.*X2.^2 + 2.7113.*X1.*X2.*X3 - 21.2411.*X1.*X2 +
19.8015.*X1.*X3.^2 - 478.7762.*X1.*X3 + 2865.7861.*X1 + 2.0908.*X2.^3 +
8.795.*X2.^2.*X3 - 164.6255.*X2.^2 + 64.2498.*X2.*X3.^2 - 1716.2114.*X2.*X3 +
11283.8959.*X2 + 90.3893.*X3.^3 - 4564.7024.*X3.^2 + 68723.6701.*X3 -
317756.4546
mdl = fitlm(x,y)
yp2 = 9.9974.*X1 + 137.46.*X2 - 412.45.*X3 + 5377.9; %linear regression
plot(y,'linewidth',2);
hold on;
plot(yp1,'-or');
hold on;
plot(yp2,'-.c','linewidth',2);
grid on;
rmse1 = sqrt(immse(y,yp1));
r1 = rsquare(y,yp1);
rmse2 = sqrt(immse(y,yp2));
r2 = rsquare(y,yp2);
title('Fuel Consumption Estimator');
legend(['actual'],['polynomial regression with RMSE: ' num2str(rmse1) ' and R^2: '
num2str(r1)],['linear regression with RMSE: ' num2str(rmse2) ' and R^2: ' num2str(r2)],'loca-
tion','northeast');
xlabel('number of data');
ylabel('Fuel Consumption (Liter)');
```

3.4 Comparison of Linear and Non-linear Regression in Fuel Consumption Analysis

Figure 4 presents a comparative graph of actual data versus expected data derived from two statistical methodologies. The picture indicates that the third-order polynomial regression achieved the lowest RMSE value, with an R-squared value of 0.939, as an R-squared value nearing 1 indicates the real data.

Linear regression is representative and suitable for straightforward and linear patterns, whereas non-linear regression is more successful for more complicated patterns. The characteristics of the data and the requirements of the analysis both play a role in determining which model to use. Non-linear regression can capture differences in patterns that cannot be reached by linear regression, although linear regression is easier to apply and faster in processing. Linear regression, on the other hand, is more straightforward.

4 Data Analytics Technique Using Machine Learning (ML)

The reliability value pattern, derived by component failure times, will be calibrated to match the data distribution, after which machine learning (ML) will be employed for approximation. Research conducted by [24] has shown that the ML approach is more efficient and reliable for estimating continuous data. In machine learning, data is essential for training before output generation. The training results are assessed upon generating output using either identical data or data distinct from

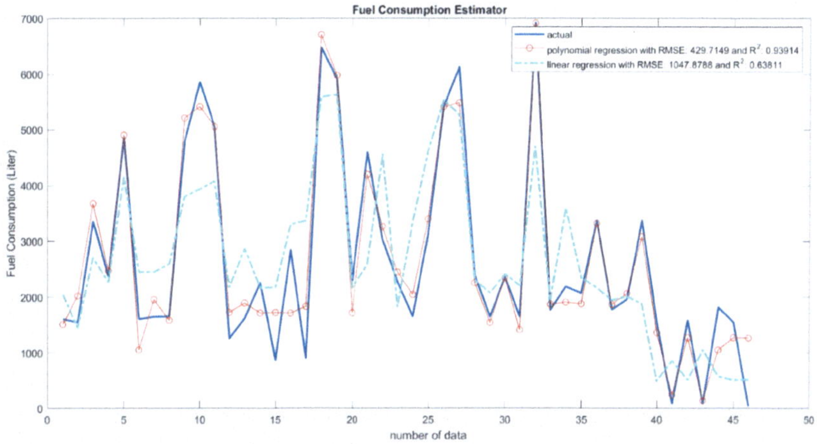

Fig. 4 Fuel consumption predictor using Matlab file (script)

the training set. Categorization and prediction constitute the fundamental aspects of our comprehension of machine learning.

Researchers across several research interests employ a multitude of forecasting methodologies. The conventional time series method is frequently used to deliver precise predictions concerning a component's lifespan. Nonetheless, categorizing failure kinds constitutes a significant weakness of the methodology, hindering the assessment of existing assumptions [25]. Machine learning, a subject of computer science under the domain of artificial intelligence (AI), empowers computer systems to perform tasks autonomously without explicit programming. The learning process involves data observation or pattern recognition to facilitate improved decision-making.

The algorithms employed in machine learning have existed for a considerable duration. They can independently do intricate mathematical calculations on vast datasets while concurrently enhancing their speed. Machine learning approaches are often categorized into three types: supervised, unsupervised, and semi-supervised [26]. Trends in fuel usage can be predicted using supervised learning, which considers many operational characteristics of vessels. Supervised learning encompasses many classification techniques frequently employed in data modeling. The methods encompass support vector regression (SVR) and random forest regression, both appropriate for modeling data related to fuel use.

Support Vector Regression (SVR) is the regression counterpart to the Support Vector Machine (SVM), primarily recognized as a classification model. In regression, SVR seeks to identify a function that most accurately represents the connection between independent features (inputs) and continuous dependent variables (outputs). SVR reduces error, ensures tolerance within defined limits (epsilon), and identifies the most suitable hyperplane in high-dimensional space [27] to forecast the output value with minimal error [28]. The primary parameters in SVR consist of:

1. Parameters of the SVR kernel designed for managing non-linear datasets. The kernel facilitates the transformation of initially non-linear data into a linear format by projecting it into higher dimensions.
2. Epsilon (ε) serves as a parameter that defines the margin within which no penalty is applied for the errors predicted by the model. Any predictions that fall within epsilon of the actual value are regarded as accurate.
3. The Regularisation Parameter (C) balances the fit to the training data with the complexity of the model. An increased C value indicates that the model is more effective at preventing misclassification, whereas a decreased value results in a broader margin.

Random Forest Regressor (RFR) is a method that combines multiple models to tackle regression tasks, providing predictions on numerical scales. The technique operates by integrating multiple decision trees to create a forest. Every tree is built using a random sample of data, and each division is evaluated based on a random selection of features. Every subset is utilized to train an individual decision tree. The random selection of data and features enhances the model's robustness while minimizing variance. The ultimate prediction is typically derived from the mean of all tree predictions in regression tasks. This collection enhances precision [29] and mitigates overfitting [30]. In contrast to traditional decision trees that determine the optimal split feature from the complete set of features at each node, RF opts for the best feature from a randomly selected subset of k features. The parameter k influences the degree of randomness: when k equals d, the process mirrors that of conventional decision trees, and the split feature is chosen randomly. This ability to regulate the degree of randomness enables the individual to customize the model according to their particular requirements [31].

Regarding predicting fuel consumption, Support Vector Regression (SVR) and Random Forest Regression (RFR) each present distinct advantages. SVR demonstrates remarkable proficiency in managing high-dimensional data and exhibits robust generalization capabilities. Simultaneously, RFR is recognized for its ability to handle extensive and intricate datasets that may contain numerous interrelated features. In fuel consumption prediction, it is anticipated that RFR will yield superior trends due to its capability to handle complex features and deliver precise predictions despite the model's tendency to be less interpretable than SVR [32, 33]. In the realm of inquiry, in a study, RFR hybridized with conventional techniques demonstrates a significant improvement in prediction accuracy, effectively adapting to variations in data concerning fuel consumption [34].

In practical scenarios, examining fuel consumption data through these two regressor methods can assist vessels in optimizing their routes and speeds to achieve fuel savings. Predictive models created using SVR or RFR enable shipping companies.

4.1 Support Vector Regression (SVR) in Fuel Consumption Analysis

1. **Definition and Uses**

The Support Vector Machine (SVM) regression version, more commonly referred to as a classification model, is called Support Vector Regression (SVR). SVR looks for the function that best describes the relationship between continuous independent variables (inputs) and dependent variables (outputs) in regression tasks. SVR finds the best hyperplane in high-dimensional space to forecast the output value with the least error. It also allows for an error tolerance within predetermined bounds (epsilon).

2. **Models**

The general formula of Support Vector Regression (SVR) is (Stalhane 2012) [10] is as follows $f(X) = W^T\varphi(X) + b,$, where w is the weight vector, φ(X) is the mapping function from x to the dimension indicating the error. Through the quadratic function and Lagrange function, the following SVR last model is obtained $f(X) = \sum_{i=1}^{N}(\beta_i^* - \beta_i)K(X_i, X) + b$ with limitations $\sum_{i=1}^{N}(\beta_i^* - \beta_i) = 0; 0 \leq \beta_i \leq C$ and $K(X_i, X)$ is a kernel function.

3. **Analysis Steps** [6]

   a. Dataset collection. Select the independent variable and dependent variable for the analysis.
   b. Data pre-processing consists of:

      - Data Cleaning
      - Data Normalization or Standardization to ensure that the model is not biased towards features on a larger scale.
      - Data Splitting: Split the data into a training set and a testing set to train the model and test the accuracy of its predictions (in this case, use 80% training and 20% testing).

   c. The selection of appropriate models and parameters (linear, polynomial, RBF, sigmoid, etc.) depends on the type and distribution of data and the value of C parameters and epsilon (error tolerance). In this case, the RBF kernel is used, parameter value C is 100, and epsilon 0.1.
   d. Model evaluation.
      With a test set to assess its performance. To assess the accuracy of model predictions, use evaluation metrics such as Mean Squared Error (MSE), Mean Absolute Error (MAE), or R-squared.
   e. Model tuning.
      Through adjusting model parameters or selecting a different kernel. Perform hyperparameter optimization with grid or random search techniques to find the best parameter combination.
   f. Model Implementation: applying the trained and tuned model to a production system or using it to make predictions based on new data.

4. **Advantages and disadvantages**
   SVR is excellent at handling high-dimensional data and has strong generalization capabilities. Still, it is more challenging to interpret directly than linear regression models because it requires a clear understanding of the model and high interpretability.

4.2 Random Forest Regressor (RFR) in Fuel Consumption Analysis

1. **Definition and Uses**

Random Forest Regressor is an ensemble-based machine learning algorithm used for regression. This algorithm consists of a collection of decision trees that are trained independently, and the regression prediction results from each tree are averaged to produce a final prediction [35]. This method was proposed by Breiman [36] and developed by Liaw and Wiener [37], adding a layer of randomness. Operating as a classic machine learning algorithm, RF is used for classification and regression. It utilizes a bagging algorithm to collect data from the original data, which is then followed by individual training for each combined group using a decision tree model.

2. **Models**

RF selects the best feature from a randomly chosen subset of k features. The parameter k determines the level of randomness: if $k = d$, the process is identical to that of traditional decision trees and if the split feature is selected randomly. This control over the level of randomness empowers the user to fine-tune the model to their specific needs. Typically, the recommended value of k is $k = log_2 d$.

3. **Analysis Steps**

   a. Dataset collection. Select the independent variable and target variable for the analysis.
   b. Data pre-processing consists of:
      - Data Cleaning
      - Data Splitting: Split the data into a training set and a testing set to train the model and test the accuracy of its predictions (in this case, use 80% training and 20% testing).
   c. Determine model parameters namely n_estimators, max_depth, min_samples_split, max_features, random_state.
   d. Initialize the model according to the parameters set using training data.
   e. Use models to predict new data or test data.
   f. Perform model evaluation by calculating evaluation metrics to measure model performance.
   g. Hyperparameter tuning for parameter optimization can be done using a grid or random search.

4. **Advantages and Disadvantages**

RFR is known for its ability to manage large and complex datasets with many possibly interrelated features. RFR will likely provide better trends because it manages complex features and produces accurate predictions, even though the model is often less interpretable and more complex computing than SVR. RFR can improve prediction accuracy significantly, showing good adaptation to variations in data related to fuel consumption.

4.3 Case Study in Fuel Consumption Analysis

Suppose there is a dataset containing information about Fuel Consumption (liter) (Target Variable), Distance (nm), wind speed (knots), and vessel speed (knots), which is explained above. By using SVR and RFR, analysis can be done to predict fuel consumption under certain conditions. All data was normalized first before analysis with SVR because of the difference in scale. Data analysis was done using Python.

Support Vector Regression (SVR) Pseudocode

```
START
Import pandas library as pd
Import the CSV file "N_Fuel Consumption_Data.csv" into a DataFrame called
'data'
Print the DataFrame 'data'
 Import train_test_split from sklearn.model_selection
Import SVR from sklearn.svm
Import StandardScaler from sklearn.preprocessing
Import mean_squared_error and r2_score from sklearn.metrics
Import matplotlib.pyplot as plt
Import numpy as np

Select features 'distance (nm)', 'wind_speed (knots)', and 'vessel_speed (knots)'
into X
Select target 'fuel_consumption (liter)' into y

Split data into training and testing sets with 80% training and 20% testing, set
random state as 42

Initialize a StandardScaler called 'scaler'
Fit and transform the training data features using 'scaler' and store in
X_train_scaled
Transform the testing data features using the pre-fitted 'scaler' and store in
X_test_scaled

Initialize SVR with RBF kernel, C=100, gamma=0.1, and epsilon=0.1
Train the SVR model with X_train_scaled and y_train

Predict fuel consumption on X_test_scaled and store in y_pred
Calculate mean squared error, and R squared between y_test and y_pred, print
these metrics

Plot setup: Create a subplot for training data comparison
Plot training data predictions vs. actual data, label appropriately, and draw a per-
fect prediction line
Set title, labels, legend, and grid for training plot
```

```
Plot setup: Create a subplot for testing data comparison
Plot testing data predictions vs. actual data, label appropriately and draw a perfect
prediction line
Set title, labels, legend, and grid for testing plot

Adjust the layout and display the plot.
END
```

Figure 5 shows the MSE and R-square values from training data, which are 0.0016 and 0.971 using SVM, respectively. Meanwhile, the MSE and R-square values from testing data are 0.0015 and 0.969. The petite MSE figure shows minimal error between the actual Fuel Consumption data and the predicted Fuel Consumption. This indicates that the resulting SVR model effectively interprets Fuel Consumption data. The orange dots (predicted values from testing data) show a very close distribution to the green dots (actual values). This indicates that the SVR model makes predictions very accurately. The SVR model has succeeded in modeling the relationship between the selected input variables, namely distance, wind speeds, vessel speeds, and fuel consumption, with a very high level of accuracy, namely 0.971. This makes the SVR model an effective tool for fuel consumption prediction based on given operational conditions.

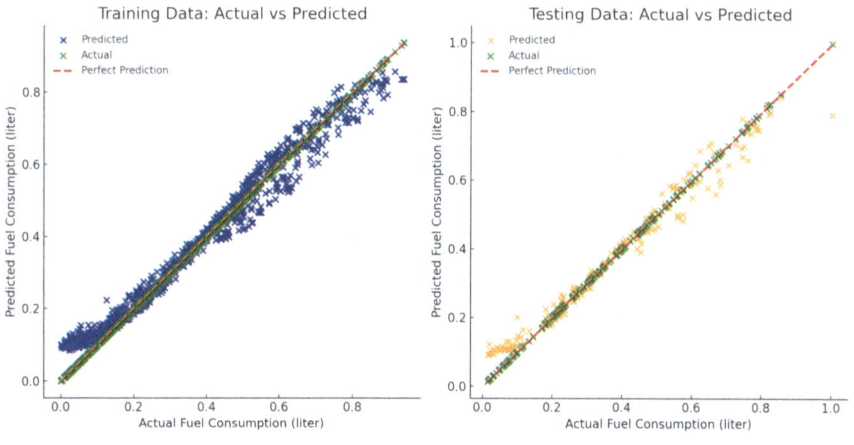

Fig. 5 SVR training and predicted result

Random Forest Regressor (RFR) Pseudocode

```
START
Import the pandas library as pd
Load data from "Fuel Consumption_Data.csv" into DataFrame 'data'
Display the contents of 'data'

Import train_test_split from sklearn.model_selection
Import RandomForestRegressor from sklearn.ensemble
Import mean_squared_error and r2_score from sklearn.metrics
Import matplotlib.pyplot as plt

Define features as all columns except 'fuel_consumption (liter)'
Define target variable 'fuel_consumption (liter)'

Split data into training and testing sets with 20% of the data as test data, set random state to 42
Create a RandomForestRegressor with 100 trees and a random state set to 42
Fit the RandomForest model using the training data

Use the trained model to predict fuel consumption for the test data and store these in y_pred
Calculate the mean squared error and R squared value between y_test and y_pred
Print the mean squared error and R squared value

Create a figure for plotting, set figure size to 14x6

Setup the first subplot for training data
Scatter plot of predicted vs. actual training data
Add a line representing perfect prediction from minimum to maximum of y_train
Set title, labels, legend, and layout for the training data plot

Setup the second subplot for testing data
Scatter plot of predicted vs. actual testing data
Add a line representing perfect prediction from minimum to maximum of y_test
Set title, labels, legend, and layout for the testing data plot

Adjust the overall plot layout and display the plot

END
```

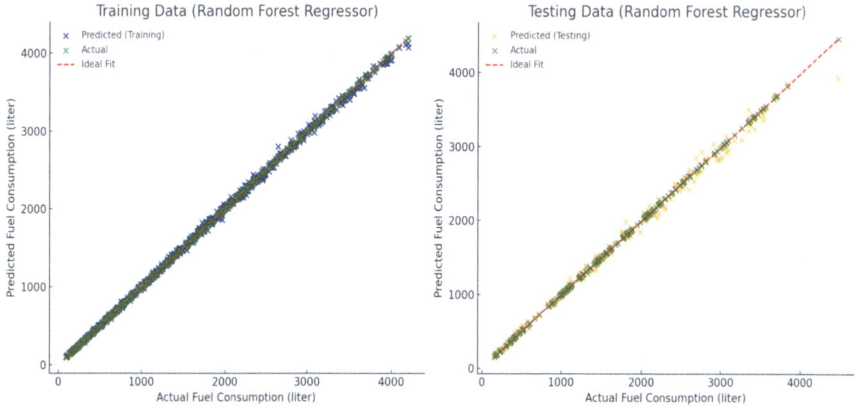

Fig. 6 RFR training and predicted result

Figure 6 illustrates the Mean Squared Error (MSE) and R-square values from training data, which are 851.461 and 0.999 from the RFR method, respectively. Meanwhile, the MSE and R-square values from testing data are 5442.51 and 0.994. A low MSE value signifies slight discrepancies between the actual and estimated Fuel Consumption, suggesting that the Random Forest Regressor (RFR) model proficiently represents Fuel Consumption data. From the testing data plot, aligning the orange dots (predicted values) with the purple dots (actual values) illustrates the accuracy of the model's predictions. The RFR model demonstrates a strong correlation between the chosen input variables and fuel consumption, achieving an impressive accuracy of 0.994. This indicates that the RFR method is dependable for estimating fuel consumption in specific operational scenarios.

4.4 Comparison Between SVR and RFR

Analysis of the data using the SVR and RFR methods indicates that the RFR method yields superior predictions compared to SVR when the data is not normalized beforehand. The R-square value of RFR, which exceeds that of SVR, effectively illustrates the comparison of values. Figure 7 illustrates a comparison between the two methods through visualization. The blue dots represent the anticipated fuel consumption using SVR, whereas the orange dots illustrate the predictions made with RFR. The orange dots align more closely with the straight line (actual value) than the blue dots.

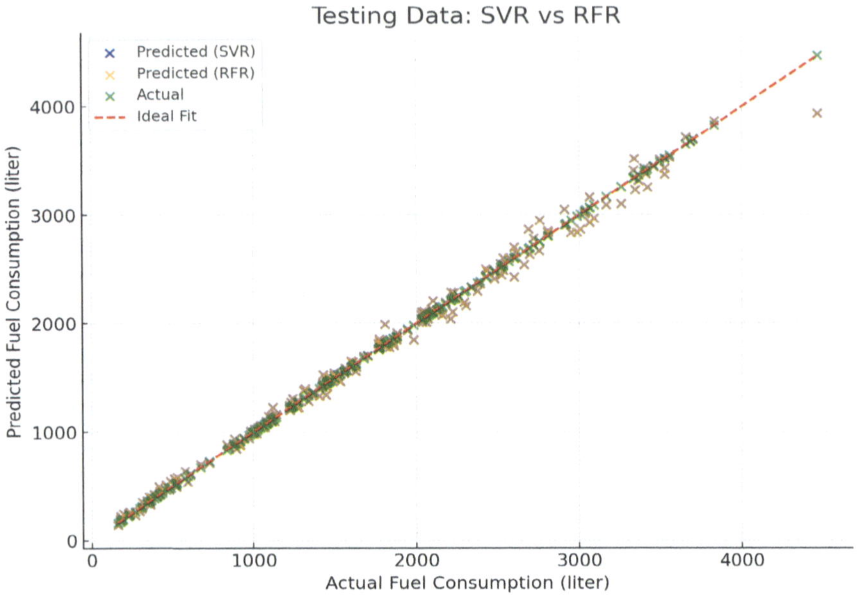

Fig. 7 Comparison of SVR and RFR predicted result

5 Conclusion

The international shipping industry focuses on improving energy efficiency and reducing CO2 emissions to mitigate environmental impacts. With high fuel costs and stricter regulations, the industry has adopted energy-efficient measures such as the Energy Efficiency Design Index (EEDI), Ship Energy Efficiency Management Plan (SEEMP), and Energy Efficiency Operational Indicators (EEOI) to lower fuel consumption and CO2 emissions. Fuel consumption prediction models are divided into physics-based, simulation-based, and data-driven. Data-driven approaches, leveraging machine learning techniques like artificial neural networks (ANNs) and regression models, offer accurate and efficient predictions. Hybrid models combining multiple methodologies are particularly effective.

The shipping industry faces increased sustainability concerns due to emission control regulations and the need for cost-effective operations while maintaining global trade demands. Bunker fuel, primarily from heavy fuel oil (HFO) and liquefied natural gas (LNG), is a significant source of emissions and operational costs. Key models for fuel consumption modeling include the White Box Model (WBM) and Black Box Model (BBM). The Grey Box Model (GBM) combines the strengths of both models, offering improved performance while using less data.

Understanding fuel consumption is crucial for budget planning and minimizing carbon emissions. Statistical techniques, such as linear and non-linear regression, are commonly used to predict fuel consumption trends based on operational factors. Linear regression works well for more straightforward relationships but has limitations in capturing complex, non-linear relationships.

SVR and RFR are powerful machine-learning techniques for fuel consumption prediction in the shipping industry. The choice between these models depends on the complexity of the data, interpretability needs, and computational resources available.

References

1. Yan R, Wang S, Psaraftis H (2021) Data analytics for fuel consumption management in maritime transportation: status and perspectives. Transp Res Part E 155
2. Yan R, Wang S, Psaraftis HN (2021) Data analytic for fuel consumption management in maritime transportation: status and perspectives. Transp Res Part E, 155:102489
3. Hu Z, Zhou T, Zhen R, Jin Y, Li X, Osman MT (2022) A two-step strategy for fuel consumption prediction and optimization of ocean-going ships. Ocean Eng 249
4. Golias MM, Saharidis GK, Boile M, Theofanis S, Ierapetritou MG (2009) The berth allocation problem: optimizing vessel arrival time. Marit Econ Logist
5. Shao W, Zhou P (2012) Development of a 3D dynamic programming method for weather routing. Int J Mar Navig Saf Sea Transp 6(1):79–85
6. Agarwal R, Ergun O (2008) Ship scheduling and network design for cargo routing in liner shipping. Transp Sci 42(2):175–196
7. Stålhane M, Andersson H, Christiansen M (2015) A branch-and-price method for a ship routing and scheduling problem with cargo coupling and synchronization constraints. EURO J Transp Logist 4(4):421–443
8. Armstrong V (2013) Vessel optimization for low-carbon shipping. https://www.sciencedirect.com/science/article/pii/S0029801813002643. Accessed 9 Aug 2013
9. ABB (2020) Shipping 4.0 navigating the 4th industrial revolution. ABB
10. Fan A, Yang J, Yang L, Wu D, Vladimir N (2022) A review of ship fuel consumption models. Ocean Eng 264:112405
11. Coraddu A, Oneto L, Baldi F, Anguita D (2017) Vessels fuel consumption forecast and trim optimization: a data analytic perspective. Ocean Eng 130:351–370
12. Wang X (2020) Prediction model of inland ship fuel consumption considering influence of navigation status and environmental factors. IEEE Xplore
13. Liu F (2019) Fuel consumption prediction for enroute ships using machine learning approaches. In: Proceedings of IEEE conference on intelligent transportation systems
14. Chen R (2022) The role of regression analysis in predicting ship energy usage. IEEE Access
15. Park H (2020) Fuel efficiency optimization for inland shipping using regression techniques. IEEE Trans Sustain Energy
16. Tan K et al (2022) Analysis of fuel consumption influencing factors in smart ships. IEEE Internet Things J
17. Rao S (2020) A regression-based approach to predicting energy efficiency in shipping. IEEE Trans Ind Appl
18. Zhang C (2021) machine learning and regression models in ship emission control. IEEE Trans Veh Technol
19. Tak N (2021) Factors affecting fuel efficiency in modern marine operations. IEEE Access
20. Liu X (2020) Regression-based analysis of fuel consumption in high-seas cargo vessels. In: IEEE conference on sustainable maritime transport
21. Silva L (2021) Optimization strategies for ship speed and fuel consumption. In: Proceedings of IEEE power energy conference
22. Torres B (2020) Multi-variable regression models in maritime fuel use prediction. IEEE Trans Autom Sci Eng

23. Brown D (2021) Analysis of fuel consumption patterns in international shipping. IEEE Trans Transp Electrif
24. Adhitya EK, Satria R, Subagyo H (2015) Komparasi Metode Machine Learning dan Metode Non Machine Learning untuk Estimasi Usaha Perangkat Lunak. J Softw Eng 1(2):109–113
25. Liang YH (2011) Analyzing and forecasting the reliability for repairable systems using the time series decomposition method. Int J Qual Reliab Manag
26. Fan GF, Peng LL, Hong WC, Sun F (2016) Electric load forecasting by the SVR model with differential empirical mode decomposition and autoregression. Neurocomputing
27. Smola AJ, Schölkopf B (2004) A tutorial on support vector regression. statistics and computing archive. Stat Comput
28. Arumsari N, Pamuji FA, Puspita BD (2021) Application of machine learning method to predict reliability in lubricating oil system components. IOP Conf Ser: Mater Sci Eng 1175(1):012012. https://doi.org/10.1088/1757-899X/1175/1/012012
29. Dudek G (2022) A comprehensive study of random forest for short-term load forecasting. Energies 15(20):7547. https://doi.org/10.3390/en15207547
30. Pradhananga P, ElZomor M (2023) Revamping sustainability efforts post-disaster by adopting circular economy resilience practices. Sustainability 15(22):15870. https://doi.org/10.3390/su152215870
31. Kim D, Lee S, Lee J (2020) An ensemble-based approach to anomaly detection in marine engine sensor streams for efficient condition monitoring and analysis. Sensors 20(24):7285. https://doi.org/10.3390/s20247285
32. Zhou ZH (2021) Machine learning. Springer Nature. https://doi.org/10.1007/978-981-15-1967-3
33. Michałowska K, Riemer-Sørensen S, Sterud C, Hjellset OM (2021) Anomaly detection with unknown anomalies: application to maritime machinery. IFAC-PapersOnLine 54(16):105–111. https://doi.org/10.1016/j.ifacol.2021.10.080
34. Zhao D, Li H, Hou J, Gong P, Zhong Y, He W, Fu Z (2023) A review of the data-driven prediction method of vehicle fuel consumption. Energies 16(14):5258. https://doi.org/10.3390/en16145258
35. Doz D, Cotič M, Felda D (2023) Random forest regression in predicting students' achievements and fuzzy grades. Mathematics 11(19):4129. https://doi.org/10.3390/math11194129
36. Breiman L (2001) Random forests. Mach Learn 45:5–32
37. Liaw A, Wiener M (2002) Classification and regression by randomforest. R News 2(3):18–22. https://www.researchgate.net/publication/228451484

Chapter 5
Digital Twin Applications in Maritime Systems

Syamsiar Kautsar, Aulia Siti Aisjah, Mat Syai'in, and Syamsul Arifin

Abstract Digital Twin (DT) is a technology that creates a digital replica of a physical object or system, enabling virtual monitoring, simulation, and testing. DT technology integrates historical and real-time data to predict future situations, proving advantageous in areas with high-value assets. In the maritime industry, DT plays a crucial role in the development of autonomous vessels, predictive maintenance, energy management, and port operations. DT facilitates the simulation of maritime environmental conditions and the monitoring of vessel health, hence improving operating efficiency and mitigating hazards. The main challenges in DT implementation encompass sufficient information technology infrastructure, data integrity, security, and interoperability. By tackling these difficulties, DT possesses significant potential to transform the marine sector, enhancing data-driven decision-making and promoting more efficient and environmentally sustainable vessel management.

Keywords Digital Twin (DT) · Maritime industry · Predictive maintenance · Energy management · Sustainability

---

S. Kautsar (✉)
Engineering Departement, Politeknik Negeri Jember, Jember, Indonesia
e-mail: syamsiar_kautsar@polije.ac.id

S. Kautsar · A. S. Aisjah · S. Arifin
Department of Engineering Physics, Sepuluh Nopember Institute of Technology, Surabaya, Indonesia
e-mail: auliasa@ep.its.ac.id

S. Arifin
e-mail: syamsul@ep.its.ac.id

M. Syai'in
Department of Marine Electrical Engineering, Shipbuilding Institute of Polytechnic Surabaya, Surabaya, Indonesia
e-mail: matt.syaiin@ppns.ac.id

M. Elsisi et al. (eds.), *Maritime Infrastructure for Energy Management and Emission Reduction Using Digital Transformation*, Studies in Infrastructure and Control,
https://doi.org/10.1007/978-981-96-4438-4_5

1 Digital Twin Framework

According to [16], a digital twin (DT) is defined as a virtual representation or twin of a real-world entity, which can be physical or conceptual. The concept of DT was first introduced in 2002 by Dr. Michael Grieves at the Society of Manufacturing Engineering (SME) conference in Troy, Michigan. Dr. Grieves introduced the concept of a virtual system that encompasses information pertinent to its physical counterpart, highlighting the seamless integration of digital and physical entities [9]. This DT facilitates virtual oversight, simulation, and evaluation. This technology employs real-time and historical data to illustrate past and present conditions and to model anticipated future states. DT links the physical object to its digital duplicate via a data communication channel, guaranteeing information synchronization between them. DT covers all at-tributes, states, and behaviors of the physical item via models and data. This concept enables the simulation of the object's behavior within its operational environment and acts as a virtual companion to the physical object throughout its lifecycle. DT technology offers substantial benefits in predictive maintenance and fault tolerance, crucial for industries that employ high-value assets and operate continuously, as downtime can greatly affect costs and productivity.

The key components of the DT concept consist of:

(a) Physical twin—The physical object in the real world.
(b) Virtual twin—The digital replica of the physical object.
(c) Communication channel—The data channel that links the physical object and its DT to ensure data synchronization between them.

In terms of the physical twin, DT is also integrated with controller and sensor da-ta, as seen in Cyber-Physical Systems (CPS). However, although it encompasses physical items and embedded system architecture, DT possesses a unique idea that diverges from Cyber-Physical Systems (CPS) in both purpose and methodology [14]. CPS denotes a close integration of physical systems and computer processes, wherein physical components and software engage in a complex interaction necessitating real-time responsiveness. CPS emphasizes the direct supervision and regulation of the physical system using computers linked to sensors and actuators. In CPS, embedded devices not only collect data but also provide direct management of the physical system based on computational inputs. In intelligent manufacturing facilities, Cyber-Physical Systems (CPS) are employed to regulate and oversee machinery and robotics inside the production process [4]. The control system incorporates sensors to operate actuators that manage machine performance according to the data collected. Through CPS, factories can function autonomously with limited human involvement. The primary distinction between DT and CPS is that DT functions as a digital representation for comprehensive simulation and modeling, whereas CPS focuses on the integration between physical and digital components for direct control and monitoring of the physical system.

In the context of the DT, the communication channel plays a crucial role in maintaining synchronization between the physical twin (physical object) and the virtual twin (digital replica). An illustration of a communication channel utilized in DT is the IoT network. The IoT establishes the technical framework for physical things to interact with their digital counterparts in real-time and online. IoT sensors facilitate data acquisition from physical objects, which is subsequently communicated to the DT to correctly represent the state and condition of the physical object. Moreover, IoT guarantees ongoing synchronization between physical and digital entities via a communication architecture that facilitates swift data transmission. This synchronization enables the DT to update the status, properties, and behaviors of the physical object in real-time, markedly improving system efficiency and reliability.

Through IoT middleware architecture, DT can retrieve data from several devices interconnected within the network and manipulate physical entities via IoT actuators. Figure 1 depicts an architectural paradigm that underpins the concept of a DT within an IoT framework. This paradigm comprises several layers that oversee different components of DT, including physical devices, edge computing, cloud computing, and other virtualization activities. The foundational layer directly engages with devices and computing resources, whereas the superior layer oversees the lifetime of logical objects (LO), including modeling, self-management, orchestration, and entanglement. Furthermore, there exists a data layer that aggregates, contextualizes, and analyzes data, incorporating a crawling function for the acquisition of external information. The simulation layer facilitates DT simulation and visualization, allowing users to model and delineate DTs using accessible, user-centric tools and APIs. This guarantees adaptability in the deployment and management of DTs across diverse IoT applications and settings.

The potential for DT implementation is considerable and encompasses multiple industrial areas. This technology's primary potential lies in its capacity to improve interoperability among systems and applications across several industry sectors. DTs enable diverse businesses to communicate data and knowledge more effectively, facilitating the reuse of models and applications without intricate data conversion or manual involvement. In the manufacturing sector, DT has the capacity to enhance product lifetime management, encompassing design, production processes, maintenance, and disposal methods. This technology can be utilized for operational process simulation and modeling, facilitating the anticipation of problems prior to their emergence, thus mitigating risks and expenses. In the context of manufacturing and Industry 4.0, (M. [8] explains the function of DT in facilitating the product lifecycle, encompassing design to disposal, and in enhancing production processes via simulation and prediction. The DT offers a more effective, efficient, and coordinated method for attaining operational excellence in production and lifecycle management. In platform-based IoT system monitoring, [2] define the benefits of employing DTs for simulation, prediction, and historical data aggregation in smart factories, and elucidate a DT architecture model for IoT systems that incorporates integration with edge and cloud computing. Performance evaluation data demonstrate that DT can decrease typical execution time by as much as 20% and substantially diminish quality inspection batches for diverse structural components.

The DT concept is applicable in the aerospace sector. NASA utilized DT principles during the Apollo missions to replicate the state of a spacecraft, enabling engineers to

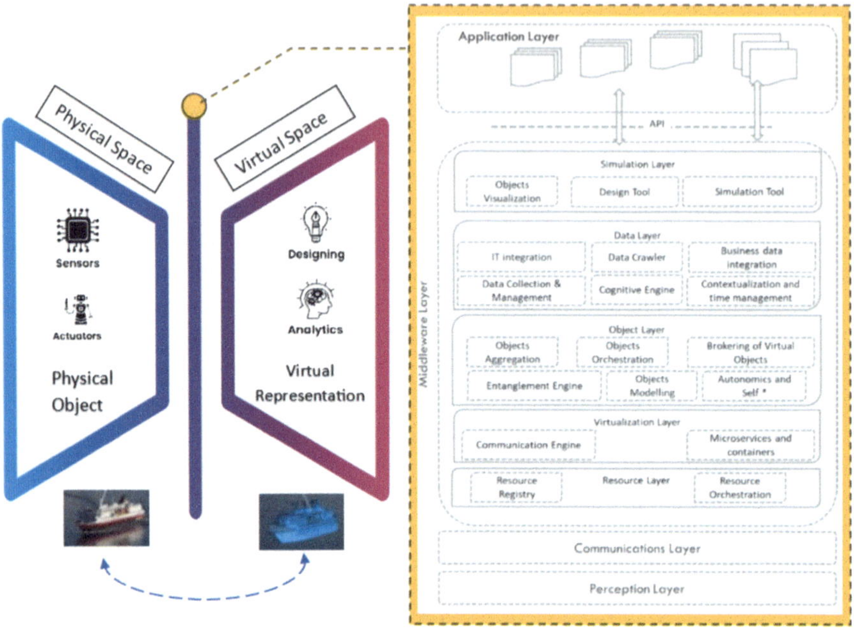

Fig. 1 Digital twin system (*Source* [14])

perform risk management throughout operations [7]. Moreover, DT can be applied in the healthcare sector, facilitating enhanced information interchange among hospitals, laboratories, and other healthcare providers by diminishing reliance on closed and proprietary systems. In a military environment, DT can enhance interoperability among diverse weapon systems and military equipment across multiple service branches [17]. Additionally, in the context of smart cities, DTs can be utilized in city management systems for infrastructure monitoring and urban scenario modeling, with the objective of improving efficiency and sustainability in urban settings [15]. Furthermore, in the context of information technology, DTs can be employed for forecasting product failures and simulating the behavior of actual entities under diverse environmental conditions within a virtual environment [12]. These references offer an extensive understanding of DT application across several sectors and its advantages in propelling enterprises toward Industry 5.0. The primary advantage of DT adoption is its capacity to facilitate automation and operational efficiency, mitigate risks, and expedite data-driven decision-making across diverse sectors.

2 Application of Digital Twin in the Maritime Sector

According to [11], the initial application of DT technology in the maritime industry, focusing on the testing and development of unmanned autonomous ships, began with the ReVolt vessel by DNV GL. ReVolt is a 60-m battery-operated cargo vessel engineered for unmanned operation, wherein DT was integral to the ship's design and testing processes. Evaluation of ReVolt's essential systems was performed utilizing a DT model, facilitating cost and time efficiency. Moreover, DT technology facilitated the modeling of essential trials under many situations with less risk, allowing for safer and more effective plan-ning. The ReVolt DT model employed previous data in the operating phase to forecast component longevity under various environmental variables, such as speed, position, and load. The implementation of autonomous systems and DT is projected to yield savings of up to USD 34 million in operational expenses throughout ReVolt's 30-year lifespan.

DT technology is applicable in training simulators, including Full Mission Simulators, Part Task Simulators, and computer-based training systems. Part Task Simulators frequently employ Hardware in the Loop (HIL) solutions, enabling physical components to function concurrently with the virtual platform. This guarantees the safety of staff (instructors and trainees alike) and facilitates participation in essential tasks without jeopardizing the real equipment. Organizations like as DNV GL employ Hardware-in-the-Loop (HIL) testing for the integration and optimization of novel control systems on DT simulation platforms prior to mass production, thereby mitigating errors during ship construction. In initiatives like as "Nerves of Steel," hydrodynamic data simulations across diverse maritime circumstances facilitate the design of safer and more efficient hulls.

A Digital Twin-based fault diagnosis approach is also applied to Autonomous Surface Vehicles (ASV) by [3]. This approach uses graphical models and the Adaptive Extended Kalman Filter (AEKF) algorithm to detect and estimate actuator faults in vehicles. DT technology incorporates real-time sensor systems that leverage data from global navigation satellite systems (GNSS) to perpetually assess the condition of actuator systems. The case study pertains to the Otter ASV, created by Maritime Robotics, which had a malfunction in one of its propellers. Both simulation and experimental findings indicate that this method can precisely identify and quantify flaws. Furthermore, [3] emphasize the significance of DT in enhancing operating efficiency, decreasing maintenance expenses, and prolonging asset lifespan. The research findings indicate that the DT-based fault diagnosis method utilizing the Adaptive Extended Kalman Filter (AEKF) algorithm effectively identified and assessed the actuator failure on the Otter ASV with considerable precision. In the simulation, the AEKF identified faults in less than one second and precisely monitored variations in fault parameters. Field studies revealed a propeller flaw with a damage magnitude of 12%. Despite the increased difficulty of S-maneuvers for defect detection, the algorithm yielded accurate estimations, indicating that this method can improve ASV reliability and safety. This study demonstrates the application of DT technology for problem diagnosis and fault tolerance control in Autonomous

Surface Vehicles (ASVs), which can markedly enhance the safety and dependability of autonomous vehicles.

Surface vessels comprise intricate components, including communication, navigation, cargo, and propulsion systems, as well as structural frameworks, power generators, auxiliary systems, and additional elements. These components are amalgamated and interlinked. Figure 2 depicts the primary functions of surface vessels during their operational cycle. A malfunction in any component of a ship can impact the vessel's overall performance. DT can contribute to all these functions, including design, production, operation, maintenance, and decommissioning. DT facilitates Condition-Based Predictive Maintenance (CBPM), forecasting the necessity for component maintenance, hence minimizing downtime, prolonging equipment lifespan, and reducing maintenance expenditures. Operators can monitor the ship's health, engine performance, and other important aspects using real-time data synchronized with DT. This ensures the vessel runs under optimal circumstances and mitigates the chance of abrupt breakdowns that could be hazardous. The operational data gathered and processed within the DT offers critical insights for operators and shipowners to enhance decision-making for asset management and operations. DT can assist ship management firms in adhering to environmental requirements and minimizing their carbon footprint by monitoring and optimizing fuel efficiency and energy consumption trends. In the high-risk, continuous-operation marine sector, DT enhances safety, optimizes costs, and fulfills environmental standards, making it a crucial element in the modernization of ships and ports.

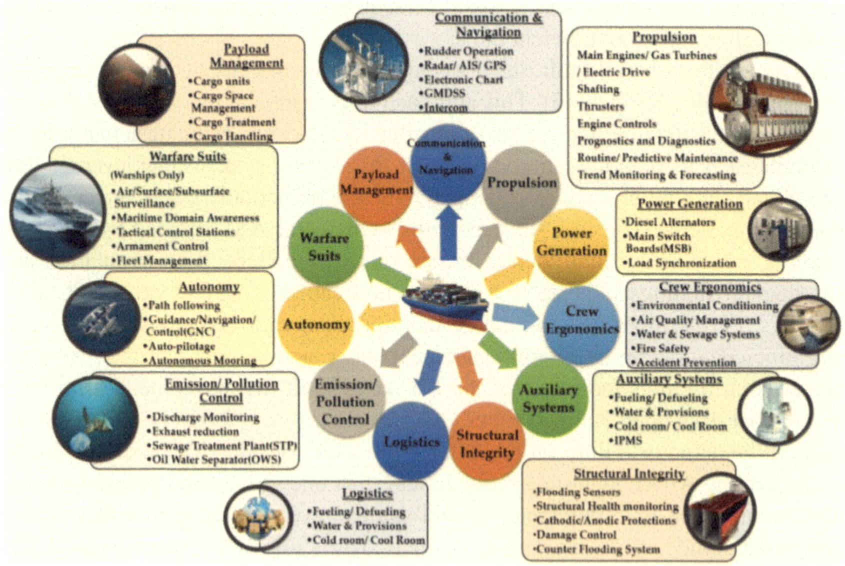

Fig. 2 General functions on ships (*Source* [11])

In the maritime context, weather substantially affects ship performance, as environmental factors such as elevated waves, intense winds, and powerful currents augment the resistance encountered by vessels during navigation. This additional resistance compels the ship's engine to exert greater effort to sustain speed, hence augmenting fuel consumption. In the study by [13] this additional resistance reduces operational efficiency, escalates carbon emissions, and inflates ship operating expenses. Oceanic waves and wind are the primary meteorological factors influencing a vessel's fuel use. When a ship encounters strong waves, the force imparted by the waves might impede the vessel's speed or compromise its stability, necessitating increased energy to sustain the same velocity. Intense winds can increase resistance against the ship, necessitating greater power expenditure to surmount this impediment.

In addition to fuel consumption, inclement weather can also impact the vessel's safety. Intense waves and powerful winds elevate the likelihood of marine incidents, including collisions or vessel capsizing. Unpredictable weather may postpone voyages or necessitate docking if conditions are hazardous [1]. In maritime weather forecasting, DT can accurately predict marine and atmospheric conditions, hence improving forecasting capabilities in maritime areas. An instance of digital technology application in maritime weather forecasting is the creation of the Indonesian Weather Information for Shipping (INA-WIS) system by Indonesia's Meteorology, Climatology, and Geophysics Agency (BMKG). This technology offers interactive marine weather data that can assist fishermen and other maritime transport users in mitigating the danger of sea accidents due to sudden weather fluctuations.

3 Digital Twin for for Marine Operational Optimization

The application of DT technology in maritime logistics management in the Industry 5.0 era, driven by advancements in IoT and smart sensors, is discussed by [18]. This technology enables digitalization and information management to enhance operational efficiency, safety, and sustainability in the maritime logistics industry. According to [18] DT technology plays a critical role in:

(a) Real-time monitoring and simulation of physical assets.

DT technology enables real-time monitoring of physical assets via precise digital mapping and simulation. This enables shipping businesses and manufacturers to optimize asset management, mitigate risks, and enhance operational performance by forecasting changing conditions. An application on ships involves simulating the marine environment and vessel construction to forecast the ship's response to forthcoming waves. This mitigates accident risk and enhances ship fuel efficiency.

(b) Optimization of smart port operations.

Constructing DT-based smart ports can enhance operational efficiency and diminish logistics expenses. Smart ports can offer feedback to port managers for operational modifications through enhanced and real-time data integration. This is especially beneficial in overseeing the intricate port supply chain.

The technology facilitates thorough port position identification and intelligent interactions among facilities within the port.

(c) Improving Ship Efficiency by Predictive Maintenance and Simulation of Operational Conditions.

The DT system facilitates simulations that assist in forecasting maintenance requirements, hence markedly improving ship operational efficiency. Real-time DT-based monitoring of vessel conditions facilitates the early detection of probable failures, hence averting more significant difficulties and minimizing unexpected repair duration. Vessels equipped with this technology can autonomously modify speed and navigation in response to simulated environmental conditions.

(d) Mitigating carbon emissions throughout the maritime supply chain.

DT plays a role in reducing carbon emissions in the supply chain by enabling real-time monitoring and simulation of operations. Through DT, companies can visualize the entire supply chain process and identify areas with high emission potential, such as transportation or energy use in facilities. DT enables scenario simulations to optimize delivery routes, fuel consumption, and energy usage. Thus, DT can help companies make more environmentally friendly decisions, improve operational efficiency, and minimize overall carbon emissions throughout the supply chain.

It can be stated that the implementation of DT technology in maritime logistics systems enhances operational efficiency and environmental sustainability, enabling data-driven decision-making. However, challenges remain concerning the widespread integration of this technology within the maritime industry, particularly regarding costs and the complexity of implementation.

4 Digital Twins for Marine Energy and Environmental Efficiency

Monitoring ship emissions in ports is essential for several key reasons. First, ship emissions, such as CO2, NOx, SOx, and other particulates, significantly contribute to air pollution and global warming. CO2 Carbon dioxide is a significant contribution to the greenhouse effect, resulting in climate change. Consequently, emission monitoring is essential for diminishing the carbon footprint of the maritime sector. Furthermore, regulations established by the International Maritime Organization (IMO), including the greenhouse gas (GHG) reduction strategy, aim to decrease ship carbon intensity by 40% by 2030. These regulations promote cooperation between ports and the maritime industry to reduce greenhouse gas emissions. Efficient monitoring is crucial for maintaining adherence to stringent regulations in Emission Control Areas (ECAs), where vessels are required to comply with designated emission requirements. This surveillance may prevent infractions by vessels that do not comply with these rules. Moreover, emission monitoring facilitates the implementation of alternative technologies and low-carbon fuels that the maritime sector advocates to reduce environmental effect. This technology is anticipated to enable ships to function more sustainably, thus ensuring adherence to emission reduction objectives. These factors highlight the necessity of monitoring ship emissions at ports to facilitate environmental policies, guarantee adherence to international standards, and mitigate adverse effects on the environment and public health in port regions.

An IoT-based drone system equipped with deep learning technology to address ship emission issues at ports is proposed by [5]. This technology employs drones to monitor maritime pollution in real time within port regions. The drones are outfitted with sensors to identify emissions, including carbon dioxide (CO2), generated by vessels. The emission data gathered by the drones is analyzed using a deep learning model founded on Convolutional Neural Networks (CNN) augmented with Continuous Wavelet Transform (CWT) to precisely identify and assess ship emissions. The system is engineered to endure cyber assaults, including False Data Injection (FDI) attacks and adversarial attempts that may distort emission data. The technology utilizes a deep learning algorithm to identify and categorize both authentic emission data and data altered by cyber assaults. The integration of IoT and deep learning technologies facilitates enhanced and secure emission monitoring, hence efficiently aiding emission control decisions at ports. The device can also notify the port control center about the status of identified emission data, indicating its validity or the possibility of a cyber attack.

The drone and deep learning-based IoT system presented in this research possesses significant potential for advancement into DT technology. This system for monitoring ship emissions at ports includes essential components for DT development, including real-time emission data collection via drones, data processing and analysis utilizing deep learning models, and direct response capabilities through notifications to the port control center. To transform this system into a comprehensive DT, it necessitates the incorporation of additional features, including continuous synchronization between physical and virtual assets, operational scenario simulation for the virtual testing of emission reduction strategies, and the optimization of vessel performance regarding fuel efficiency and environmentally sustainable navigation routes. Through advanced data synchronization, simulation, and optimization, this system could transform into a DT that aids ports in improving operating efficiency and markedly diminishing environmental impact.

The development of DT and IoT-based maritime transportation systems to enhance security and digitalization of maritime transportation is also discussed by [10]. This study aims to investigate the security efficacy of maritime transportation systems by implementing a DT model integrated with relays in the data transmission pathway and confirming the security and data transmission performance via simulation. This method involves processing historical data from the marine Silk Road to construct a marine transportation DT model. Relay nodes are incorporated into the data transmission pathway to enhance communication security and efficiency. Simulation results demonstrate that interference information is essential for preserving information secrecy. The created approach may extract energy from interference data to augment transmission power, consequently improving communication efficacy and data security. Simulation results indicate that the likelihood of outage (transmission failure) diminishes as the number of relays and the fading index grow. An increased number of relays in the transmission path enhances system performance and reduces the probability of outages. Furthermore, power distribution can be improved to attain a balance that enhances system security performance, particularly when nodes are situated in diverse places. The relay cooperation IoT-based Maritime DT model achieves a data transmission success rate of 100% and maintains transmission delays under 700 ms. The developed model provides an experimental foundation for the future development of smart and secure maritime transportation.

5 Challenges and Future Directions of Digital Twin in the Maritime Sector

In the development of DT technology, several primary challenges arise, as discussed by [6]. The first challenge pertains to IT infrastructure. DT necessitates an infrastructure that facilitates data analytics and IoT for optimal functionality. In the absence of a robust infrastructure, DT cannot realize its complete potential. Moreover, data quality constitutes a significant challenge. Data transformation relies significantly on continuous data flow and superior data quality. Inadequate data quality might hinder DT performance, resulting in inferior decision-making. A notable difficulty is security and privacy. The extensive data employed by DT poses a risk to sensitive information within industrial systems. To address this, it is imperative to use the most recent security and privacy protocols in DT-supporting technologies, including data analytics and the IoT. Trust is a pivotal concern for both users and businesses. Model validation enhances confidence in decision trees by confirming their performance aligns with anticipated criteria. Ultimately, standardization and interdisciplinarity pose significant challenges in the development of digital technologies. The lack of defined DT modeling methodologies leads to variations among projects, potentially obstructing the progress of this technology. Interdisciplinarity poses challenges as it involves multiple research fields with differing objectives, potentially slowing down research outcomes.

To ensure the effective operation of DT, it is essential to establish standards, frameworks, and ecosystems that facilitate interoperability among systems and applications [16]. This includes the capacity to disseminate and employ data, models, and algorithms across systems without the necessity for manual conversion or modification. Consequently, "Making Digital Twin Work" involves dismantling current information silos, fostering an atmosphere conducive to seamless automation and data sharing, and guaranteeing the efficient operation of DT technology across many domains and industries. According to [16], some necessary precautions are required to improve DT efficacy:

(a) First, establishing a consensus on the concept and scope of DT applications is crucial. There is presently ambiguity on the definition of a DT and the contexts in which this technology is most suitably utilized. A consensus would facilitate the establishment of essential norms and structures.

(b) Furthermore, the establishment of standards and interoperability is essential. DT will operate optimally only if systems and applications spanning many domains, sectors, and markets can seamlessly exchange data, models, and algorithms without user intervention or intricate data conversion. This standardization is essential for facilitating workflow automation and reducing human error.

(c) A supportive ecosystem for the integration of DT across sectors is essential. This ecosystem must encompass a glossary, data dictionary, extensible meta-model definitions, accessible and reusable DT definition catalogs, and certification authority to guarantee quality and consistency.

(d) Fourthly, procedures and frameworks must be established to facilitate the extensive development and implementation of DT. These frameworks should enable data and information management in a platform-independent manner, allowing data and models to be utilized across systems and applications without alteration or customization for each new instance.

(e) Fifthly, system automation emerges as a crucial component. Robust interoperability enables automation to minimize human involvement in data processing, component sourcing, or product design validation, hence enhancing efficiency across the product or system lifecycle.

A primary challenge in implementing DT in the maritime sector is that data quality is often inconsistent and not real-time due to the remote locations of ships, far from communication infrastructure. Moreover, dependence on satellite networks for transoceanic data transmission can result in significant latency, indicating that the received data may not consistently be up-to-date. Furthermore, some systems within the maritime sector have been functioning for years on obsolete technology. The Digital Twin must seamlessly interact with current infrastructure without interrupting everyday operations, posing a significant barrier as some legacy systems were not engineered for compatibility with contemporary technologies like IoT and Artificial Intelligence (AI). Consequently, the implementation of DT in the maritime industry necessitates a multidisciplinary strategy, including the utilization of hybrid networks that integrate satellite, LTE, and 5G technologies to guarantee stable, real-time access in remote places. Technologies such as Edge Computing can process data nearer to its origin prior to transmission to the cloud, thereby lowering latency. The application of machine learning methods and adaptive AI enables the DT model to perpetually learn from real-world settings, hence enhancing model accuracy over time. To enable the proper functioning of DT, cross-industry collaboration and the advancement of technology that facilitates integration, automation, and interoperability across many applications and technology domains are essential.

6 Conclusion

This article highlights the transformative role of Digital Twin (DT) technology in the maritime industry. DT creates digital replicas of physical assets, enabling real-time monitoring, simulation, and predictive maintenance. By improving operational efficiency, optimizing energy management, and reducing environmental impacts, DT supports sustainability and regulatory compliance. Despite challenges such as IT infrastructure, data quality, and integration with legacy systems, DT offers significant potential to modernize maritime operations, enhance decision-making, and drive environmental and economic benefits in the industry.

References

1. Aisjah AS, Arifin S (2011) Maritime weather prediction using fuzzy logic in java sea. In: 2011 2nd international conference on instrumentation control and automation, pp 205–208. https://doi.org/10.1109/ICA.2011.6130157
2. Bao J, Guo D, Li J, Zhang J (2019) The modelling and operations for the digital twin in the context of manufacturing. Enterp Inform Syst 13(4):534–556. https://doi.org/10.1080/17517575.2018.1526324
3. Bhagavathi R, Kwame Minde Kufoalor D, Hasan A (2023) Digital twin-driven fault diagnosis for autonomous surface vehicles. IEEE Access. https://doi.org/10.1109/ACCESS.2023.3268711
4. Carreras Guzman NH, Wied M, Kozine I, Lundteigen MA (2020) Conceptualizing the key features of cyber-physical systems in a multi-layered representation for safety and security analysis. Syst Eng 23(2):189–210. https://doi.org/10.1002/sys.21509
5. Elsisi M, Yu J-T, Lai C-C, Su C-L (2024) A drone-assisted deep learning-based IoT system for monitoring ship emissions in ports considering adversarial attacks. IEEE Trans Instrum Meas 73:1–11. https://doi.org/10.1109/TIM.2024.3374306
6. Fuller A, Fan Z, Day C, Barlow C (2020) Digital twin: enabling technologies, challenges and open research. IEEE Access 8:108952–108971. https://doi.org/10.1109/ACCESS.2020.2998358
7. Glaessgen E, Stargel D (2012) The digital twin paradigm for future NASA and US Air Force vehicles. In: 53rd AIAA/ASME/ASCE/AHS/ASC structures, structural dynamics and materials conference 20th AIAA/ASME/AHS adaptive structures conference 14th AIAA, vol 1818
8. Grieves M (2014) Digital twin: manufacturing excellence through virtual factory replication. https://www.researchgate.net/publication/275211047
9. Grieves MW (2023) Digital twins: past, present, and future. In: The digital twin. Springer International Publishing, pp 97–121. https://doi.org/10.1007/978-3-031-21343-4_4
10. Liu J, Li C, Bai J, Luo Y, Lv H, Lv Z (2021) Security in IoT-enabled digital twins of maritime transportation systems. IEEE Trans Intell Transp Syst 24:1–9. https://doi.org/10.1109/TITS.2021.3122566
11. Madusanka NS, Fan Y, Yang S, Xiang X (2023) Digital twin in the maritime domain: a review and emerging trends. J Mar Sci Eng 11(5):1021. https://doi.org/10.3390/jmse11051021
12. Marcus G, Rossi F, Veloso M (2016) Beyond the Turing test. AI Mag 37(1):3–4. https://doi.org/10.1609/aimag.v37i1.2650
13. Megawati S, Aisjah AS, Widjaja S (2023) Prediction of ship fuel consumption due to the effect of weather conditions. In: 2023 international seminar on intelligent technology and its applications (ISITIA), pp 786–791. https://doi.org/10.1109/ISITIA59021.2023.10221158

14. Minerva R, Lee GM, Crespi N (2020) Digital twin in the IoT context: a survey on technical features, scenarios, and architectural models. Proc IEEE 108(10):1785–1824. https://doi.org/10.1109/JPROC.2020.2998530
15. Mohammadi N, Taylor JE (2017) Smart city digital twins. In: 2017 IEEE symposium series on computational intelligence (SSCI), pp 1–5. https://doi.org/10.1109/SSCI.2017.8285439
16. Piroumian V (2023) Making digital twins work. Computer 56(1):42–51. https://doi.org/10.1109/MC.2022.3206101
17. Tokar JL (2017) A comparison of avionics open system architectures. ACM SIGAda Ada Lett 36(2):22–26. https://doi.org/10.1145/3092893.3092897
18. Zhou F, Yu K, Xie W, Lyu J, Zheng Z, Zhou S (2024) Digital twin-enabled smart maritime logistics management in the context of industry 5.0. IEEE Access 12:10920–10931. https://doi.org/10.1109/ACCESS.2024.3354838

Chapter 6
Cybersecurity in Digital Maritime Infrastructure

Rachmad Andri Atmoko, Afif Zuhri Arfianto, Muhammad Khoirul Hasin, Mohammad Basuki Rahmat, and Lukman Arif Kurniawan

Abstract The rapid digitalization of the maritime industry has transformed operational efficiency and global supply chain connectivity but has also introduced significant cybersecurity challenges. This study explores the critical components of cybersecurity within maritime infrastructure, highlighting vulnerabilities in vessels, port systems, logistics networks, and cargo management platforms. It addresses the increasing prevalence of cyber threats such as ransomware, phishing, and GPS spoofing, and their potential impacts on operational safety, financial stability, and data integrity. Key strategies for mitigating these risks include implementing robust network security measures, encrypting sensitive data, developing comprehensive incident response plans, and fostering cybersecurity awareness through employee training. Furthermore, adherence to international compliance standards, such as the IMO's cybersecurity guidelines, underscores the importance of a unified and proactive approach to managing cyber risks in maritime operations. Case studies, such as the NotPetya attack on Maersk, illustrate the real-world implications of cybersecurity breaches and the necessity for resilient frameworks. This research underscores the urgent need for the maritime sector to integrate advanced cybersecurity practices to safeguard critical infrastructure, maintain global trade continuity, and build a secure future for digital maritime operations.

R. A. Atmoko (✉)
Department of Creative Industry and Digital, Faculty of Vocational Studies, Universitas Brawijaya, Malang, Indonesia
e-mail: ra.atmoko@ub.ac.id

A. Z. Arfianto · M. K. Hasin · M. B. Rahmat
Department of Marine Electrical Engineering, Shipbuilding Institute of Polytechnic Surabaya, Surabaya, Indonesia
e-mail: afif@ppns.ac.id

M. K. Hasin
e-mail: khoirul.hasin@ppns.ac.id

M. B. Rahmat
e-mail: mbasuki.rahmat@ppns.ac.id

L. A. Kurniawan
Liteon Singapore, Lorong Chuan, Singapore

M. Elsisi et al. (eds.), *Maritime Infrastructure for Energy Management and Emission Reduction Using Digital Transformation*, Studies in Infrastructure and Control,
https://doi.org/10.1007/978-981-96-4438-4_6

Keywords Maritime cybersecurity · Digital transformation · Cyber threats · Network security · Data encryption · Incident response · Cyber risk management · Cyber resilience · Maritime infrastructure · Digital maritime operations

1 Introduction to Maritime Cybersecurity

The maritime industry has undergone a profound transformation due to digitalization, which has significantly enhanced operational efficiency and connectivity across global supply chains. Technologies such as advanced navigation systems, automated port operations, and integrated logistics platforms have become integral to maritime operations. However, this digital transformation has also introduced a new dimension of risks, particularly in the realm of cybersecurity. As maritime operations increasingly rely on interconnected systems, they become more vulnerable to cyber threats that can disrupt logistics, compromise sensitive data, and endanger physical safety.

This section underscores the critical importance of cybersecurity in maritime operations, focusing on essential domains such as vessels, cargo management systems, port infrastructure, and logistics networks. It provides a detailed exploration of the evolving cyber threat landscape in the maritime sector and emphasizes the need for comprehensive strategies to secure these critical systems.

Cyber security is considered one of the top three risks in Ports alongside piracy and terrorism that use submarine security [1]. Some ports are not fully prepared to counter these risks. Cyber security is urgently needed due to the large and evolving threats. The IMO has stated that there is an 'urgent need to raise awareness of cyber risk threats and vulnerabilities, to support safe and secure shipping, which is operationally resilient to cyber risks' encouraging 'administrations to ensure that cyber risks are appropriately addressed in the safety management system, no later than the first annual verification of the company's Document of Compliance after 1 January 2021' [2].

1.1 The Impact of Digital Transformation in Maritime Operations

Digital transformation has revolutionized the maritime industry by integrating advanced technologies that streamline processes, reduce operational costs, and enhance decision-making. Automated systems for cargo tracking, digital communication networks, and the Internet of Things (IoT) have redefined how ships, ports, and logistics chains operate. For instance, real-time tracking of vessels and cargo enables greater transparency and efficiency in global trade. However, this growing reliance on digital systems also exposes maritime operations to cyber threats, such as ransomware attacks [3, 4], phishing, and unauthorized access to critical systems.

A single cyber-attack can cause significant disruptions, leading to financial losses and reputational damage [5].

1.2 Key Vulnerabilities Within Maritime Infrastructure

The interconnected nature of modern maritime operations makes them susceptible to various cyber vulnerabilities. Critical areas of concern include:

1. **Vessels**: Ships are equipped with sophisticated systems such as Electronic Chart Display and Information Systems (ECDIS) and Automatic Identification Systems (AIS). These systems, while improving navigation and safety, are potential targets for cyber-attacks that can manipulate vessel positions or disable navigation controls.
2. **Cargo Systems**: Digital systems used for container tracking, loading, and unloading processes are vulnerable to attacks that could lead to cargo mismanagement or theft.
3. **Port Infrastructure**: Ports depend on integrated systems for logistics, communication, and security. Cyber-attacks on port operations can result in delays, loss of critical data, and even physical damage to equipment.
4. **Logistics Networks**: Cyber-attacks on logistics platforms can disrupt supply chains, affecting the timely delivery of goods and causing cascading effects across industries reliant on maritime trade.

1.3 The Importance of Risk Assessment in Securing Maritime Assets

Risk assessment is a cornerstone of maritime cybersecurity. Identifying and evaluating potential threats enables stakeholders to develop tailored strategies to mitigate risks effectively. Risk assessments involve analyzing vulnerabilities across vessels, port infrastructure, and logistics networks, as well as understanding the potential impact of cyber threats on operations [6]. A comprehensive risk assessment includes:

- **Threat Identification**: Recognizing the types of cyber threats, such as malware, insider attacks, or denial-of-service (DoS) attacks that could target maritime assets.
- **Impact Analysis**: Assessing the potential consequences of a cyber-attack, including operational downtime, financial losses, and reputational harm.
- **Mitigation Strategies**: Developing robust cybersecurity measures, such as encryption protocols [7], access controls, and regular system audits, to protect against identified threats.

By addressing these vulnerabilities through systematic risk assessment and mitigation, the maritime industry can enhance its resilience to cyber threats and ensure the safety and efficiency of its operations.

In conclusion, as digital transformation continues to reshape the maritime industry, the significance of cybersecurity cannot be overstated. A proactive approach that prioritizes the identification and mitigation of risks is essential to safeguard maritime infrastructure and ensure the continuity of global trade.

2 Cyber Threat Landscape in Maritime Infrastructure

The *Cyber Threat Landscape in Maritime Infrastructure* explores the myriad cyber threats that challenge the safety and continuity of maritime operations, emphasizing the risks these threats pose to both vessels and ports. Maritime infrastructure faces specific vulnerabilities that cybercriminals exploit to disrupt operations, compromise security, and impact logistical and financial stability. Among these threats are Malware and Phishing Attacks, which exploit human and technical weaknesses through infected files and deceptive emails. Such attacks can compromise critical systems, lead to the loss of sensitive information, and severely disrupt operations.

Another significant threat, GPS Spoofing and Jamming, targets the navigational integrity of maritime vessels by manipulating or blocking GPS signals [8]. This manipulation can result in route deviations or even collisions, endangering both the crew and cargo. Additionally, Ransomware poses a unique risk to maritime infrastructure by locking down essential data and systems, demanding a ransom for their release. When ransomware attacks [3, 4] occur, they can paralyze operations at ports and onboard vessels, often resulting in substantial financial losses and extended downtime.

Insider Threats add another layer of complexity, as employees may unintentionally or, in some cases, intentionally expose maritime systems to risk. Factors such as insufficient cybersecurity training or malicious intent can make insider threats particularly challenging to manage. Meanwhile, Supply Chain Attacks highlight the vulnerabilities present in the interconnected logistics networks that underpin maritime operations [6]. When third-party suppliers or partners are compromised, it can create cascading failures across the supply chain, impacting the entire network.

In summary, the cyber threat landscape in maritime infrastructure is vast and varied, with each type of attack exploiting different vulnerabilities and presenting unique challenges. From malware and phishing that prey on human error to sophisticated GPS spoofing and ransomware that target critical systems, these threats underscore the need for comprehensive cybersecurity measures. Addressing insider threats and supply chain vulnerabilities further emphasizes the importance of securing the entire operational ecosystem to ensure resilience against these evolving cyber risks.

In today's digital age, the maritime industry not only faces physical challenges, such as extreme weather or the risk of maritime accidents, but also a range of cyber threats. Among the most common are malware and phishing attacks. These types of attacks exploit both technical and human vulnerabilities, often preying on personnel who may be unaware of cybersecurity risks. This section delves into how malware and phishing can infiltrate maritime infrastructure and discusses the potential impacts of these attacks on operations and data security.

2.1 Understanding Malware

Malware, or malicious software, is designed to disrupt, damage, or gain unauthorized access to computer systems. Malware includes a variety of types such as viruses, worms, trojans, spyware, ransomware, and adware. Each form of malware has its own attack methods, yet they share similar objectives: to weaken security defenses and steal sensitive information. In the maritime sector, malware can spread through connected devices, such as onboard computers or port systems, serving as entry points for broader attacks [9].

2.2 Phishing in the Maritime Context

Phishing is a social engineering technique where attackers trick users into revealing sensitive information like passwords or financial details. In a maritime setting, phishing emails often impersonate official communications from port authorities or logistics partners. Using psychological manipulation, attackers can deceive employees into providing access that enables further breaches. Unlike purely technical threats, phishing relies heavily on human factors, exploiting gaps in awareness and training [10].

2.3 Exploiting Human Vulnerabilities

Maritime employees are often unaware of cyber threats. Phishing emails that mimic professional communications can deceive ship crew or port staff into clicking harmful links or downloading attachments that contain malware [11]. Often, a lack of training on cybersecurity leads employees to overlook signs of suspicious emails or messages. Attackers capitalize on this lack of awareness to introduce malware into larger networks.

2.4 *Exploiting Technical Vulnerabilities*

On the technical side, many maritime systems still rely on outdated software and hardware, which are susceptible to exploitation. Malware can enter through various loopholes, such as unpatched software, weak communication protocols, or physically connected devices. Once malware infiltrates a network, it can spread further, allowing attackers access to steal data, delete critical information, or even take control of systems essential to ship and port operations [12].

2.5 *Impact: The Potential for Compromised Systems, Loss of Sensitive Information, and Operational Disruption*

When malware or phishing successfully penetrates a system, the operational integrity of a vessel or port can be severely compromised. Ransomware attacks, for example, can lock down access to ship control systems or critical logistical data, preventing access and demanding a ransom [13]. Such incidents not only delay operations but also result in significant recovery costs. In many cases, malware is designed to steal sensitive information. In the maritime industry, this data might include details about ship routes, sailing schedules, crew information, or financial data used in logistics processes. If this information is leaked, the consequences could be severe, such as sabotage or theft of goods while at sea. Operational disruption is a direct and severe consequence of malware or phishing attacks. Systems locked or data damaged by malware can delay shipping schedules, disrupt port logistics management, or even cause damage to equipment. In some instances, such disruptions have a domino effect on the global supply chain, especially if major ports or large vessels are impacted by cyberattacks [5].

2.6 *Real-World Examples in the Maritime Context*

2.6.1 Ransomware Attack on Port Systems

One real-world example is the ransomware attack that affected the global shipping company Maersk in 2017. Known as NotPetya, this attack disrupted computer systems across multiple countries, including Maersk-managed ports [14]. As a result, port operations were halted for several days, causing significant financial losses and reputational damage [5] to the company. This case illustrates how malware can have a widespread impact on broader maritime operations.

2.6.2 Phishing Attack on Port Employees

In some cases, phishing attacks have successfully tricked port employees into granting unauthorized access to attackers. For instance, a phishing email posing as a request from port authorities may deceive an employee into disclosing login credentials, which attackers then use to infiltrate control systems [11]. Such incidents highlight how human vulnerabilities represent a significant cybersecurity gap within the maritime environment.

To minimize the risk posed by malware and phishing, several mitigation measures should be implemented within maritime infrastructure:

- **Employee Training and Awareness**
  Raising awareness among employees about cyber threats is crucial [11]. Regular training on identifying phishing emails, following security protocols, and handling data securely can help prevent attacks that rely on human vulnerabilities.
- **System and Software Updates**
  Ensuring that all devices and software within maritime networks are updated can reduce the risk of malware attacks. Security patches released by software vendors close vulnerabilities that attackers might otherwise exploit.
- **Implementing Strict Security Policies**
  Adopting policies such as two-factor authentication, role-based access control, and real-time network monitoring can be effective in preventing unauthorized access. These policies add an extra layer of security, making systems more resilient against cyber threats.

2.6.3 GPS Spoofing and Jamming

GPS spoofing is a method in which attackers broadcast fake GPS signals to deceive a vessel's navigation system, causing it to report incorrect location data. In a spoofing attack, the false signals overpower authentic GPS signals, tricking the system into recalculating its position based on these falsified coordinates [8]. In some cases, the spoofed signals are so convincing that the vessel's navigation systems fail to recognize the discrepancy, effectively leading the vessel off its intended route. GPS spoofing can mislead a ship's automated navigation system or even deceive crew members who rely on digital navigation displays for accurate information.

The sophistication of GPS spoofing can vary. In a basic attack, an intruder may simply attempt to alter a vessel's position, shifting it a short distance off course. However, more advanced spoofing methods can continuously manipulate the vessel's location data, simulating a gradual course deviation or even creating the illusion of staying on the correct path while actually moving in the wrong direction [15]. Such advanced techniques make spoofing attacks difficult to detect in real-time, especially for crew members who may not suspect manipulation until the vessel is far off course. While GPS spoofing manipulates location data, GPS jamming involves emitting powerful radio frequency signals to disrupt or block GPS reception altogether [16]. GPS signals are inherently weak by the time they reach Earth, which makes them vulnerable to interference from stronger signals emitted nearby. In a jamming attack, the attacker deploys a jammer device that emits noise on the same frequencies used by GPS satellites, effectively drowning out the legitimate signals and rendering the GPS receiver unable to calculate its position.

GPS jamming, unlike spoofing, doesn't provide false data; instead, it results in a complete loss of navigational information. The absence of GPS data can create confusion for both automated systems and crew members who rely on consistent GPS updates for safe navigation. Jamming is especially dangerous in busy shipping lanes, congested port areas, or narrow channels where precise positioning is crucial to prevent accidents [15].

2.6.4 Impact of GPS Spoofing and Jamming on Maritime Operations

- **Route Deviation and Delays**

One of the most immediate impacts of GPS spoofing is route deviation, where vessels unintentionally veer off their intended paths due to falsified coordinates [15]. Even small deviations can lead to significant consequences, especially when traveling in regions with restricted or high-traffic areas. For instance, a slight course shift can lead a vessel into territorial waters it didn't intend to enter, potentially causing international disputes or legal issues. In other cases, vessels may stray into environmentally sensitive areas or protected zones, risking legal penalties and environmental damage.

Delays can also arise from GPS jamming, as vessels encountering navigation issues may slow down or stop to recalibrate their positions. When GPS is disrupted, additional time is needed to identify alternative navigation methods, such as reverting to manual plotting or using backup systems like radar. These delays impact not only the vessel in question but also other ships waiting for port access or transit clearances, leading to a cascade of operational disruptions.

- **Risks of Collisions and Groundings**

GPS spoofing and jamming [8, 15] can increase the risk of vessel collisions and groundings. Spoofed GPS data may create a misleading sense of position that leads a vessel too close to another ship, creating the potential for a collision. In extreme cases, spoofing could direct a vessel onto a collision course with another, particularly in congested areas where proximity to other vessels is tightly managed.

Jamming poses a similarly high risk in terms of navigational hazards. When GPS data is unavailable, crew members may struggle to maintain accurate situational awareness, particularly in adverse weather conditions or at night. Without reliable GPS data, even minor miscalculations in position or course can lead to catastrophic outcomes, including collisions with other vessels, underwater obstacles, or unintended grounding on reefs or coastlines. The risks are amplified in narrow channels or busy harbors, where precise navigation is essential to avoid accidents.

- **Compromised Safety and Security**

GPS spoofing and jamming [8, 15] compromise not only navigational safety but also the overall security of maritime assets. GPS manipulation could facilitate hijacking by directing a vessel to a location where attackers lie in wait. Spoofing could also be used to disguise a vessel's actual position, making it vulnerable to piracy or unauthorized boarding. Additionally, because spoofing can obscure a ship's true path, vessels could be lured into areas where monitoring and emergency response capabilities are limited, reducing the likelihood of a quick response in case of an emergency.

GPS jamming can further complicate security operations, especially for vessels transporting valuable or hazardous cargo. In cases where the vessel's GPS is disabled, security teams lose access to real-time location tracking, making it challenging to ensure the vessel's safety and exposing it to potential security breaches [16].

- **Disruptions to the Global Supply Chain**

The maritime industry plays a central role in the global supply chain, with the majority of goods being transported by sea. GPS spoofing and jamming [8, 15] attacks can have a ripple effect throughout the supply chain, delaying cargo shipments and affecting inventory management for companies relying on timely deliveries. These disruptions are particularly damaging for time-sensitive goods, such as perishable items or medical supplies, which can spoil or become obsolete if not delivered on schedule.

A compromised navigational system can also lead to financial losses for shipping companies and their clients due to increased fuel consumption from extended routes, delayed arrival times, and potential legal fees or penalties. Additionally, insurance claims may rise in the event of incidents caused by GPS manipulation, further adding to the financial burden on companies and affecting global trade stability [5].

2.6.5 Mitigating the Risks of GPS Spoofing and Jamming

To address the threats posed by GPS spoofing and jamming [8, 15], several strategies and technologies have been developed:

- **Enhanced Crew Training and Awareness**
  Crew members should be trained to recognize the symptoms of GPS spoofing and jamming [8, 15], such as sudden changes in coordinates or loss of GPS signal [11]. Awareness of these threats allows for quick detection and enables crew members to take alternative navigational measures when needed. Regular drills and simulations can help crews practice responses to GPS disruptions, ensuring preparedness in real-world scenarios.
- **Use of Backup Navigational Systems**
  Integrating redundant navigational systems, such as inertial navigation systems (INS) or radar, helps vessels maintain position awareness even when GPS is compromised [16]. INS, which uses internal sensors to track movement relative to a known starting point, can be particularly useful when GPS is unavailable. Radar

can also provide real-time positional data by mapping nearby objects, giving crew members additional references for safe navigation.

- **Implementing Signal Authentication and Encryption**
  Emerging technologies in signal authentication aim to protect GPS signals by verifying their authenticity before allowing them to influence navigational systems. Encrypted signals are harder to spoof, as they require a key for decryption that attackers do not possess [7]. Although still in the development phase, these technologies hold promise for enhancing GPS security in the future.
- **Regulatory and Industry-Wide Standards**
  The International Maritime Organization (IMO) and other regulatory bodies have started to establish guidelines for cybersecurity, including GPS protection, within the maritime sector [2, 17]. Industry standards and certifications for GPS-reliant technologies help ensure that vessels are equipped with resilient systems capable of withstanding spoofing and jamming attempts.

2.6.6 Ransomware: Paralyzing Maritime Operations by Locking Critical Data and Demanding Ransoms

Ransomware is a type of malicious software that infects a computer system, encrypting critical files and data so they can no longer be accessed. Typically, attackers gain entry to systems through phishing emails, compromised websites, or vulnerabilities in network security [10]. Once inside, ransomware spreads quickly across the network, encrypting files and locking users out. In many cases, the attackers demand a ransom, often in cryptocurrency, promising to provide the decryption key once payment is received. Without this key, companies may be unable to access vital data, leading to a standstill in operations.

The maritime industry, with its reliance on centralized digital systems, is especially vulnerable to ransomware [13]. Attackers often target operational technology (OT) systems that control cargo handling, scheduling, and vessel management, as well as IT networks managing communications and data storage. A successful ransomware attack on any of these systems can halt port activities, delay shipments, and prevent vessels from maintaining safe and efficient routes.

Ransomware attackers exploit vulnerabilities in digital infrastructure through several means. Phishing emails are among the most common, where a deceptive email prompts an unsuspecting employee to click a link or download an attachment that installs malware [10]. In maritime settings, where crew members and port staff frequently handle digital communications, phishing remains a potent method. Additionally, attackers may exploit outdated software or unpatched systems within maritime infrastructure, taking advantage of the slow pace at which some companies update legacy systems. Once ransomware gains a foothold, it spreads within the network, eventually encrypting the data that drives port or vessel operations.

One of the most immediate effects of a ransomware attack is the paralysis of operations. Ports and vessels rely on continuous data flow to coordinate logistics, manage cargo, and ensure the safety of operations. When ransomware encrypts this data, the systems are effectively locked down. For instance, a ransomware attack on a port's cargo management system can prevent the loading and unloading of goods, halting trade and creating significant backlogs [3]. In some cases, vessels may be unable to depart or dock without access to critical navigational or scheduling information.

Downtime in maritime operations can have a ripple effect on the entire supply chain. Ports serve as vital links in global trade, and delays in one port can affect numerous other ports, distribution centers, and retailers relying on timely shipments. A ransomware-induced delay may extend beyond the immediate downtime, as vessels stuck at ports or delayed shipments can throw off schedules for weeks or even months, particularly for ports dealing with high volumes of cargo [18].

The financial impact of ransomware attacks [3, 4] on the maritime sector is staggering. Aside from the initial ransom demand, which can reach millions of dollars, companies face indirect costs such as lost revenue from halted operations, expenses for incident response, and the cost of restoring systems. Additionally, insurance premiums may increase following an attack, and affected companies may face legal fees, regulatory fines, or compensation claims from customers impacted by the delay [5].

In cases where companies choose to pay the ransom, the financial burden only deepens. Not only does this add a direct cost, but paying ransoms may also encourage further attacks. However, choosing not to pay can lead to extended downtime and potential data loss if attackers destroy data as punishment for non-payment. In the maritime sector, where time is money, every hour of downtime can mean thousands, if not millions, in lost revenue [13].

Ransomware also poses safety risks, especially when it affects operational technology (OT) on vessels or in ports. OT systems control physical operations such as machinery, navigation systems, and cargo cranes. When ransomware disrupts OT systems, it can create hazardous situations. For example, if the cargo crane controls in a port are encrypted and rendered inoperable, it can prevent workers from safely loading or unloading cargo, increasing the risk of accidents. Onboard vessels, compromised navigational systems may lead to errors in route planning or position tracking, placing crew and cargo in potential danger, especially in adverse weather conditions or high-traffic areas.

Additionally, ransomware attacks [3, 4] on systems used to monitor security can expose vessels and ports to unauthorized access, increasing the risk of theft, piracy, or sabotage. Ransomware attackers may threaten to release or delete sensitive data if the ransom is not paid, potentially exposing proprietary information or sensitive communications that could compromise the security of maritime assets.

Beyond operational and financial impacts, ransomware attacks [3, 4] can severely damage a maritime company's reputation. Ports and shipping companies are trusted to handle valuable goods and ensure timely delivery. A ransomware attack undermines this trust, as delays, compromised data, and disrupted operations can lead customers to question the reliability and security of the affected company. Reputational damage can result in lost business as clients seek more secure alternatives, impacting a company's long-term revenue.

In today's competitive maritime industry, where efficiency and reliability are paramount, any incident that affects service delivery can have long-lasting effects on customer loyalty. Even after systems are restored, companies may face months or even years of rebuilding trust with clients and stakeholders.

2.6.7 Real-World Examples of Ransomware in Maritime Infrastructure

One prominent example of ransomware's impact on the maritime industry is the 2017 NotPetya attack on Maersk [14, 18], a global leader in container shipping. NotPetya, a ransomware variant, spread through Maersk's global network, disrupting operations across multiple continents [4]. The attack paralyzed terminals, affected cargo loading and unloading, and led to substantial financial losses estimated at over $300 million. Maersk's response involved a complete system overhaul, highlighting the extensive damage ransomware can inflict on critical infrastructure and the time required to restore operations fully.

Another incident in 2018 targeted the Port of San Diego, where a ransomware attack disrupted administrative functions, forcing port officials to adopt manual processes [3, 18]. Although the attack did not impact physical operations, the administrative backlog underscored how even limited ransomware attacks [3, 4] could hinder port efficiency and create delays for customers relying on port services. To defend against ransomware, the maritime sector must adopt a comprehensive cybersecurity strategy that addresses both prevention and response.

- **Employee Training and Awareness**
  Since many ransomware attacks [3, 4] begin with phishing emails, training employees to recognize and report suspicious communications is essential. Regular cybersecurity training, including phishing simulations, can help build a security-conscious culture that reduces the likelihood of successful ransomware attacks [3, 4].
- **Data Backup and Recovery Plans**
  A reliable data backup system is critical for mitigating ransomware impact. By frequently backing up data and storing it offline, maritime companies can ensure access to essential information even if ransomware encrypts primary systems. Recovery plans should also be tested regularly to ensure a swift response if systems are compromised.
- **Network Segmentation and Access Control**
  Segmenting networks limits the spread of ransomware within an organization. By isolating OT systems from IT networks, companies can prevent ransomware from moving laterally across the entire infrastructure. Implementing access controls, such as multi-factor authentication, also helps restrict unauthorized access, making it more challenging for attackers to infiltrate critical systems.
- **Incident Response Planning**
  Having a well-defined incident response plan is crucial in managing ransomware attacks [3, 4]. Maritime organizations should establish a response team, identify

critical assets, and outline specific steps for isolating affected systems, restoring operations, and communicating with stakeholders. Incident response plans help companies react swiftly and reduce the operational and financial impact of ransomware.

2.6.8 Insider Threats: The Role of Human Factors in Cybersecurity Vulnerabilities in Maritime Infrastructure

- **Understanding Insider Threats in the Maritime Sector**

Insider threats are a significant cybersecurity concern in the maritime industry, stemming from individuals within an organization such as employees, contractors, or partners who unintentionally or intentionally compromise security. These threats can be malicious, where individuals act with intent to harm the organization, or unintentional, where errors or negligence expose systems to risk. In the maritime sector, insider threats are particularly impactful due to the extensive access that many employees and partners have to critical operational technology (OT) and information technology (IT) systems used in port and vessel operations [11].

Human factors play a central role in creating vulnerabilities. For example, lack of cybersecurity awareness, poor adherence to security protocols, or simple human error can open the door to cyber risks. A single click on a phishing email or failure to follow security guidelines can result in malware spreading through a network, compromising sensitive information, or disrupting essential operations.

Given the interconnected nature of maritime logistics, insider threats can have far-reaching consequences, affecting safety, operational continuity, and data security. Addressing these risks requires comprehensive training, stringent access controls, and a strong cybersecurity culture to mitigate human error and safeguard maritime infrastructure from internal vulnerabilities

- **Types of Insider Threats: Malicious versus Unintentional**
  Insider threats in the maritime industry can generally be divided into two categories: malicious and unintentional.

  - **Malicious Insiders**: These are individuals who knowingly act against their organization's interests. They may be motivated by personal gain, grievances, or financial incentives from external entities. Malicious insiders can intentionally access, steal, or sabotage data and systems, potentially causing significant operational and financial harm.
  - **Unintentional Insiders**: In contrast, unintentional insiders are those who inadvertently create security vulnerabilities, often due to a lack of awareness or negligence [11]. Common examples include employees who click on phishing emails, misconfigure systems, or mishandle sensitive information. Although their actions are not intentional, the damage caused by unintentional insiders can be just as severe as that of malicious insiders.

Both types of insiders present unique challenges. Malicious insiders are difficult to detect because they often operate under legitimate credentials, while unintentional insiders may unknowingly expose systems to risk due to insufficient training or weak security practices.

- **How Human Factors Create Cybersecurity Vulnerabilities**
  Human factors play a crucial role in creating cybersecurity vulnerabilities, especially within complex environments like maritime infrastructure. Despite technological advancements, human behaviors, attitudes, and knowledge gaps continue to contribute to security risks. The maritime sector is particularly vulnerable due to the following human-related factors:
  - **Lack of Cybersecurity Awareness**: Many maritime employees lack sufficient cybersecurity training, leaving them unaware of common attack methods like phishing, social engineering, and data breaches [11]. This lack of awareness makes employees more susceptible to falling victim to scams or inadvertently exposing sensitive information.
  - **Negligence and Mistakes**: Mistakes such as sending sensitive data to the wrong recipient, failing to secure devices, or disregarding password policies are common across all industries, including maritime. In a highly interconnected environment like a port or vessel, even minor mistakes can lead to significant vulnerabilities, providing attackers with entry points into critical systems.
  - **Access to Sensitive Systems and Data**: In the maritime sector, many employees and contractors have extensive access to operational technology (OT) and information technology (IT) systems. While this access is necessary for day-to-day operations, it also increases the risk of insider threats. Individuals with access to sensitive systems can unintentionally or deliberately misuse their privileges, impacting network security.
- **The Impact of Insider Threats on Maritime Operations**
  - Operational Disruption and Downtime: One of the most immediate impacts of insider threats in maritime operations is the disruption of essential activities. For example, an insider who accidentally installs malware on a system could cause critical equipment or software to malfunction, leading to delays in cargo handling, vessel departure schedules, or docking procedures. Such disruptions are especially costly in the maritime industry, where operational continuity is crucial for maintaining schedules and ensuring efficient trade.
  - Malicious insiders can cause even more severe operational disruptions. An employee with technical knowledge and ill intent could tamper with navigation systems, disable cargo tracking tools, or interfere with scheduling platforms, essentially halting maritime operations. These disruptions create substantial backlogs, particularly at busy ports, where any delay in processing shipments can lead to cascading effects throughout the supply chain [19].

- Financial Losses and Insider threats can lead to substantial financial losses for maritime companies. Operational downtime caused by insider actions—whether intentional or accidental—directly impacts revenue by delaying shipments, increasing costs, and incurring penalties from clients and regulatory bodies. For instance, a cargo ship delayed due to compromised systems may incur additional docking fees, extended fuel consumption, or spoilage of perishable goods.
- Financial losses may also arise from data breaches. Insiders who gain unauthorized access to proprietary data, such as shipping manifests, client records, or trade routes, expose the organization to potential data theft. If sensitive information is leaked, companies may face legal fees, compensation costs, and lost business due to damaged reputations. In some cases, insider incidents lead to long-term financial setbacks, particularly if customers lose trust and turn to competitors with stronger security measures.
- Threats to Safety and Security: The maritime industry is unique in that cybersecurity breaches can translate into physical safety risks. An insider threat targeting OT systems on a vessel could impact navigation, communication, or cargo handling systems, potentially jeopardizing the safety of the crew and cargo. For example, if a malicious insider manipulates navigational data, it could lead to a vessel deviating from its intended course, increasing the risk of collisions or grounding in dangerous areas [16].
- Unintentional insider threats can also pose safety risks. An employee who unwittingly introduces malware into the network may disrupt critical safety systems, such as automated emergency response protocols or fire suppression controls on a ship. Such vulnerabilities can lead to dangerous situations, especially in adverse weather conditions or high-traffic sea routes, where operational integrity and quick decision-making are essential for safety.
- Reputational Damage a cyber incident resulting from an insider threat can significantly damage a company's reputation. Ports, shipping companies, and other maritime stakeholders rely on trust and reliability in their operations. When insider threats expose sensitive data or disrupt services, clients may question the company's commitment to cybersecurity, leading to decreased customer loyalty and lost business opportunities. In an industry where trust is essential, the reputational impact of an insider breach can linger long after the initial incident.

- **Mitigating Insider Threats in Maritime Operations**
  - **Employee Training and Cybersecurity Awareness**
    To reduce the risk of unintentional insider threats, comprehensive cybersecurity training is essential [11]. Maritime organizations should regularly educate employees on common cyber threats, secure data handling practices, and how to recognize and report suspicious behavior. Training programs, including phishing simulations, can reinforce awareness and equip employees with the skills to avoid common security pitfalls.

- **Access Management and Role-Based Permissions**
  Implementing strict access controls helps minimize insider threat risks by limiting access to sensitive systems and data. Role-based permissions ensure that employees have access only to the resources necessary for their jobs, reducing the likelihood of misuse. Additionally, continuous monitoring of access privileges and regular audits can help identify unusual patterns that may indicate an insider threat.
- **Monitoring and Detection Systems**
  Advanced monitoring tools can detect unusual activity in real-time, providing early warning of potential insider threats. Behavior analysis software, for instance, can alert security teams to deviations from an employee's typical usage patterns, such as accessing unusual files or logging in at odd hours. By flagging these activities, monitoring systems help security teams address insider threats before they cause significant harm.
- **Creating a Positive Security Culture**
  A strong, positive security culture encourages employees to report suspicious activity, seek help when needed, and remain vigilant about cybersecurity. Organizations can foster a supportive environment by emphasizing that cybersecurity is a shared responsibility. Clear policies, open communication, and a focus on teamwork can reinforce the importance of security and make employees feel empowered to contribute to the organization's safety.

• **Supply Chain Attacks: Addressing Risks in Interconnected Maritime Logistics**

A supply chain attack occurs when cybercriminals exploit a vulnerability within an organization's network of suppliers, vendors, contractors, or other third-party partners [6]. Rather than directly targeting a well-protected organization, attackers look for weaknesses in its partners, which may have less stringent security protocols. Once they gain access to a third party, they can infiltrate the primary organization through shared systems, software, or data channels, creating a pathway to sensitive information or critical systems.

In the maritime industry, supply chain attacks often target logistics companies, port authorities, equipment suppliers, and software providers, many of whom have direct access to operational technology (OT) and information technology (IT) systems used by shipping companies and ports [20]. This complex web of relationships presents multiple entry points for attackers, making the maritime supply chain an attractive target for cybercriminals seeking to exploit interconnected systems.

Supply chain attacks in maritime logistics can take several forms, each targeting different aspects of the operational ecosystem:

- **Compromised Software Updates**: Attackers may inject malware or malicious code into legitimate software updates provided by third-party vendors. When maritime companies install these updates, the malware spreads within their systems, giving attackers access to sensitive data or control over critical systems.
- **Third-Party Service Providers**: Many maritime companies outsource functions such as cargo tracking, fleet management, or port logistics to third-party service

providers. If these service providers have inadequate cybersecurity measures, attackers can exploit their systems to access the primary company's network and data [9].

- **Hardware Compromise**: Attackers can also tamper with hardware components used in maritime infrastructure, such as sensors, GPS units, or communications equipment. By infiltrating hardware through the supply chain, attackers can gain access to onboard systems or port management networks, potentially compromising navigation, cargo handling, and safety operations [6].
- **Impact of Supply Chain Attacks on Maritime Operations**

  (a) **Operational Disruptions and Downtime**
  Supply chain attacks have the potential to disrupt entire operations by compromising critical systems that maritime companies rely on for their daily functions. For instance, if a compromised software update contains malware that disables a port's cargo management system, the loading and unloading of goods may be delayed, creating a backlog that impacts vessel schedules and port efficiency. Additionally, because supply chain attacks often remain undetected until they have caused significant damage, organizations may face prolonged downtime as they work to identify and remove the threat. Operational disruptions caused by supply chain attacks can cascade throughout the logistics network. A delay in one port, for example, can affect the timing of shipments at other ports, delay downstream deliveries, and create a ripple effect that impacts the global supply chain. For time-sensitive or perishable goods, such delays can lead to spoilage, financial losses, and reputational damage.

  (b) **Financial Losses**
  The financial repercussions of a supply chain attack in the maritime sector are substantial. Beyond the costs associated with downtime, companies face expenses related to containment, remediation, and system restoration. They may also need to invest in cybersecurity audits, implement more robust security measures, and provide compensation to clients affected by the delays. Additionally, a supply chain attack that results in data theft or intellectual property loss can create long-term financial setbacks. Competitors or other malicious actors can exploit stolen trade secrets, client data, or operational details, giving them a competitive advantage or enabling further attacks. In some cases, companies may face legal liabilities or regulatory fines if the attack exposes sensitive customer information, particularly if data protection laws such as GDPR or CCPA apply.

  (c) **Compromised Security and Safety**
  Supply chain attacks pose a significant risk to security and safety in the maritime industry. In a scenario where attackers gain access to operational technology systems on a vessel or at a port, they could manipulate critical functions such as navigation, cargo handling, or engine control, leading to dangerous situations. A compromised navigation system could send a vessel off course, potentially causing collisions or grounding in restricted

or hazardous waters. Similarly, manipulated cargo handling systems could result in unsafe loading practices, increasing the risk of accidents at the port. The security risk extends to cybersecurity measures as well. If attackers can access confidential information, such as shipping routes, cargo contents, or personnel records, they can use this information to target specific vessels or shipments, increasing the likelihood of piracy or unauthorized boarding. The maritime sector's reliance on third-party logistics and service providers creates numerous entry points for these security threats, necessitating stringent cybersecurity protocols throughout the supply chain.

(d) **Reputational Damage**
A successful supply chain attack can severely damage the reputation of a maritime organization. Clients, partners, and stakeholders depend on reliable, secure services, and any breach of trust can lead to a loss of confidence. When clients or regulatory bodies learn of compromised data, delayed deliveries, or unsafe practices due to a supply chain attack, they may seek alternative service providers with stronger security measures. Reputational damage has long-term financial implications, as affected companies may lose existing clients, experience reduced demand for services, and face challenges in forming new partnerships. In an industry driven by reputation and reliability, a compromised supply chain can result in a prolonged recovery period as companies work to rebuild trust and reassure stakeholders of their commitment to security.

- **Strategies to Mitigate Supply Chain Attack Risks in Maritime Logistics**
- **Comprehensive Third-Party Risk Management** To reduce the risk of supply chain attacks, maritime companies must implement rigorous third-party risk management practices. This process involves assessing and vetting the cybersecurity practices of suppliers, vendors, and contractors before establishing a partnership. Organizations should evaluate the cybersecurity protocols, incident response plans, and compliance with security standards of each third party to ensure that their partners meet minimum security requirements. Regular audits of third-party cybersecurity practices are also essential. By continuously monitoring the security posture of suppliers and vendors, maritime companies can detect potential vulnerabilities early and take steps to mitigate risks. Establishing cybersecurity clauses in contracts, such as requiring timely disclosure of security incidents and adherence to industry security standards, further strengthens third-party risk management.
- **Implementing Multi-Layered Security Measures** A multi-layered security approach, often referred to as defense-in-depth, adds layers of protection to reduce the risk of supply chain attacks. This approach includes network segmentation, encryption, access controls, and continuous monitoring. Network segmentation restricts access to critical systems, preventing attackers who breach one area of the network from moving laterally into sensitive systems. Data encryption protects sensitive information during transmission, ensuring that intercepted data

remains unreadable to unauthorized individuals. Access controls, such as role-based permissions and multi-factor authentication, limit access to critical systems and data, making it more challenging for attackers to exploit third-party vulnerabilities. By continuously monitoring network activity, maritime organizations can detect unusual patterns and take immediate action if suspicious behavior is identified.

- **Supply Chain Cybersecurity Awareness and Training** Educating employees about supply chain cybersecurity risks is crucial for mitigating attacks. Training programs that focus on recognizing phishing attempts, securing data, and understanding the importance of cybersecurity within the supply chain can help reduce human error, a common entry point for supply chain attacks. Employees should understand the risks of interacting with third-party systems and be aware of the company's protocols for safely handling vendor interactions and data exchanges. Regular cybersecurity training ensures that employees remain vigilant and proactive, particularly in high-stakes environments like maritime logistics, where human error can lead to substantial security vulnerabilities.
- **Collaborating with Industry Partners and Regulatory Bodies** Finally, maritime organizations can enhance supply chain security by collaborating with industry partners, cybersecurity experts, and regulatory bodies. Industry groups and maritime associations often provide resources, guidelines, and best practices for securing supply chains. Additionally, regulatory bodies may issue cybersecurity standards and frameworks that maritime companies can adopt to ensure compliance and improve resilience against supply chain attacks. Collaborative initiatives, such as sharing threat intelligence and participating in joint cybersecurity exercises, help organizations prepare for supply chain risks. By participating in these collaborative efforts, maritime companies contribute to a safer and more secure industry-wide approach to cybersecurity.

3 Essential Components of Maritime Cybersecurity

The maritime industry is rapidly transforming through digitalization, integrating advanced technologies such as Industrial Internet of Things (IIoT), artificial intelligence (AI), and data analytics to improve efficiency, safety, and environmental performance. However, this digital transformation brings significant cybersecurity risks, as these technologies increase connectivity between vessels, shore-based facilities, and various control systems, creating a larger attack surface for cyber threats. Critical maritime operations, including cargo management, vessel navigation, and energy management, rely on interconnected systems that are vulnerable to unauthorized access, data breaches, and cyber-physical attacks.

Cyber threats in the maritime sector range from ransomware and malware attacks to espionage and state-sponsored cyber-attacks aimed at disrupting global supply chains. Incidents in recent years have highlighted the urgency of securing maritime infrastructure; for instance, attacks on major shipping companies have led to operational downtime, data theft, and significant financial losses. The potential risks extend beyond financial damage, posing threats to safety, environmental sustainability, and even national security, especially as maritime infrastructure plays a pivotal role in international trade.

Addressing these cybersecurity challenges [9] is complex due to the unique structure of the maritime industry. Vessels often operate in remote areas with limited connectivity, and port infrastructures house a mix of legacy and modern systems, each with different security requirements. This diversity makes implementing uniform cybersecurity measures challenging, requiring tailored solutions to address both shore-based and offshore environments. Regulatory bodies like the International Maritime Organization (IMO) have responded by establishing cybersecurity guidelines that emphasize risk management and resilience-building in maritime systems.

In light of these challenges, a robust cybersecurity framework must encompass multiple components: network security, data protection, incident response, workforce training, and regulatory compliance. Each of these components addresses specific aspects of cybersecurity within maritime operations, from preventing unauthorized access to preparing for and responding to cyber incidents. The "Essential Components of Maritime Cybersecurity" highlights these critical areas, providing an approach for securing digital maritime infrastructure amidst evolving threats.

3.1 Network Security: Implementing Firewalls and Intrusion Detection Systems

Network security is fundamental to protecting maritime infrastructure against a growing array of cyber threats. Implementing firewalls and intrusion detection systems (IDS) forms the first layer of defense, designed to regulate and monitor traffic between trusted and untrusted zones. Firewalls act as gatekeepers, blocking unauthorized access to networked systems while allowing safe data flow for critical operations. Advanced IDS, including both Network-Based IDS (NIDS) and Host-Based IDS (HIDS), provide deeper inspection and identify suspicious activity in real time. In maritime environments, IDS technologies tailored to handle the complexities of Operational Technology (OT) and Industrial Control Systems (ICS) are essential. Firewalls and IDS configurations must be regularly updated to adapt to evolving threats and vulnerabilities unique to maritime systems, such as those related to Vessel Traffic Services (VTS) and cargo management systems. Together, these solutions contribute to a robust network security framework, reducing the likelihood of unauthorized breaches and data leaks.

Network security serves as the first line of defense in protecting maritime digital infrastructure, which includes essential components such as vessel control systems, cargo management databases, and port operational networks. With the proliferation of interconnected systems, the maritime industry has become increasingly vulnerable to cyber threats that target these critical network elements. A robust network security approach in maritime infrastructure aims to ensure safe, secure, and reliable communication between vessels, port facilities, and shore-based operational centers, which handle the exchange of sensitive data. At its core, network security focuses on preventing unauthorized access, ensuring data integrity, and maintaining the confidentiality of information circulating within and between maritime systems.

Implementing firewalls is one of the fundamental measures for achieving effective network security. Firewalls act as gatekeepers, controlling data flow between trusted and untrusted networks by defining what type of traffic is allowed or blocked based on a set of security rules. These rules often consider various parameters, such as IP addresses, protocols, and port numbers, to determine the legitimacy of incoming and outgoing data packets. Firewalls can be configured in several ways—network-based or host-based—and are often deployed as a combination of both types in maritime networks to create layered security that shields critical systems from direct exposure to the open internet.

In maritime environments, firewalls play a crucial role in protecting operational technology (OT) systems, which are often integrated with information technology (IT) systems in a single network. OT systems control physical processes within vessels and port facilities, such as propulsion systems and cargo handling equipment. Firewalls help isolate OT from IT systems, ensuring that any vulnerabilities in the IT domain do not impact OT operations. This isolation is vital because OT systems are usually less adaptable to frequent updates or security patches, making them potentially susceptible to cyber threats if exposed to open network connections.

Beyond firewalls, intrusion detection systems (IDS) add another critical layer of defense by monitoring network activity and alerting administrators to suspicious behavior. Unlike firewalls, which focus on blocking or allowing traffic based on predefined rules, IDS actively scans network traffic for abnormal patterns or known attack signatures that may indicate malicious activity. There are two main types of IDS: Network-Based Intrusion Detection Systems (NIDS), which monitor network traffic across multiple devices, and Host-Based Intrusion Detection Systems (HIDS), which focus on individual devices. The integration of both types within maritime networks provides comprehensive visibility into the activities occurring across the entire infrastructure.

One of the key challenges in deploying IDS in maritime networks is balancing between performance and accuracy. IDS can be resource-intensive, especially when deployed on older network infrastructures that are common in the maritime sector. For instance, NIDS need substantial processing power to analyze network traffic in real time, which may be challenging for vessels operating with limited bandwidth or outdated systems. To address this, many organizations in the maritime industry utilize IDS solutions optimized for low-latency environments, enabling real-time threat detection without significantly impacting system performance.

In modern maritime networks, intrusion prevention systems (IPS) often work alongside IDS to proactively respond to detected threats. While IDS typically generate alerts, IPS takes further action to mitigate threats by blocking malicious traffic, quarantining affected devices, or even triggering automatic shutdown procedures. IPS technology is particularly beneficial in maritime operations where real-time response is essential to prevent potential disruptions. For example, if an intrusion attempt targets a vessel's navigation or propulsion systems, the IPS can intervene immediately, preventing the attacker from executing harmful actions or gaining unauthorized access.

To maximize the effectiveness of firewalls and IDS, maritime organizations must establish well-defined security policies and protocols that guide their configuration and use. Security policies outline acceptable use standards, access control measures, and protocols for incident response. These policies are tailored to address specific risks within maritime operations, such as unauthorized access to vessel control systems or data interception during shore-to-ship communications. Clear protocols enable operators to adapt firewall and IDS configurations based on evolving security needs, such as increasing restrictions on network access when transiting through high-risk regions.

Regular updates and maintenance are critical to keeping firewalls and IDS effective against emerging threats. Cyber threats constantly evolve, and attackers frequently develop techniques to bypass standard network defenses. Therefore, firewalls and IDS must be continuously updated with the latest threat signatures, security patches, and firmware. This process can be challenging in maritime environments due to operational constraints, including the remote location of vessels and limited access to reliable internet connectivity. Maritime cybersecurity teams often coordinate with vendors to schedule regular maintenance and update cycles that align with vessel docking times or other scheduled downtimes.

Despite their importance, firewalls and IDS alone are not sufficient to ensure complete network security. Effective network security requires a layered approach, with firewalls and IDS supported by additional measures such as encryption, access control, and regular network audits. Encryption protects data in transit, ensuring that sensitive information remains secure even if intercepted, while access control protocols restrict network access to authorized users only. Additionally, network audits provide valuable insights into the performance and effectiveness of security measures, allowing organizations to identify vulnerabilities and take corrective actions as needed.

The future of network security in maritime infrastructure will likely involve increased integration of artificial intelligence (AI) and machine learning (ML) technologies to enhance threat detection and response capabilities. AI and ML can improve the accuracy and efficiency of IDS by analyzing vast amounts of network traffic and learning to recognize subtle anomalies that may indicate emerging threats. This technology holds significant potential for maritime operations, enabling adaptive and proactive network security measures that can address complex threats, such as sophisticated cyber-attacks that target multiple systems simultaneously.

In summary, network security, supported by firewalls and intrusion detection systems, is essential to safeguarding maritime infrastructure against cyber threats. These tools work together to create a layered defense that controls access to critical systems, monitors network activity for malicious behavior, and responds quickly to threats. By continuously improving and adapting network security measures, the maritime industry can strengthen its resilience against cyber-attacks, ensuring that critical operations and assets remain secure in an increasingly interconnected digital world.

3.2 Data Encryption and Communication Security: Safeguarding Sensitive Data Across Networks

In maritime cybersecurity, data encryption and secure communication protocols are critical to protecting sensitive information in transit [7]. Maritime operations involve the exchange of confidential data, from navigational instructions to cargo manifests, which are valuable to both the industry and malicious actors. Employing data encryption ensures that information, even if intercepted, remains unreadable without the appropriate decryption keys. Encryption protocols such as AES (Advanced Encryption Standard) and TLS (Transport Layer Security) play crucial roles in safeguarding data within the maritime ecosystem. Beyond encryption, secure communication practices like Virtual Private Networks (VPNs) provide an additional layer by creating secure, encrypted channels for remote communication. This is particularly important for shore-to-ship data exchanges and cloud-based management systems in maritime infrastructure. Ensuring encrypted data flow between vessels, port facilities, and centralized control centers not only preserves data integrity but also builds resilience against attacks aiming to disrupt maritime operations.

Data encryption and communication security play a critical role in protecting sensitive information within maritime networks, which involve extensive communication between vessels, ports, and shore-based systems. The maritime industry handles vast amounts of sensitive data, from cargo manifests and vessel locations to crew information and financial transactions, making it an attractive target for cyber attackers. The goal of data encryption and secure communication practices is to ensure that, even if data is intercepted, it remains unreadable and unusable by unauthorized entities. This foundational element of cybersecurity mitigates the risk of data breaches, prevents unauthorized data manipulation, and maintains operational integrity across maritime systems.

Encryption transforms data into a coded form that can only be deciphered with a specific decryption key, making it highly effective in securing information both at rest and in transit. For maritime operations, encryption standards such as Advanced Encryption Standard (AES) and RSA (Rivest–Shamir–Adleman) are commonly used. AES, for instance, provides a robust level of encryption suited for securing large volumes of data, such as those exchanged in vessel tracking systems or port management software. By using these encryption methods, maritime organizations protect confidential information from being accessed or altered during transmission, thereby enhancing overall data security.

One of the primary areas where data encryption is crucial is in ship-to-shore communication. This type of communication is vital for exchanging navigational updates, cargo information, and safety protocols between vessels and their shore-based operators. However, due to the high levels of radio frequency (RF) transmission involved in such communication, it is vulnerable to interception. Implementing end-to-end encryption (E2EE) ensures that data remains secure as it moves from the vessel to the port facility. E2EE encrypts data at the origin point, and it only becomes readable once it reaches its intended recipient, preventing third parties from accessing sensitive information even if they intercept the transmission.

In addition to encryption, Virtual Private Networks (VPNs) are frequently used in maritime operations to create secure, encrypted tunnels for data transmission. VPNs provide an additional layer of security by masking the data's point of origin and adding an extra encryption layer as it travels across potentially unsecured networks. This is particularly beneficial in maritime environments where satellite-based internet connections are often required, such as during long voyages. VPNs enable vessels to securely access shore-based systems and services without exposing data to interception or unauthorized access, thereby protecting against man-in-the-middle (MITM) attacks and similar threats.

Public Key Infrastructure (PKI) is another crucial technology for secure communication in maritime operations. PKI systems utilize digital certificates to verify the identity of communicating parties, which prevents malicious actors from impersonating legitimate users within the network. PKI is commonly used for secure email, access control, and ensuring the authenticity of software updates within the maritime industry. By incorporating PKI into communication protocols, maritime organizations can establish trust between vessels, ports, and shore-based systems, ensuring that only verified entities can initiate data exchanges or execute critical commands.

Challenges to data encryption in maritime environments include limited bandwidth, legacy systems, and complex network infrastructures. Many vessels rely on satellite communication systems with limited data transfer rates, making it challenging to implement high-strength encryption without impacting performance. Legacy equipment, which may not support modern encryption protocols, further complicates the integration of data security measures. As a result, maritime cybersecurity strategies often involve a balance between implementing robust encryption methods and ensuring compatibility and functionality within the constraints of existing infrastructure.

Regular key management practices are essential to maintaining the effectiveness of encryption systems. Secure key management involves generating, distributing, storing, and revoking encryption keys as necessary to prevent unauthorized access. In a maritime setting, this includes securely handling keys on vessels, ensuring safe transmission between ships and shore facilities, and enforcing strict access controls. Poor key management can weaken encryption efforts, leaving data vulnerable to unauthorized access even if it is technically encrypted. For example, if encryption keys are improperly stored or shared, malicious actors could potentially gain access to encrypted data, undermining the benefits of encryption.

Data encryption and communication security also extend to protecting data in storage, such as within onboard databases or cloud-based platforms used by maritime companies. Encrypting data at rest ensures that, even if physical access to storage devices is obtained such as through hardware theft or device loss—sensitive information remains protected. Encryption at rest is especially critical for portable storage devices commonly used on ships, like USB drives and laptops, as well as data stored on cloud services accessed by maritime operators. These precautions help maintain data security even outside of active network transmissions, providing a comprehensive approach to data protection.

Security protocols such as Transport Layer Security (TLS) and Secure Sockets Layer (SSL) are also central to securing data in transit in maritime operations. TLS and SSL create encrypted links between networked devices, ensuring the security and confidentiality of data being exchanged. For instance, TLS is widely used in secure web browsing and remote management tools, which are commonly employed by maritime operators to monitor and manage vessels from remote locations. By securing these connections, TLS and SSL help protect against data interception and unauthorized access, providing a secure foundation for the remote management of maritime assets.

Compliance with international data protection standards, such as the General Data Protection Regulation (GDPR), is also essential to maritime communication security. GDPR and other regulations mandate strict measures to protect personal data and ensure data privacy, imposing penalties for data breaches. For the maritime industry, this means implementing data encryption and communication security measures that align with regulatory requirements, as sensitive crew and passenger data is often transmitted across maritime networks. Compliance not only safeguards data privacy but also strengthens trust with stakeholders and enhances the organization's overall cybersecurity posture.

In summary, data encryption and communication security are essential to safeguarding sensitive information across maritime networks. By implementing robust encryption methods, secure communication protocols, and stringent key management practices, the maritime industry can protect against data breaches, unauthorized access, and other cyber threats. These measures help ensure that critical data remains confidential and secure, thereby supporting safe and efficient maritime operations in an increasingly digitalized environment. As maritime organizations continue to embrace digital transformation, data encryption and communication security will remain fundamental pillars of maritime cybersecurity, protecting against evolving threats and ensuring the resilience of maritime infrastructure.

3.3 Incident Response Planning: Minimizing Impact Through Prompt Recovery Measures

Incident response planning is essential for mitigating the consequences of cyber incidents in maritime environments. A robust incident response plan (IRP) outlines the steps necessary to detect, contain, and recover from cyber threats quickly and effectively. It typically involves threat assessment, notification protocols, containment procedures, and post-incident evaluation. Key stakeholders—such as vessel operators, port authorities, and IT teams—must have well-defined roles within the IRP to ensure cohesive and swift action. Regular drills and simulation exercises help prepare these teams, allowing them to respond efficiently during actual incidents. Additionally, effective incident response involves establishing communication with relevant authorities, such as the International Maritime Organization (IMO) and regional cybersecurity centers, to comply with reporting requirements. As digital infrastructure in maritime operations expands, IRPs are critical for reducing operational downtime and safeguarding assets against prolonged cyber impacts.

Incident response planning is a critical component of maritime cybersecurity, designed to minimize the impact of cyber incidents through structured and timely recovery measures. As maritime systems grow increasingly interconnected, the potential risks of cyber threats—ranging from ransomware and malware attacks to system intrusions—continue to escalate. Given the critical role that maritime infrastructure plays in global trade and transportation, even minor disruptions can have significant operational and financial impacts. A well-prepared incident response plan (IRP) helps maritime organizations respond quickly and effectively, ensuring that essential operations remain resilient and any impacts are mitigated.

An effective IRP involves a comprehensive set of protocols to guide the organization through the detection, containment, eradication, and recovery stages of a cyber incident. The **detection** phase emphasizes identifying unusual activities or potential threats within the network, often through intrusion detection systems (IDS), security information, and event management (SIEM) solutions, or routine monitoring by security personnel. In maritime settings, early detection is crucial because of the unique operating environments—ships at sea may have limited bandwidth and communication capabilities, which can slow response times. A strong detection mechanism enables swift action, allowing organizations to contain the threat before it spreads further into critical systems.

The containment phase is essential to prevent the incident from affecting additional systems or escalating in severity. Containment strategies may include isolating compromised devices from the network, restricting user access, and applying network segmentation. For example, if a vessel's navigational system is affected by malware, immediate containment would involve disconnecting it from the broader vessel network to prevent further disruption or data corruption. Containment efforts often balance quick intervention with minimizing operational disruptions, as certain systems cannot be taken offline without impacting safety. Properly trained personnel and contingency plans are essential for making containment decisions that prioritize both security and operational continuity.

In the eradication phase, the focus shifts to completely removing the threat from the affected systems. This involves eliminating any malware, closing security vulnerabilities, and conducting a thorough analysis to ensure that no remnants of the threat remain. In maritime operations, eradication can be complex due to the diversity of connected systems, including both operational technology (OT) and information technology (IT) components. It often requires collaboration with IT teams, OT specialists, and possibly third-party cybersecurity experts. Eradication measures may also include reconfiguring system permissions, updating firewall rules, and applying patches to mitigate vulnerabilities identified during the incident.

The recovery phase is aimed at restoring normal operations and involves the careful reactivation of systems, verifying their functionality, and monitoring for any signs of residual issues. Recovery plans in maritime environments must consider the possibility of limited internet connectivity and other logistical challenges unique to vessels at sea. For instance, updating security patches or restoring data may need to be coordinated with shore-based facilities to ensure the integrity of restored systems. This phase may also include testing system performance and functionality to verify that all threats have been fully addressed and that systems are ready to resume full operations.

One of the critical components of incident response planning is defining clear roles and responsibilities for all personnel involved in the process. In a maritime organization, this might include shipboard IT teams, shore-based support teams, vessel operators, and management. Assigning specific responsibilities ensures that each team member understands their role in the response process, whether it's coordinating communication, executing containment strategies, or liaising with external cybersecurity experts. Having clear lines of authority helps avoid confusion and delays during an incident, which can be especially important for crews operating under high-stress conditions on vessels.

Communication is also a fundamental aspect of incident response planning, as effective communication with internal teams, external partners, regulatory bodies, and affected stakeholders is essential during and after an incident. Organizations should establish pre-defined communication protocols that outline how information is shared across teams and with external entities, including the International Maritime Organization (IMO) or other regulatory bodies. Clear communication not only helps in coordinating response efforts but also ensures transparency with stakeholders who may be affected by the incident. For instance, if a data breach exposes sensitive customer or partner information, timely and clear communication is critical to maintaining trust and complying with regulatory requirements.

Regular training and simulation exercises are essential for preparing personnel to respond effectively in the event of a cyber incident. Incident response drills enable team members to practice their roles, familiarize themselves with protocols, and learn how to handle simulated incidents. In the maritime sector, these simulations may include mock scenarios like malware affecting navigation systems or a ransomware attack on cargo data. Training should be conducted periodically and include different types of incidents to ensure a well-rounded preparedness. By regularly testing the IRP, organizations can identify potential weaknesses, refine protocols, and ensure that personnel are ready to respond efficiently when real incidents occur.

Documentation and post-incident analysis are equally important steps in the incident response process, enabling organizations to learn from each event and enhance their IRP over time. Post-incident analysis involves reviewing the response process, identifying what worked well, and pinpointing areas that require improvement. Documentation of each incident, including timelines, actions taken, and outcomes, provides valuable insights that help strengthen future responses. This process is particularly important in the maritime industry, where understanding the root cause of an incident and adapting the IRP can significantly reduce the likelihood of recurrence.

Compliance with regulatory standards such as those set by the IMO or other international organizations is another vital aspect of incident response planning in maritime cybersecurity. Compliance ensures that response strategies align with industry best practices and regulatory expectations, which is especially important for multinational maritime organizations. For example, the IMO's guidelines on maritime cybersecurity call for robust response plans and reporting protocols to handle cyber incidents effectively. Adhering to these standards helps organizations meet legal obligations, reduces potential liabilities, and demonstrates a commitment to cybersecurity resilience.

In summary, incident response planning is indispensable for maritime organizations aiming to mitigate the impact of cyber incidents and ensure rapid recovery. By establishing clear response protocols, conducting regular training, and fostering strong communication, maritime organizations can create a resilient framework capable of handling a wide range of cyber threats. As maritime operations continue to embrace digital transformation, the importance of a well-structured IRP will only grow, enabling the industry to maintain secure, reliable, and safe operations in the face of an evolving cybersecurity landscape.

3.4 Employee Training: Building Cybersecurity Awareness Within the Workforce

Human error remains one of the primary contributors to cyber vulnerabilities, making employee training an essential component of maritime cybersecurity [11]. Comprehensive cybersecurity training equips maritime professionals with the knowledge to identify, prevent, and respond to cyber threats. Training programs should address phishing scams, password security, software updates, and specific threats pertinent to maritime technology, such as GPS spoofing and unauthorized access to ICS. Frequent refreshers and hands-on practice enhance employees' understanding of both cybersecurity best practices and potential attack vectors they might encounter. Building a culture of cybersecurity awareness involves fostering a proactive mindset, where employees are vigilant and mindful of cybersecurity risks. Additionally, organizations can implement role-based training to ensure that personnel with higher access privileges, such as captains and chief engineers, receive more specialized cybersecurity instruction.

Employee training is a fundamental element of maritime cybersecurity, focused on building awareness and empowering the workforce to recognize and respond to cyber threats. In the maritime industry, personnel across various roles from crew members on ships to IT staff in port facilities handle critical systems and sensitive data. With rising incidents of phishing attacks, malware, and other cyber threats, building cybersecurity awareness ensures that employees can act as a first line of defense, minimizing risks and enhancing the organization's overall security posture. Training employees to recognize potential threats and respond appropriately can significantly reduce vulnerabilities in maritime operations.

One of the primary goals of cybersecurity training is to educate employees on recognizing phishing attacks, which are among the most common forms of cyber threats. Phishing attempts often come in the form of emails or messages disguised as legitimate communications, attempting to trick recipients into revealing sensitive information or downloading malicious software. In a maritime context, where crew members may be handling multiple roles, phishing attempts could target anyone in the organization. Training sessions that teach employees how to identify suspicious messages and verify the authenticity of links or attachments play a crucial role in preventing successful phishing attacks.

Password security is another key component of employee cybersecurity training. Weak or easily guessed passwords are a common vulnerability that cybercriminals exploit to gain unauthorized access to systems. Training should emphasize the importance of strong passwords and encourage practices such as using complex combinations of letters, numbers, and symbols, as well as changing passwords regularly. Additionally, training employees on multi-factor authentication (MFA) methods can strengthen access security, making it more challenging for attackers to compromise sensitive systems and data.

In the maritime industry, secure handling of USB drives and external devices is essential, as malware can easily be introduced through infected USBs. Vessels at sea frequently use USBs to transfer data between onboard and shore-based systems, creating an entry point for malicious software. Employees should be trained to handle USBs with caution, including scanning them for malware and refraining from using personal or unknown USB devices on work systems. Raising awareness about this risk can help mitigate the chances of introducing malware into critical systems, particularly on vessels with limited network security.

Training also emphasizes the importance of secure communication practices. Maritime employees, including vessel crew and port operators, frequently exchange sensitive information over email, messaging apps, or radio communications. Employees should understand the importance of using secure channels, encryption, and VPNs for transmitting sensitive data. Additionally, they should be aware of the risks of transmitting unencrypted data, which can be intercepted and exploited by cybercriminals. Cybersecurity training reinforces the use of secure communication tools and protocols to maintain confidentiality in maritime operations.

An effective cybersecurity training program also covers social engineering tactics used by cybercriminals to manipulate employees into divulging confidential information or granting unauthorized access. In social engineering attacks, attackers may impersonate trusted individuals, such as managers, coworkers, or IT staff, to persuade employees into making security compromises. By training employees to verify requests and identify unusual behavior, maritime organizations can reduce the risk of social engineering attacks and foster a culture of vigilance against potential scams.

Physical security practices are also integral to cybersecurity training in the maritime sector. Many cyber threats can originate from physical security lapses, such as unauthorized individuals accessing sensitive areas of a vessel or port. Employees should be trained to follow physical security protocols, including locking computer terminals, securing sensitive documents, and reporting any suspicious individuals. Ensuring that employees are aware of physical security measures reinforces cybersecurity efforts, as securing physical access points is essential to preventing unauthorized access to networks and systems.

Continuous learning through regular cybersecurity drills and simulations enhances the effectiveness of training programs. These drills allow employees to practice responding to simulated cyber incidents, such as ransomware attacks or system breaches, in a controlled environment [3, 4]. By simulating real-life scenarios, employees can familiarize themselves with incident response protocols and gain confidence in handling cyber threats. Regular drills also enable organizations to assess the effectiveness of their training programs, identify areas for improvement, and ensure employees remain vigilant against emerging threats.

Training should also include an understanding of compliance standards and regulatory requirements relevant to maritime cybersecurity, such as the International Maritime Organization (IMO) guidelines. Compliance with these standards often requires strict adherence to security protocols, incident reporting procedures, and documentation practices. Educating employees on these compliance requirements not only reduces the risk of regulatory penalties but also ensures that cybersecurity practices align with industry standards, enhancing the organization's resilience to cyber threats.

Another key benefit of cybersecurity training is the creation of a cybersecurity-aware culture within the maritime organization. When employees at all levels are knowledgeable about cybersecurity risks and understand their role in mitigating those risks, they are more likely to take proactive steps to secure systems and data. A strong cybersecurity culture fosters collaboration, encourages reporting of suspicious activities, and empowers employees to prioritize security in their daily tasks. This shared sense of responsibility strengthens the organization's defenses and creates a workforce that is actively engaged in cybersecurity efforts.

In summary, employee training in cybersecurity is a cornerstone of maritime cyber resilience, equipping the workforce with the knowledge and skills necessary to protect critical infrastructure and sensitive data [11]. By addressing key areas such as phishing awareness, password security, secure communication, and social engineering, maritime organizations can significantly reduce vulnerabilities. As digitalization in maritime operations continues to advance, ongoing training and a cybersecurity-aware culture will remain essential in countering evolving threats and ensuring the safe, secure, and efficient functioning of maritime assets.

3.5 Compliance Standards: Aligning with International Regulations, Like IMO Guidelines

Compliance with international cybersecurity standards and guidelines is critical to ensuring maritime cybersecurity aligns with industry best practices and regulatory requirements. The IMO's International Ship and Port Facility Security (ISPS) Code and its cybersecurity guidelines serve as foundational standards, outlining protocols for risk assessment, vulnerability management, and incident reporting. Compliance helps organizations to meet legal and operational standards and reinforces their credibility and security posture. Other frameworks, such as the National Institute of Standards and Technology (NIST) Cybersecurity Framework and ISO/IEC 27,001, provide additional guidance for implementing comprehensive cybersecurity measures in maritime settings. Compliance with these frameworks requires regular audits, documentation, and risk assessments, which help organizations identify areas for improvement and maintain security readiness. By adhering to these standards, maritime organizations can foster a safer operational environment, mitigate legal liabilities, and demonstrate their commitment to cybersecurity integrity.

Compliance standards are a fundamental aspect of maritime cybersecurity, aimed at aligning the industry's practices with international regulations and guidelines to ensure safe, secure operations across the global maritime sector. As cyber threats continue to evolve, regulatory bodies like the International Maritime Organization (IMO) have established guidelines to help maritime operators build resilient cybersecurity frameworks. Compliance with these standards not only protects against cyber risks but also strengthens operational integrity, fosters trust with stakeholders, and reduces the potential for costly disruptions. IMO guidelines, along with other regulatory frameworks, are essential tools for achieving consistent, industry-wide cybersecurity practices.

The IMO's Resolution MSC.428(98) is one of the primary regulatory frameworks guiding cybersecurity practices in maritime operations. This resolution requires shipowners and operators to incorporate cybersecurity risk management into their existing safety management systems (SMS) by January 1, 2021. The resolution emphasizes a proactive approach to managing cyber risks, urging organizations to identify, assess, and mitigate potential vulnerabilities in both information technology (IT) and operational technology (OT) systems on ships. By adhering to IMO standards, maritime organizations can establish cybersecurity practices that meet global expectations and reduce vulnerabilities across vessels and shore-based facilities.

To support compliance with Resolution MSC.428(98), the IMO has developed guidelines on maritime cyber risk management. These guidelines provide a framework for assessing cyber risks and implementing appropriate safeguards, focusing on five functional elements: identify, protect, detect, respond, and recover. Each element addresses a critical aspect of cybersecurity, from recognizing potential risks and protecting systems to detecting incidents and ensuring rapid recovery. Maritime organizations are encouraged to follow these guidelines to ensure comprehensive risk management and maintain alignment with IMO's cybersecurity goals.

Compliance standards extend beyond IMO guidelines to include other international regulations such as the International Organization for Standardization's ISO/IEC 27,001, a widely recognized standard for information security management. ISO/IEC 27,001 establishes requirements for creating, implementing, maintaining, and continually improving an information security management system (ISMS). For maritime organizations, certification to ISO/IEC 27,001 demonstrates a commitment to cybersecurity, providing a structured approach to securing data and systems against threats. Compliance with ISO standards complements IMO guidelines, enhancing the robustness of cybersecurity efforts.

Another essential standard is the General Data Protection Regulation (GDPR), which impacts any maritime company that handles data related to European Union citizens. GDPR mandates stringent data privacy and protection measures, requiring organizations to implement data security practices that safeguard personal information. This includes measures such as data encryption, regular audits, and breach notification protocols [7]. Compliance with GDPR is especially critical for cruise lines, shipping companies, and port operators who collect and process data on passengers, employees, and business partners, helping to build trust and avoid penalties associated with data breaches.

Compliance standards also involve continuous monitoring and auditing to ensure that cybersecurity practices remain effective over time. Audits assess adherence to cybersecurity standards, identifying any gaps in policies, procedures, or technologies that could pose risks. By conducting regular cybersecurity audits, maritime organizations can validate their compliance with regulatory requirements and adapt their cybersecurity frameworks to address new vulnerabilities. Auditing also provides a benchmark for measuring cybersecurity maturity, helping organizations to track progress and make informed decisions about future security investments.

Employee awareness and training are critical for ensuring compliance with cybersecurity standards. Many regulations require that maritime personnel understand the basics of cybersecurity and adhere to best practices in their daily tasks. Compliance-focused training programs help employees recognize threats, follow secure practices, and respond appropriately in the event of a cyber incident. For example, IMO guidelines emphasize the need for cybersecurity awareness across all levels of an organization, from the bridge crew to the port office staff. By investing in cybersecurity training, maritime organizations can meet regulatory expectations while building a security-aware workforce.

Compliance with cybersecurity standards also fosters collaboration with external partners and stakeholders, including regulatory bodies, cybersecurity experts, and other industry players. Adhering to shared standards facilitates communication and coordination among entities within the maritime ecosystem, allowing for a united approach to combating cyber threats. For instance, ports, shipping companies, and regulatory agencies can work together to share information about emerging threats, best practices, and response strategies. Compliance with industry standards creates a common language and framework for addressing cybersecurity risks, enhancing the resilience of the entire maritime supply chain.

The financial implications of non-compliance are another crucial consideration. Cyber incidents can lead to substantial financial losses through direct operational disruptions, reputational damage, and regulatory fines. Non-compliance with regulations such as the IMO's cybersecurity guidelines [17] or GDPR can result in penalties that further compound these costs. By proactively aligning with compliance standards, maritime organizations not only reduce the risk of incidents but also protect their financial stability. Compliance serves as an investment in risk management, enabling maritime operators to avoid the steep costs associated with cyber breaches.

In summary, compliance with international cybersecurity standards is essential for maritime organizations aiming to operate securely and efficiently in an increasingly digital landscape. IMO guidelines, along with complementary frameworks like ISO/IEC 27,001 and GDPR, provide a structured approach to addressing cyber risks, promoting security across vessels, ports, and shore-based facilities. Adherence to these standards strengthens the maritime industry's resilience against cyber threats, builds stakeholder trust, and ensures that organizations meet their regulatory responsibilities. As digitalization advances in maritime operations, compliance with cybersecurity standards will remain a cornerstone of sustainable and secure maritime practices.

In conclusion, a robust approach to maritime cybersecurity must integrate essential components—network security, data encryption [7], incident response, employee training [11], and adherence to compliance standards—to create a resilient framework against evolving cyber threats. Network security measures such as firewalls and intrusion detection systems protect critical systems from unauthorized access, while data encryption safeguards sensitive information in transit and storage. These measures, in conjunction with an established incident response plan, ensure that maritime organizations can detect, contain, and recover from cyber incidents swiftly, minimizing operational and financial impacts.

Employee training is another fundamental pillar, equipping personnel with the awareness and skills to recognize and respond to cyber threats, thereby reducing human error—a common vulnerability in cybersecurity. By fostering a cybersecurity-aware culture, maritime organizations can empower their workforce to become an active line of defense. Additionally, compliance with international standards, including IMO guidelines and other regulations, aligns maritime operations with industry best practices, creating a consistent approach to managing cyber risks across the global maritime network. Compliance not only mitigates legal and financial risks but also strengthens trust with stakeholders and regulatory bodies.

As the maritime industry continues to embrace digital transformation, these cybersecurity components are crucial in safeguarding complex networks of vessels, ports, and shore facilities. By prioritizing these essential elements, maritime organizations can build a sustainable cybersecurity framework that ensures operational continuity, protects sensitive data, and upholds the safety of maritime infrastructure. This holistic approach to cybersecurity not only shields the maritime sector from cyber threats but also supports the growth and resilience of global maritime trade.

4 Case Studies in Maritime Cybersecurity

As the maritime industry becomes increasingly dependent on digital and automated systems, the threat landscape evolves, making cybersecurity an essential focus for safeguarding operations. This section examines prominent real-world cyber incidents that illustrate the complexity and potential devastation of cybersecurity attacks in the maritime sector. These incidents not only disrupted operations but also led to significant financial losses and damage to the reputation of the affected companies. We analyze cases of attacks on Maersk, the Port of San Diego, and GPS spoofing incidents in the Black Sea. Each case reveals unique vulnerabilities, consequences, and crucial lessons, providing a foundation for developing resilient cybersecurity strategies in the maritime industry.

4.1 The Maersk NotPetya Attack: Impact of Malware on Global Shipping Operations

One of the most notable cyber incidents in the maritime industry was the NotPetya malware attack on Maersk in 2017. NotPetya, a type of malware, rapidly spread globally due to its worm-like properties, encrypting data and damaging systems indiscriminately. Maersk, one of the largest shipping companies in the world, was severely impacted, with operations disrupted across 76 of its ports worldwide, costing the company an estimated $300 million. This attack exploited a vulnerability in outdated software and spread via networked systems, demonstrating the risks of unpatched systems and outdated software dependencies in critical infrastructure.

4.1.1 Methodology and Attack Vector

The NotPetya malware spread through the EternalBlue vulnerability, which exploited weaknesses in Microsoft's Server Message Block (SMB) protocol. NotPetya, unlike typical ransomware, had no mechanism to decrypt the files after encryption, making it a "wiper" rather than traditional ransomware. A third-party software provider compromised Maersk's IT network to facilitate Ukrainian tax compliance, indicating the risk of supply chain vulnerabilities [6, 14].

4.1.2 Industry Response and Lessons Learned

Maersk's response involved a massive recovery operation. Maersk IT teams rebuilt over 4,000 servers and 45,000 computers within ten days. This case underscored the importance of network segmentation, supply chain security, and maintaining updated systems. In the aftermath, Maersk implemented enhanced cybersecurity protocols, including network isolation for critical systems and regular vulnerability assessments. This incident also raised awareness industry-wide about the importance of rapid incident response and robust data recovery strategies. The company also conducted extensive employee training [11] to increase awareness about cybersecurity threats and how to prevent them.

4.2 Port of San Diego Ransomware Attack: Financial and Operational Effects of Ransomware

In 2018, the Port of San Diego fell victim to a ransomware attack that temporarily crippled its IT infrastructure, impacting operations and causing delays. This attack was significant due to its financial and operational consequences on port activities, including administrative disruptions, port fee collection, and service delays. The port reported being unable to access its billing system, leading to revenue loss, operational delays, and diminished public trust [18].

4.2.1 Methodology and Attack Vector

This ransomware attack targeted critical infrastructure through phishing emails containing malware-laden attachments. The attackers demanded a ransom to restore the system's functionality, but the port managed the situation without paying. The ransomware in this attack encrypted files and systems, making data inaccessible and spreading through compromised user accounts and permissions [18].

4.2.2 Industry Response and Lessons Learned

The Port of San Diego's incident response team coordinated with local and federal agencies to mitigate the attack's impact. They restored operations by isolating affected systems and conducting an extensive recovery process. The key lessons from this attack were not just for the port, but for the entire industry. They include the importance of frequent data backups, employee cybersecurity training to reduce phishing risks, and collaboration with external agencies for rapid response. This case illustrates the importance of preparing for ransomware incidents with clear recovery strategies, isolated backups, and strong access controls [3].

4.3 GPS Spoofing in the Black Sea: Risks to Vessel Navigation

GPS spoofing, an emerging threat in the maritime domain, has been observed in regions such as the Black Sea, where vessels reported strange navigational data indicating that their positions were miles off course. Over 20 ships experienced GPS manipulation in one incident, which falsely located them inland. This GPS spoofing event demonstrated the vulnerability of navigation systems to signal interference and the potential risks to vessel safety and situational awareness [15].

4.3.1 Methodology and Attack Vector

GPS spoofing involves intentionally broadcasting false GPS signals to manipulate the location data received by a GPS receiver. In the Black Sea incidents, unauthorized signals deceived ships' navigation systems, disrupting their actual course and positioning. This attack highlights the vulnerability of GPS-based navigation in open water and the need for alternative navigation support to prevent reliance on potentially compromised signals [8].

4.3.2 Industry Response and Lessons Learned

This GPS spoofing case underscores the critical need for maritime operations to diversify navigational methods, incorporating traditional techniques and alternative technologies like eLoran as a GPS backup. These incidents serve as a reminder to continuously monitor the integrity of positioning signals and validate position data using multiple sources. This case study emphasizes the importance of redundancy in navigational systems to mitigate the risks of spoofing and other interference-based attacks [16].

4.4 Key Takeaways for Maritime Cybersecurity Strategy

Each case study reveals specific areas where cybersecurity practices can be improved to mitigate the risks of similar incidents. The Maersk NotPetya attack highlights the importance of securing supply chains and maintaining software updates, the Port of San Diego ransomware incident emphasizes employee training [11] and ransomware response planning, and GPS spoofing in the Black Sea demonstrates the need for resilient and diversified navigation systems. Together, these examples underscore the importance of proactive measures, including:

1. **Network Segmentation and Isolation**: Separating critical systems from general networks can prevent malware from spreading.

2. **Employee Training and Awareness**: Educating employees about phishing and social engineering attacks is essential.
3. **Incident Response and Backup Strategies**: Ensuring all critical data and systems have regular backups and are equipped with tested recovery processes.
4. **Redundant Navigation Systems**: Employing alternative navigation methods or backup systems to counter GPS spoofing or signal jamming.

These case studies provide essential insights into the types of cybersecurity threats that maritime organizations face. More importantly, they offer valuable lessons for crafting robust cybersecurity strategies. By implementing proactive measures such as network segmentation, employee training [11], incident response planning, and redundant navigation systems, the industry can empower itself to prevent or mitigate the risks of similar incidents.

5 Future Directions and Best Practices for Maritime Cybersecurity

As digital technologies continue to shape the future of maritime operations, the industry must adapt to counter increasingly sophisticated cyber threats. This section discusses future directions and best practices designed to safeguard maritime infrastructure, enhance resilience, and meet emerging global cybersecurity standards. By employing advanced strategies and integrating innovative technologies, the maritime sector can develop robust defenses to address both current and future cyber risks.

One foundational approach is Layered Security Architectures or defense-in-depth, which applies multiple layers of security to protect each aspect of maritime operations, from port IT systems to onboard navigation tools. This multi-layered defense makes it harder for attackers to penetrate deeper into critical systems. Key practices include using firewalls, intrusion detection systems, and network segmentation. Segmentation is especially critical, as it restricts access to sensitive areas, isolating them from other network sections, thereby limiting the spread of any potential breach and providing redundancy.

Equally important are Routine Vulnerability Assessments, which involve identifying and fixing weaknesses before they can be exploited. Regular assessments of IT and OT systems, coupled with simulations of potential attack scenarios, allow maritime organizations to proactively improve their cybersecurity. Routine testing of systems and timely patching help in identifying vulnerabilities early, reducing the risk of unexpected disruptions.

Advanced Access Controls like Multi-Factor Authentication (MFA) and role-based permissions add an essential layer of protection by limiting unauthorized access. Implementing MFA at every access point, including port and vessel control systems, and regularly reviewing access rights ensures that only essential personnel interact with critical systems. These practices minimize the exposure of sensitive information and reinforce security.

The use of AI and Machine Learning is transforming cybersecurity through predictive analytics and real-time anomaly detection. These technologies can analyze large volumes of data to identify unusual patterns, helping detect cyber threats before they escalate. For instance, machine learning models can monitor network activity and flag irregular login times or unexpected data flows, enabling cybersecurity teams to act swiftly, thus preventing potential breaches and improving response times.

In terms of data security, Blockchain and Quantum Encryption offer promising advancements. Blockchain creates tamper-proof data records that enhance the security of data shared among trusted entities, such as ports and logistics partners. Quantum encryption, still emerging, holds potential for unbreakable data protection, adding an advanced layer of security as cyber threats evolve. Implementing these technologies can secure sensitive maritime data and guard against unauthorized alterations.

Finally, adapting to Global Regulations is vital as cybersecurity standards are expected to become more stringent. Staying informed about international regulations, such as those from the IMO, and aligning with industry standards ensure compliance and prepare maritime organizations for future requirements. By proactively adopting these guidelines, maritime companies can bolster their cybersecurity resilience and maintain adherence to evolving regulations.

In conclusion, securing maritime infrastructure against future cyber threats requires a blend of best practices and readiness for technological and regulatory shifts. Implementing layered security, conducting regular vulnerability assessments, applying advanced access controls, integrating AI-driven analytics, leveraging blockchain and quantum encryption, and preparing for regulatory changes collectively strengthen the sector's ability to navigate the complex cybersecurity landscape. Through these measures, the maritime industry can future-proof its operations and ensure safe, resilient, and secure maritime logistics in an increasingly digital world.

5.1 Layered Security Architectures: Adopting Defense-In-Depth for Ships and Ports

As the maritime industry increasingly relies on digital systems to optimize port and ship operations, it faces escalating cyber threats targeting these critical infrastructures. A successful cyberattack can lead to severe consequences, such as operational disruptions, financial losses, or compromised safety. To counter these risks, layered security architectures, also known as defense-in-depth, have become essential for maritime cybersecurity. This approach involves creating multiple layers of defense to protect every aspect of maritime infrastructure, from IT systems in ports to navigation and control systems onboard ships. By securing each layer, maritime organizations can build a comprehensive defense that minimizes the likelihood of intrusions and limits the spread of any potential breaches. This section explains the concept of layered security and explores best practices, including firewalls, intrusion detection systems, and network segmentation, that help create a robust defense framework.

5.2 Understanding Layered Security Architecture in Maritime Cybersecurity

Layered security, or defense-in-depth, is a cybersecurity strategy that deploys multiple layers of protection to secure each component of an organization's infrastructure. Rather than relying on a single security solution, layered security incorporates diverse security measures at different levels. Each layer serves as a barrier, forcing attackers to navigate through multiple defenses, which reduces their chances of reaching critical systems or sensitive data.

In the context of maritime operations, layered security applies to a range of systems and networks, from port IT systems that manage cargo and logistics to OT (operational technology) systems onboard ships that control navigation, engine functions, and communication channels. With numerous stakeholders, interconnected systems, and remote operations involved, layered security is crucial for maintaining resilience in the face of increasingly complex cyber threats.

5.3 Why Layered Security Matters for Ships and Ports

Ships and ports have become the prime targets for cyberattacks due to the high-stakes nature of their operations and their reliance on technology. From logistical software and digital communication channels to navigation systems, these interconnected technologies create multiple points of vulnerability. If a single layer of security fails, an attacker could potentially exploit the vulnerability to access sensitive information, disrupt operations, or gain control over critical functions.

Layered security not only prevents direct breaches but also contains the impact of a successful attack. By implementing distinct security measures across different layers, maritime organizations can slow attackers' progress, allowing time for detection and response before the attack reaches core systems. Moreover, layered security creates redundancy, meaning that if one defense fails, other layers remain intact, minimizing the potential impact of a breach.

5.4 *Best Practices for Layered Security in Maritime Infrastructure*

- **Firewalls: The First Line of Defense** Firewalls are an essential component of layered security, acting as the first barrier between internal networks and external threats. They monitor and filter incoming and outgoing network traffic based on established security rules, blocking unauthorized access and protecting against various types of attacks. In a maritime setting, firewalls can be used at multiple points within the infrastructure. For example, port IT systems can be protected by perimeter firewalls that block malicious traffic from entering the network. Onboard ships, firewalls safeguard connections between the vessel's systems and external networks, especially crucial for systems that rely on satellite communication or internet access. By controlling traffic at these entry points, firewalls prevent attackers from gaining initial access to sensitive systems.
  Best practices for firewalls in maritime infrastructure include regularly updating firewall configurations, setting strict rules to limit traffic from unknown sources, and using next-generation firewalls (NGFWs) that offer advanced features such as threat intelligence and application awareness. Firewalls should also be configured to allow only essential communications between different segments of the network, reducing potential access points for attackers.
- **Intrusion Detection and Prevention Systems (IDPS): Monitoring for Anomalies** Intrusion Detection and Prevention Systems (IDPS) add another layer of defense by monitoring network activity to detect and respond to suspicious behavior. IDPS tools use predefined rules, threat signatures, and machine learning algorithms to identify potential intrusions. When a suspicious activity is detected, the system generates alerts, and in some cases, automatically takes action to block the activity.
  In maritime operations, IDPS is essential for real-time monitoring of both port and vessel networks. For example, an IDPS can flag unusual login attempts or unexpected data transfers from ship systems to external networks, allowing cybersecurity teams to investigate and respond to potential breaches quickly. By continuously monitoring network activity, IDPS provides an added layer of security that detects attacks that may bypass the firewall or occur from within the network.
  To enhance the effectiveness of IDPS, maritime organizations should regularly update their threat databases, review alert thresholds to reduce false positives, and integrate IDPS with other cybersecurity tools for comprehensive visibility. Regular audits of IDPS logs and automated responses can also help identify gaps in coverage and improve the system's ability to detect advanced threats.
- **Network Segmentation: Restricting Access to Sensitive Areas**
  Network segmentation is a crucial component of layered security that involves dividing a network into distinct segments, each with its own security protocols. By isolating sensitive areas from other network sections, segmentation prevents attackers from moving freely across systems if they gain access to one area. For instance, if a breach occurs in a less critical system, segmentation limits the

attacker's access to more sensitive systems, such as navigation controls or cargo handling systems.

For ports and vessels, network segmentation can be achieved by creating separate networks for different functions. For example, a port may establish individual segments for cargo logistics, administrative data, and public Wi-Fi. Onboard ships, segmentation can separate navigational systems, communication tools, and crew internet access. This way, if an attacker compromises one segment, they face additional barriers to accessing other parts of the network.

Best practices for network segmentation in the maritime sector include defining strict access controls, using virtual local area networks (VLANs) to separate functions, and enforcing role-based access permissions. Regularly reviewing network architecture and access permissions ensures that sensitive areas remain isolated, minimizing potential attack paths.

- **Multi-Factor Authentication (MFA) and Role-Based Access Controls** Access control is another critical layer in a defense-in-depth strategy. Multi-factor authentication (MFA) adds an extra layer of security by requiring users to verify their identities through two or more factors—such as a password and a mobile verification code—before accessing critical systems. Role-based access control (RBAC) limits access based on an individual's role within the organization, ensuring that only authorized personnel can interact with sensitive systems.

  In maritime operations, MFA can be applied to all remote logins, especially for users accessing systems from off-site locations or over public networks. Role-based access ensures that only essential personnel have access to critical systems, reducing the risk of insider threats or accidental misconfigurations. Best practices for access control include enforcing MFA for all sensitive systems, conducting regular access audits, and assigning permissions based on the principle of least privilege. By limiting access to only those who need it, maritime organizations reduce the number of potential entry points for attackers.

5.5 Redundancy and Response Preparedness

Layered security also emphasizes redundancy, ensuring that if one security measure fails, others remain effective. Redundancy allows maritime organizations to maintain a stable security posture even in the event of a breach, buying time for incident response teams to act.

In addition to building redundant layers, maritime organizations should establish a well-prepared response plan for handling incidents. Regular training, including drills and simulations, ensures that personnel know how to respond quickly to threats, isolate affected systems, and minimize the impact of a breach. Combining layered security with a proactive incident response plan strengthens maritime defenses and prepares organizations to face evolving cyber threats.

5.6 Routine Vulnerability Assessments: Identifying and Remediating Security Weaknesses

As the maritime industry increasingly adopts digital technologies to improve operational efficiency, it also becomes more vulnerable to cybersecurity threats. With port management, cargo handling, navigation systems, and communication channels now relying on interconnected digital systems, cybersecurity risks are higher than ever. Routine vulnerability assessments play a critical role in identifying and mitigating these risks, helping maritime organizations proactively secure their infrastructure. Through regular assessments, organizations can spot weaknesses in their IT and OT (Operational Technology) systems before they are exploited by malicious actors. This section explains the importance of vulnerability assessments, outlines best practices, and emphasizes how a proactive approach can significantly enhance cybersecurity resilience within the maritime sector.

5.7 Understanding Vulnerability Assessments and Their Role in Maritime Security

Vulnerability assessment is a systematic process that evaluates digital systems, networks, and applications for security weaknesses that could be exploited by attackers. These assessments identify, classify, and prioritize vulnerabilities, allowing cybersecurity teams to address them before they become security incidents. Vulnerability assessments differ from penetration testing, which involves attempting to exploit vulnerabilities; instead, they focus on identifying and analyzing potential weaknesses without actually exploiting them.

In the maritime industry, vulnerability assessments are crucial for both IT systems (such as data management, scheduling, and communications) and OT systems (including navigation, propulsion, and cargo control). Regular assessments ensure that security measures remain robust as systems evolve and new threats emerge, making vulnerability assessments a foundational practice for maritime cybersecurity. Routine vulnerability assessments provide maritime organizations with the insights needed to strengthen cybersecurity continuously. Without these assessments, organizations may unknowingly expose critical systems to cyber risks, potentially resulting in breaches, operational disruptions, or financial losses. Given the interconnected nature of maritime operations, where ports, ships, and logistics partners are digitally linked, a single vulnerability can compromise the entire ecosystem.

By identifying vulnerabilities early, maritime organizations can implement timely fixes and prevent attackers from exploiting weaknesses. This proactive approach reduces the risk of unexpected disruptions, minimizes downtime, and ultimately protects the integrity and safety of maritime operations. Additionally, vulnerability assessments support regulatory compliance, as many cybersecurity standards and guidelines (such as those from the International Maritime Organization) mandate regular security checks.

5.8 Best Practices for Conducting Vulnerability Assessments

- **Frequent Assessments of IT and OT Systems**
  To maintain strong cybersecurity, maritime organizations should conduct vulnerability assessments on both IT and OT systems. IT systems, such as those managing cargo tracking, port administration, and communication networks, are frequent targets for cyberattacks. These systems often handle sensitive data and rely on network connectivity, making them vulnerable to threats like data breaches and ransomware. OT systems, including navigation controls, power systems, and cargo handling equipment, also require regular assessments. Although OT systems were traditionally isolated from external networks, increasing connectivity and IoT (Internet of Things) integration have made them accessible to cyber threats. Compromised OT systems can lead to physical risks, such as navigation errors, machinery malfunctions, or safety system failures.
  To ensure comprehensive coverage, assessments should be scheduled regularly, such as quarterly or biannually, and supplemented with additional checks when significant changes or updates are made to systems. Frequent assessments allow organizations to monitor evolving vulnerabilities and keep pace with emerging threats.
- **Simulating Potential Attack Scenarios**
  Simulating cyberattacks, often known as "red teaming" or "tabletop exercises," is an effective way to test vulnerabilities under realistic conditions. These exercises mimic potential attack scenarios, allowing cybersecurity teams to evaluate how vulnerabilities might be exploited and to identify weaknesses in existing defenses. For example, a simulated ransomware attack on a port's IT system could reveal how well systems are protected and how quickly they can be restored if compromised. Similarly, an attack simulation on a vessel's navigation system can assess the resilience of OT defenses. Such exercises expose gaps that may not be apparent in regular vulnerability scans and highlight areas for improvement. By simulating attacks, maritime organizations can refine their response strategies, optimize incident handling procedures, and train personnel to react effectively under pressure. Regular simulations foster a culture of cybersecurity awareness and ensure that maritime infrastructure is prepared to withstand real-world cyber threats.
- **Regular Patching and System Updates**
  One of the simplest yet most critical steps in vulnerability management is ensuring that systems are patched and updated regularly. Cyber attackers often exploit known vulnerabilities in outdated software, taking advantage of unpatched systems to gain access to networks and data. Routine patching closes these gaps and strengthens the overall security posture. In maritime settings, both IT and OT systems require timely updates. However, applying patches can be challenging,

especially for OT systems that operate continuously and cannot be taken offline easily. To address this challenge, organizations can implement a patch management strategy, scheduling updates during maintenance windows or low-traffic periods to minimize disruptions.
Best practices for patching include maintaining a patch management schedule, automating updates where possible, and prioritizing critical patches for systems with known vulnerabilities. Additionally, organizations should verify that patches are successfully applied by performing post-update vulnerability scans, ensuring that the patching process is effective.

- **Comprehensive Network and System Testing**
Vulnerability assessments are not limited to scanning individual components; they also involve comprehensive testing of network infrastructure and system configurations. Network testing identifies weaknesses in network security, such as unsecured communication channels, outdated protocols, or poorly configured access controls.
For maritime organizations, network testing should cover both local networks (within ships or ports) and external connections (such as satellite links, internet access, and remote data exchanges). By identifying and addressing network vulnerabilities, organizations can secure communication channels, prevent unauthorized access, and limit the spread of potential breaches. System testing is equally important, focusing on ensuring that configurations, permissions, and authentication mechanisms are properly set. For instance, access control policies should be reviewed to confirm that only authorized personnel can access critical systems. Testing also ensures that backups are functioning correctly, providing a reliable method for restoring systems in case of an incident.
- **Prioritizing and Addressing Identified Vulnerabilities** Not all vulnerabilities are equal, and maritime organizations must prioritize those that pose the greatest risk. After conducting a vulnerability assessment, cybersecurity teams should classify vulnerabilities based on their severity, likelihood of exploitation, and potential impact on operations.
Critical vulnerabilities affecting essential systems, such as navigation or cargo handling, should be addressed immediately, while less critical issues can be scheduled for future remediation. By prioritizing vulnerabilities, organizations can allocate resources efficiently and ensure that high-risk areas receive the necessary attention. Addressing vulnerabilities requires a collaborative approach, involving coordination between IT teams, OT personnel, and management. Clear communication channels and defined workflows help streamline the remediation process, enabling timely resolution of security weaknesses.

5.9 Enhancing Cybersecurity Posture with a Proactive Approach

Routine vulnerability assessments offer maritime organizations a proactive approach to cybersecurity. By regularly identifying and addressing security weaknesses, organizations can stay ahead of emerging threats and adapt to the evolving cyber landscape. A proactive stance enables maritime companies to enhance their cybersecurity posture continuously, building resilience against both internal and external cyber risks.

Additionally, vulnerability assessments contribute to a culture of cybersecurity awareness. When assessments are conducted regularly, employees across the organization become more attuned to cybersecurity issues, fostering a shared commitment to safeguarding maritime infrastructure. This proactive approach strengthens cybersecurity from within, reducing reliance on reactive measures and emphasizing prevention.

5.10 Advanced Access Controls: Multi-factor Authentication and Limiting Access

In today's digitally interconnected maritime industry, protecting critical systems from unauthorized access is essential to maintaining security and operational integrity. As ports, vessels, and logistics networks increasingly rely on digital systems, controlling access to these systems has become a priority for maritime cybersecurity. Implementing advanced access controls, such as Multi-Factor Authentication (MFA) and role-based access permissions, is crucial for reducing the risk of unauthorized entry and limiting exposure to sensitive information. This section explores the importance of access control, how MFA and role-based permissions can strengthen cybersecurity, and best practices for implementing these measures effectively across maritime infrastructure.

- **Understanding Access Control and Its Importance in Maritime Cybersecurity**

Access control is a cybersecurity measure that restricts who can access certain systems, files, or data within an organization. By controlling access, organizations can ensure that only authorized users can interact with critical systems, reducing the likelihood of unauthorized access and data breaches. In the maritime sector, access control is especially important because sensitive data—such as cargo information, navigational details, and operational commands—must be protected to prevent disruption, data theft, or even malicious manipulation.

Access control consists of two key aspects: identification and authorization. Identification verifies the user's identity, while authorization determines the level of access granted based on the user's role. Together, these processes prevent unauthorized individuals from accessing sensitive systems and restrict the activities that authorized users can perform based on their responsibilities.

The maritime industry faces unique cybersecurity challenges due to its reliance on a vast, interconnected network of digital systems, including onboard control systems, port management software, and communications networks [9]. A breach in access control could expose these critical systems to cyber threats, potentially causing significant operational disruption, financial loss, or even physical harm to crew members and port staff.

For instance, if a hacker gains unauthorized access to a vessel's navigation system, they could manipulate the ship's course, endangering safety and risking environmental damage. Similarly, unauthorized access to a port's logistics system could lead to tampering with cargo data, causing delays or loss of valuable goods. By implementing advanced access controls, maritime organizations can limit access to critical systems, ensuring that sensitive information and operational tools are available only to essential personnel.

Multi-Factor Authentication (MFA) is an authentication method that requires users to provide two or more verification factors to gain access to a system. MFA typically involves a combination of something the user knows (such as a password), something the user has (like a smartphone or security token), and something the user is (such as biometric data like a fingerprint). By requiring multiple forms of verification, MFA adds an extra layer of security, making it more difficult for unauthorized users to access sensitive systems even if they obtain a password.

In the maritime industry, MFA is particularly useful for securing remote access points, such as those connecting vessels to shore-based systems or allowing port staff to access centralized databases from different terminals. With the widespread use of remote access in maritime operations, MFA reduces the risk of unauthorized access by ensuring that only authenticated users can access critical systems.

5.11 Best Practices for Implementing MFA in Maritime Infrastructure

Implementing MFA across all access points is essential for enhancing security in maritime operations. Here are some best practices for effective MFA implementation:

- **Apply MFA for All Access Points**: MFA should be mandatory for all entry points into critical systems, including port and vessel control systems, cargo management tools, and administrative systems. By securing every access point, organizations can minimize unauthorized access and protect sensitive areas from potential breaches.
- **Utilize a Variety of Authentication Factors**: To strengthen MFA, use a combination of authentication factors. For example, requiring a password along with a biometric scan or one-time code sent to a mobile device ensures that even if one

factor is compromised, the system remains protected. Biometric authentication, such as fingerprint or facial recognition, adds a physical verification layer that is difficult for attackers to bypass.

- **Regularly Review MFA Requirements**: As technology advances, so do authentication methods. Periodically reviewing and updating MFA requirements can improve security by incorporating the latest technologies, such as hardware-based security keys or location-based authentication.
- **Educate Employees on MFA Importance**: Since MFA adds an extra step to the login process, some users may find it inconvenient. Providing training on the importance of MFA helps employees understand the role it plays in cybersecurity and encourages compliance.

5.12 Role-Based Access Control (RBAC): Limiting Access Based on Roles

Role-Based Access Control (RBAC) is an access control mechanism that restricts system access based on a user's role within the organization. Instead of assigning permissions individually, RBAC assigns permissions to roles (such as "captain," "engineer," or "port administrator") and then assigns users to these roles. This approach simplifies access management, reduces the risk of human error in permission assignments, and ensures that users only have access to the information and tools necessary for their duties.

RBAC is essential in maritime operations, where different personnel require access to distinct systems. For instance, a port manager may need access to cargo tracking systems but not navigational controls, while a ship's engineer may need access to machinery data but not administrative records. By restricting access based on role, RBAC minimizes the exposure of sensitive systems and prevents users from accessing data irrelevant to their responsibilities.

5.13 Best Practices for Implementing RBAC in Maritime Infrastructure

Effective RBAC implementation ensures that maritime organizations limit access appropriately and protect sensitive systems. Key best practices for RBAC include:

- **Define Roles Clearly**: Clearly defining roles and responsibilities is the foundation of effective RBAC. Each role should be associated with specific tasks and access privileges, ensuring that users only have access to what they need for their work. For example, roles could include "Vessel Operator," "Port Administrator," or "Cargo Manager," each with distinct permissions.

- **Follow the Principle of Least Privilege**: The principle of least privilege restricts users to the minimum access required for their duties. By limiting permissions based on this principle, RBAC minimizes the risk of accidental or intentional misuse of data. Only essential personnel should have access to critical systems, and privileges should be adjusted if roles change.
- **Conduct Regular Access Reviews**: Regularly reviewing access permissions ensures that users' roles and access rights remain aligned with their current responsibilities. This practice is especially important when employees change roles, leave the organization, or if security requirements change. Periodic audits help identify unnecessary permissions and remove them to enhance security.
- **Implement Temporary Access When Necessary**: In some cases, personnel may need temporary access to specific systems or data for special projects or situations. Rather than granting permanent access, temporary access permissions can be assigned and then removed once the task is complete, preventing long-term access to sensitive systems.

5.14 Combining MFA and RBAC for Enhanced Security

By integrating MFA and RBAC, maritime organizations can create a layered access control system that both verifies user identity and restricts access based on role. This combination ensures that even if an unauthorized user gains access credentials, MFA provides an additional barrier, while RBAC ensures that only essential personnel interact with critical systems. Together, these controls offer a comprehensive approach to access management that significantly reduces cybersecurity risks.

For example, implementing MFA on top of RBAC means that only port administrators who authenticate using MFA can access administrative systems, while cargo managers are limited to cargo-related systems. Such separation reduces the risk of internal threats, as employees are restricted to their specific areas, preventing unintended access to other systems.

5.15 The Benefits of Advanced Access Controls in Maritime Cybersecurity

Advanced access controls provide several benefits for the maritime sector, including:

- **Enhanced Security**: By ensuring that only authorized users access sensitive systems, MFA and RBAC reduce the risk of data breaches and unauthorized manipulation of critical systems.

- **Improved Operational Efficiency**: With clearly defined roles and permissions, personnel can access the tools they need quickly without unnecessary access requests, enhancing productivity.
- **Reduced Risk of Insider Threats**: Limiting access based on roles and requiring multiple verification factors for access points makes it harder for malicious insiders to exploit their access for personal gain.

5.16 AI and Machine Learning: Utilizing Predictive Analytics for Anomaly Detection

As cyber threats grow more sophisticated, the maritime industry faces increasing challenges in safeguarding its digital infrastructure. Traditional cybersecurity measures alone are no longer sufficient to protect against rapidly evolving attacks. Artificial Intelligence (AI) and Machine Learning (ML) have emerged as powerful tools in the cybersecurity landscape, offering predictive analytics and anomaly detection that can transform how maritime organizations detect and respond to threats. By leveraging these technologies, cybersecurity teams can identify unusual patterns in real-time, enabling a proactive approach to threat detection and response. This section explores how AI and ML strengthen cybersecurity through predictive analytics, explains the significance of anomaly detection, and outlines best practices for implementing these technologies in maritime operations.

AI and ML use complex algorithms to analyze large volumes of data, enabling cybersecurity systems to predict potential threats based on historical patterns. Predictive analytics allows AI-driven systems to "learn" from past incidents, improving their ability to recognize signs of malicious activity and anticipate similar threats in the future. In the context of maritime cybersecurity, AI and ML models can process data from multiple sources, such as network traffic, login records, and system activity logs, identifying patterns that may indicate a cyber threat.

For example, predictive models can learn the typical behavior of users accessing port or vessel control systems. When AI detects deviations from these patterns—such as an unusual login location or an unexpected spike in data transfers—it flags the activity as suspicious. Predictive analytics not only improves the speed of detection but also enhances the accuracy of threat identification, reducing false positives and ensuring that cybersecurity teams focus on genuine threats.

5.17 Anomaly Detection and Its Importance in Maritime Cybersecurity

Anomaly detection is a core function of AI and ML in cybersecurity, focusing on identifying deviations from established norms. By constantly analyzing real-time data, AI systems detect anomalies that could signify potential threats, such as unusual login times, unexpected data flows, or unauthorized access attempts. In maritime operations, where numerous systems are interconnected—from port IT systems to onboard OT systems—AI-powered anomaly detection helps maintain security by quickly identifying unusual activities.

Anomaly detection is particularly valuable because cyber threats often begin with subtle changes in behavior that traditional security tools may overlook. For instance, a slow data exfiltration over time might go unnoticed, but AI algorithms can detect these anomalies, alerting cybersecurity teams before the threat escalates. This capability is crucial in the maritime sector, where undetected threats can disrupt operations, lead to data breaches, or even compromise physical safety.

5.18 Best Practices for Implementing AI and Machine Learning in Anomaly Detection

- **Training Machine Learning Models with High-Quality Data**
  The effectiveness of ML models relies on the quality and volume of data used for training. In cybersecurity, ML algorithms need access to diverse datasets to learn the typical behavior patterns within maritime systems. By training models with historical data from network logs, access records, and system activity, cybersecurity teams create models capable of accurately distinguishing between normal and suspicious behaviors.
- Maritime organizations should prioritize gathering clean, representative data from all relevant sources. For example, network logs from various systems, including vessel control, cargo management, and communication tools, should be collected to give ML algorithms a comprehensive understanding of standard operations. Regularly updating training data ensures that models stay current with changing patterns, further improving their accuracy in identifying anomalies.
- **Real-Time Monitoring for Rapid Detection and Response**
  One of the most significant advantages of AI and ML in cybersecurity is their ability to operate in real-time. By monitoring network activity continuously, AI-driven systems provide instant feedback on unusual behaviors, enabling immediate response. For example, if an employee's login credentials are used outside of typical hours or from an unusual location, the system can instantly alert cybersecurity teams or even automatically restrict access until further verification.
- In maritime infrastructure, where any delay in detecting a threat can lead to severe operational consequences, real-time monitoring is essential. AI's rapid detection capabilities allow organizations to act quickly, isolating affected systems, restricting access, or implementing countermeasures to prevent a breach. Automated alerts based on anomaly detection reduce the burden on security teams, enabling them to focus on high-priority incidents rather than manually monitoring data flows.

- **Combining AI with Traditional Security Measures**
  While AI and ML enhance cybersecurity, they work most effectively when integrated with traditional security measures, such as firewalls, intrusion detection systems (IDS), and role-based access controls. Combining AI-driven anomaly detection with these foundational defenses creates a layered security architecture, reducing the chances of threats bypassing protective barriers. For example, a firewall may block unauthorized access attempts, while AI monitors internal network activity for irregular patterns. This combination enhances security by ensuring that even if an external attack evades one layer, AI can detect unusual behaviors within the network. Maritime organizations benefit from this defense-in-depth approach, as it provides comprehensive coverage against both external and internal threats.
- **Using AI for Behavioral Analysis and Insider Threat Detection** One of AI's unique strengths is its ability to analyze user behavior, making it highly effective for detecting insider threats. Insider threats, whether intentional or accidental, are challenging to identify because they involve individuals with legitimate access to systems. AI-powered behavioral analysis models learn typical user patterns, flagging any deviations that might indicate malicious activity or accidental errors. In the maritime sector, where sensitive data and critical systems are accessed by multiple stakeholders, AI's behavioral analysis capability can significantly enhance security. For instance, if a user begins accessing data that is outside their typical responsibilities or exhibits abnormal login behavior, AI algorithms can detect this anomaly, alerting cybersecurity teams to a potential insider threat. This real-time monitoring of user behavior helps maritime organizations proactively prevent unauthorized access or data misuse.
- **Reducing False Positives through AI-Enhanced Filtering** One common challenge in cybersecurity is the high rate of false positives—alerts triggered by normal activity that security systems mistakenly flag as threats. False positives create unnecessary work for security teams, leading to "alert fatigue," which can result in genuine threats being overlooked. AI and ML help address this issue by refining anomaly detection models, allowing systems to more accurately distinguish between legitimate activities and potential threats. By analyzing a large volume of data, AI can "learn" what constitutes normal activity in the context of maritime operations. This ability allows AI to filter out routine variations, reducing false positives and enabling security teams to focus on real threats. In maritime infrastructure, where minor deviations in behavior may frequently occur, AI's capability to minimize false positives is invaluable for maintaining a streamlined and effective cybersecurity response.
- **Developing a Continuous Learning Framework**
  Cyber threats are constantly evolving, and AI models must adapt to new attack vectors and tactics to remain effective. A continuous learning framework allows AI and ML models to evolve based on real-world data, improving their ability to detect new types of threats. For maritime organizations, implementing this framework involves regularly updating models with the latest threat intelligence and

incorporating feedback from detected incidents. A continuous learning framework is particularly beneficial for anomaly detection, as it enables models to stay current with the latest behaviors, trends, and attack patterns. By continuously refining their understanding of normal and abnormal activity, AI-driven systems can detect emerging threats faster and with greater accuracy.

5.19 Benefits of AI and Machine Learning for Maritime Cybersecurity

Implementing AI and ML for anomaly detection provides several key benefits for maritime cybersecurity, including:

- **Enhanced Threat Detection**: AI and ML detect threats more accurately than traditional methods, enabling real-time identification of cyber threats and reducing the risk of breaches.
- **Proactive Response**: Predictive analytics allow maritime organizations to anticipate potential threats, enabling proactive measures to prevent incidents before they escalate.
- **Improved Incident Response**: Real-time alerts help cybersecurity teams respond to threats swiftly, isolating affected systems or taking preventive actions to limit the spread of a breach.
- **Reduced False Positives**: AI's refined filtering reduces the frequency of false alarms, allowing security teams to focus on high-priority incidents and reducing alert fatigue.
- **Stronger Insider Threat Detection**: AI's behavioral analysis capabilities enable organizations to identify insider threats by detecting deviations in user behavior, safeguarding critical systems from unauthorized access.

5.20 Global Regulations: Preparing for Future Changes in Cybersecurity Standards

As digital technologies become indispensable to the maritime industry, so too do the risks posed by cyber threats. With operations now deeply integrated with digital systems—from navigation tools on ships to logistics software at ports—cybersecurity has become critical for maintaining operational integrity and safety. Recognizing the need to address these growing risks, international bodies and governments are implementing increasingly stringent cybersecurity regulations. These global standards require maritime organizations to adopt high levels of protection to secure their infrastructure against cyber threats. This section discusses the evolution of cybersecurity regulations, the importance of compliance in maritime operations, and best practices for aligning with global standards to enhance cybersecurity resilience.

The past decade has seen a rapid increase in cyber regulations across various industries, driven by high-profile cyber incidents and the proliferation of digital threats. The maritime industry, which plays a vital role in global trade, is no exception. Ports, vessels, and logistics networks are now seen as critical infrastructure, and disruptions within these systems can have widespread impacts on economies and supply chains. Consequently, regulatory bodies have introduced cybersecurity requirements specifically tailored to maritime operations to prevent potential cyber incidents and ensure a secure operational environment.

In response to these risks, the International Maritime Organization (IMO), an influential global regulatory body, has issued guidelines such as the IMO Resolution MSC.428(98), which mandates that shipping companies incorporate cybersecurity into their Safety Management Systems by 2021. This regulation represents a significant step toward formalizing cybersecurity requirements within the maritime sector and sets a precedent for future regulatory advancements. The IMO's approach underscores the urgency of cyber preparedness, as digital vulnerabilities now represent a real and present danger to maritime operations.

Global standards organizations, such as the IMO and the International Organization for Standardization (ISO), have established frameworks that outline best practices for cybersecurity within maritime infrastructure. For instance, the ISO/IEC 27,001 standard provides guidelines for information security management systems, covering everything from risk assessment to incident response. These standards give maritime organizations a foundation upon which to build robust cybersecurity programs.

By adhering to these frameworks, maritime companies can ensure their cybersecurity measures are not only comprehensive but also consistent with international best practices. These standards also help organizations remain resilient in the face of evolving cyber threats by providing a structured approach to security. In a globalized industry like maritime, aligning with international standards allows companies to meet regulatory requirements across multiple jurisdictions, reducing the risk of penalties and enhancing cross-border operational consistency.

5.21 *Importance of Compliance in Maritime Cybersecurity*

Compliance with global cybersecurity regulations is essential for maritime organizations aiming to mitigate cyber risks effectively. Adhering to these regulations ensures that companies implement the minimum necessary security measures, protecting their infrastructure from various threats. For example, regulations may require the use of encryption for data in transit, secure communication protocols, and robust access control mechanisms. Each of these measures reduces the likelihood of unauthorized access, data breaches, and system disruptions.

Regulatory compliance also supports maritime organizations in maintaining trust with partners and stakeholders. In an industry heavily dependent on collaborative networks—such as shipping companies, port authorities, logistics providers, and customs agencies—compliance with security standards demonstrates a commitment to safeguarding shared data and systems. This commitment reassures partners and helps establish a secure and reliable operational environment.

Failing to comply with cybersecurity regulations can result in significant financial and reputational penalties. Many jurisdictions impose fines for non-compliance, and in some cases, authorities may suspend or restrict an organization's operational licenses until it achieves compliance. For instance, under the European Union's General Data Protection Regulation (GDPR), non-compliant organizations can face fines of up to 4% of their annual global turnover. Similar penalties are expected as cybersecurity regulations evolve within the maritime sector, especially for companies handling large volumes of data or operating critical infrastructure.

Beyond financial penalties, non-compliance can damage a company's reputation, affecting its relationships with clients and partners. In a highly competitive industry like maritime, reputational damage can lead to a loss of business and may deter potential partners from establishing collaborations. Thus, aligning with regulatory standards not only minimizes legal risks but also strengthens a company's standing within the industry.

5.22 Best Practices for Staying Compliant with Global Cybersecurity Regulations

- **Staying Updated on International Regulations**
  With cybersecurity regulations constantly evolving, it is essential for maritime organizations to stay informed about the latest developments. This can be challenging, given the variety of regulations issued by different regulatory bodies, such as the IMO, ISO, and various regional authorities. Regularly reviewing updates from these organizations allows companies to remain aware of changes in compliance requirements and prepare accordingly.
  To facilitate this, maritime companies can appoint dedicated compliance officers responsible for tracking regulatory changes and ensuring that the organization aligns with evolving standards. Additionally, participating in industry associations, cybersecurity forums, and regulatory consultations helps companies gain insights into upcoming regulations, allowing them to proactively adjust their cybersecurity measures to meet future requirements.
- **Conducting Regular Compliance Audits**
  Regular audits are essential for identifying gaps in an organization's compliance with cybersecurity standards. These audits should assess both the effectiveness of existing security controls and the organization's adherence to regulatory guidelines. By conducting audits, companies can identify areas where their cybersecurity measures may fall short of compliance standards, giving them the opportunity to make necessary adjustments before facing penalties.

Audits should cover all aspects of cybersecurity, from access controls and network security to data protection and incident response protocols. Additionally, maritime companies should consider conducting third-party audits, as external auditors bring objective perspectives that can help uncover vulnerabilities or oversights that internal teams may miss. By auditing compliance on a routine basis, companies ensure that their cybersecurity measures remain up-to-date and aligned with international standards.

- **Implementing a Cybersecurity Framework Based on Best Practices** Adopting a structured cybersecurity framework helps maritime organizations build a consistent and resilient approach to security that aligns with regulatory standards. Frameworks such as the NIST Cybersecurity Framework or ISO/IEC 27,001 provide a roadmap for implementing security controls, managing risks, and responding to incidents effectively. By using these frameworks as a foundation, companies can ensure that their cybersecurity programs are both comprehensive and adaptable to changing regulatory requirements.
  For example, a maritime company could use the ISO/IEC 27001 framework to establish an information security management system (ISMS) that covers policies, procedures, and technologies for data protection. Implementing such a framework promotes consistency across all departments and helps ensure that cybersecurity practices are deeply integrated into the organization's operations. This integration is critical for meeting regulatory standards and maintaining a strong security posture.
- **Proactively Aligning with Future Regulatory Trends** To prepare for future changes in cybersecurity standards, maritime organizations should monitor emerging regulatory trends and incorporate them into their long-term cybersecurity strategies. For instance, many regulators are placing greater emphasis on data privacy and resilience, suggesting that future standards may include stricter requirements for data encryption [7], access control, and backup solutions.

One way to proactively align with these trends is to adopt advanced technologies such as multi-factor authentication, encryption, and anomaly detection systems powered by AI. By implementing these measures now, maritime organizations can ensure that they are prepared for upcoming regulations and avoid the need for sudden adjustments when new standards take effect. Staying ahead of regulatory trends not only simplifies compliance but also enhances overall cybersecurity resilience.

5.23 The Benefits of Regulatory Compliance for Maritime Organizations

Compliance with global cybersecurity standards offers multiple benefits for maritime organizations, including:

- **Enhanced Security and Resilience**: Adhering to regulatory standards strengthens an organization's defenses against cyber threats, reducing the risk of incidents that could disrupt operations or compromise safety.
- **Improved Operational Efficiency**: A well-structured cybersecurity program based on compliance standards streamlines processes, minimizes security gaps, and ensures that best practices are followed consistently across the organization.
- **Increased Trust and Credibility**: Compliance demonstrates a commitment to security, building trust with partners, clients, and stakeholders and improving the organization's reputation in the industry.

Reduced Legal and Financial Risks: By staying compliant, maritime companies minimize the risk of fines, legal challenges, and operational restrictions, ensuring continued smooth operations and profitability.

6 Conclusion

It can be concluded that along with significant digital transformation, the maritime industry faces increasing cybersecurity challenges due to its reliance on interconnected systems [9]. To mitigate risks such as ransomware, phishing, GPS spoofing, and internal threats, robust cybersecurity measures are essential. These measures include implementing encryption, network security, incident response plans, and comprehensive employee training [11]. By adopting these strategies and adhering to international standards, the maritime sector can protect critical infrastructure, ensure operational continuity, and maintain a safe and resilient global trading environment.

References

1. Deep Trekker (2020) Top 3 risks at our ports. https://www.deeptrekker.com/resources/maritime-port-security-risks. Accessed 07 July 2020
2. International Maritime Organization (IMO) (2017) Resolution MSC.428(98) on maritime cyber risk management in safety management systems. IMO. https://www.imo.org/
3. Olmstead K (2018). Port of San Diego suffers ransomware attack, delays port fees. Port Technology International. https://www.porttechnology.org/news/port_of_san_diego_suffers_ransomware_attack/
4. Ranger S (2018) NotPetya ransomware: what it is, how it spreads, and what it has done. ZDNet. https://www.zdnet.com/article/notpetya-ransomware-what-it-is-and-how-it-spreads/
5. Allianz Global Corporate & Specialty (2021) Safety and shipping review 2021: an annual review of trends and developments in shipping losses and safety. Allianz
6. Schauer S, Polemi N, Mouratidis H (2019) MITIGATE: a dynamic supply chain cyber risk assessment methodology. J Transp Secur 12:1–35. https://doi.org/10.1007/S12198-018-0195-Z
7. ISO/IEC (2020) ISO/IEC 27002: code of practice for information security controls. ISO
8. Bhatti J, Humphreys TE (2017) Hostile control of ships via false GPS signals: demonstration and detection. Navigation 64(1):51–66

9. Drougkas A, Sarri A, Kyranoudi P, Zisi A (2019) Port cybersecurity: good practices for cybersecurity in the maritime sector. Technical report, European Network and Information Security Agency (ENISA). https://www.enisa.europa.eu/publications/port-cybersecurity-good-practices-for-cybersecurity-in-the-maritime-sector. Accessed 07 July 2020
10. Verizon (2020) Data breach investigations report 2020. Verizon. https://enterprise.verizon.com/resources/reports/dbir/
11. BIMCO (2021). Guidelines on Cyber Security Onboard Ships. Version 4.0. BIMCO.
12. Cyber Keel (2014) Maritime cyber risk. Cyber Keel. https://www.cyberkeel.com/
13. Lloyd's Register (2020) Cybersecurity in the marine industry: guidelines for cybersecurity implementation on ships and ports. Lloyd's Register
14. Greenberg A (2018) The untold story of NotPetya, the most devastating cyberattack in history. Wired. https://www.wired.com/story/notpetya-cyberattack-ukraine-russia-code-crashed-the-world/
15. Humphreys TE (2019) Toward a robust maritime navigation strategy: countering GPS spoofing. J Appl Navig 12(3):98–102
16. McCallie D, Butts J, Weeks D (2011) Security analysis of the ADS-B implementation in the next generation air transportation system. Int J Crit Infrastruct Prot 4(2):78–87
17. International Maritime Organization (2021) Guidelines on maritime cyber risk management (MSC-FAL.1/Circ.3). International maritime organization
18. Port of San Diego (2018) Ransomware attack at the Port of San Diego. https://www.portofsandiego.org/press-releases
19. United Nations Conference on Trade and Development (UNCTAD) (2021) Review of maritime transport 2021. UNCTAD
20. European Union Agency for Cybersecurity (ENISA) (2020) Port cybersecurity: good practices for cybersecurity in the maritime sector. ENISA

Chapter 7
Integration of Renewable Energy Sources in Maritime Operations

Edy Setiawan, Galih Anindita, Anggara Trisna Nugraha, and Lely Pramesti

Abstract The maritime industry, a cornerstone of globalization and international trade, is responsible for roughly 80% of global trade via over 2,000 ports worldwide. However, it significantly contributes to global pollution and CO_2 emissions, primarily through diesel-powered port operations. This study explores the potential of renewable energy solutions to mitigate environmental impacts in maritime transport, with a focus on Indonesia's unique position due to its extensive maritime routes and rich renewable energy resources. Indonesia aims to meet its goal of 23% renewable energy by 2025, reflecting a broader global trend towards renewable integration, as seen in countries like Norway. Various renewable technologies, including wind energy (soft sails, Flettner rotors, kite sails), solar photovoltaics, and hydrogen fuel cells, are being tested and implemented in maritime operations. Economic analyses indicate that the Levelized Cost of Energy (LCOE) for renewable projects is decreasing, with offshore wind projected to reach $60/MWh by 2030. While the initial costs for renewable energy systems are substantial, they can yield significant fuel savings—up to 60%—and reduce long-term operational costs, making them economically viable with short payback periods. A comparative analysis shows that hydrogen fuel cells and integrated renewable systems present promising alternatives that offer both economic and environmental advantages. Ultimately, market-driven policies and global investments are essential for fostering the adoption of

E. Setiawan (✉) · A. T. Nugraha
Department of Marine Electrical Engineering, Shipbuilding Institute of Polytechnic Surabaya, Surabaya, Indonesia
e-mail: edy_setiawan@ppns.ac.id

A. T. Nugraha
e-mail: anggaranugraha@ppns.ac.id

G. Anindita · L. Pramesti
Department of Marine Engineering, Shipbuilding Institute of Polytechnic Surabaya, Surabaya, Indonesia
e-mail: galih.talnabnof@ppns.ac.id

L. Pramesti
e-mail: lelypramesti@ppns.ac.id

M. Elsisi et al. (eds.), *Maritime Infrastructure for Energy Management and Emission Reduction Using Digital Transformation*, Studies in Infrastructure and Control,
https://doi.org/10.1007/978-981-96-4438-4_7

renewable energy technologies, enhancing energy resilience, reducing emissions, and promoting sustainability in the maritime industry.

Keywords Maritime industry · Renewable energy · Emissions · Economic viability · Sustainability

1 Introduction

1.1 Renewable Energy

One of the main pillars of globalization and the pillars of world trade is maritime transportation. About 80% of global trade takes place in all ports in the world. More than 2,000 ports around the world, economic growth, and pollution emissions are the main problems [1]. Indonesia has great potential in the marine sector because the country has a large sea area. Many of its ports serve as international trade routes and play an important role in its economic growth [2]. Ports contribute greatly to the increase in global warming and emissions, as ship loading and unloading operations use diesel generator engines, which produce ship exhaust emissions, mainly CO_2, SO_2, and NOx [3]. The Fig. 1shows the total CO_2 contribution of the economy and transport sectors. Therefore, countries in the world, including Indonesia, must have considered the application of technology that utilizes new and renewable energy in ship operations.

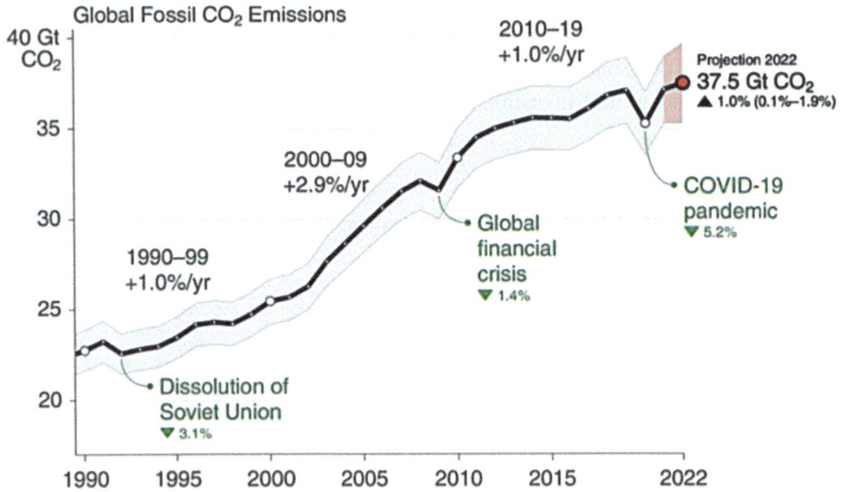

Fig. 1 Global fossil CO_2 emissions [3]

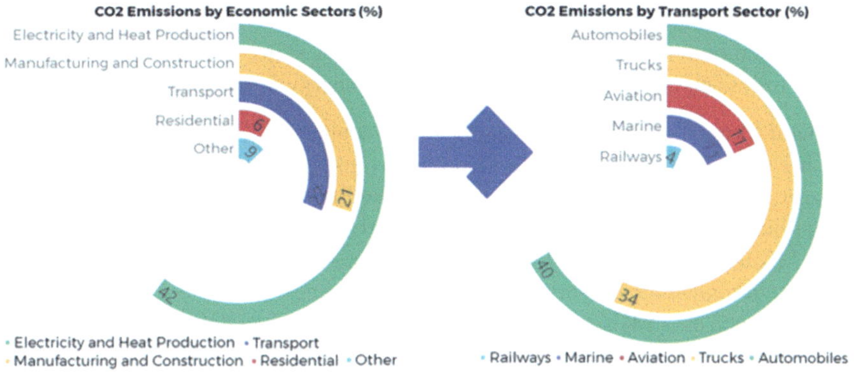

Fig. 2 Percentage of CO_2 emissions generated from economic and transportation [6]

Since 1990, data presented by Friedlingstein et al. in 2022 shows an increasing trend in CO_2 emissions. One of the main greenhouse gases that contribute to global warming and the greenhouse effect is CO_2, which is one of the big problems. To reduce CO_2 emissions, we must switch to renewable energy sources [3].

The use of new and renewable energy in maritime operations can help reduce carbon dioxide emissions and overcome global warming. Government Regulation No.79/2014 concerning Indonesia's National Energy Policy sets the level of new and renewable energy use in Indonesia to be at least 23% by 2025 and 31% by 2050 [4]. The Nordic region is an area of the European Union that always pays attention to the integration of the use of new and renewable energy. Denmark, Norway, Sweden, and Finland, among the Nordic countries, have very ambitious plans to increase the use of renewable energy in electricity generation. The Norwegian government has made policies to conserve natural resources so that they can be used sustainably since the 1990s in response to the problem of global warming. Norway aims to reduce greenhouse gas emissions by thirty percent in its energy policy by 2020. Therefore, the Norwegian Ministry of Oil and Gas and Energy uses renewable energy from water, solar, and wind to generate electricity [5] (Fig. 2).

1.2 Technology Status and Process

Currently, the use of renewable energy has been widely applied to maritime transportation operations globally at all levels, including domestic and international transportation [7]. Transportation of goods and people, as well as services for fisheries operations, tourism, and other maritime operations. Renewable energy can be applied to ships of any size to provide primary, hybrid, or auxiliary energy. It can also be applied as an offshore energy source for port infrastructure and operations. Utilization applications use renewable energy as integrated efficiency measures. These clean energy solutions are integrated through the use of energy already on current ships or incorporated into the design and development of new ships.

The following is the integration of renewable energy in maritime operations.

1. ***Wind Energy***: using ***soft-sails***, applied to Greenheart's 75 dwt freighter, B9 Shipping's 3,000 dwt bulker and Dykstra/Fair Transport's 7,000 dwt Ecoliner; ***fixed-sails***, applied to UT Wind Challenger and the EffShip's project; ***Flettner Rotors,*** applied to the Alcyone and Enercon's 12,800 dwt E-Ship 1; ***kite sails*** are applied to MS Beluga Skysails; ***Wind turbines*** (no successful prototypes until now) [7].
2. ***Solar photovoltaics***, hybrid models with other energy sources applied to small boats, such as NYK's Auriga Leader and SolarSailor by OCIUS Technology [7].
3. ***Biofuels***, such as the Meri cargo ship which claims to be the first ship to use 100% bio-oil [7] (Table 1).

In addition, hydrogen fuel cells have been used as a clean energy technology for shipping. This is seen on the FCS Alsterwasser, a passenger ship based in the Port of Hamburg in Germany, which is powered by hydrogen fuel. As part of the FellowSHIP project, in 2012, fuel cells with a power of 330 kW were tested on the Viking Lady, which operated for more than 7,000 h with an offshore supply. With electrical efficiency without emissions of NOx, SOx, and other particles, it is the first fuel cell unit to operate on a merchant ship. Germanischer Lloyd designed the zero-emission ferry "Scandlines" in 2012, as well as a 1,000 TEU (Twenty-foot equivalent unit) container ship with a speed of 15 knots that runs on hydrogen fuel [7].

Wind (soft sails, fixed wings, rotors, kites, and conventional wind turbines), solar photovoltaics, and biofuels are potential renewable energy sources that can be used in the shipping industry [7–10]. This eco-friendly solution can be integrated through retrofits of ships already in service or can be incorporated into the design and construction of new ships.

The Organisation for Economic Co-operation and Development (OECD) is making geopolitical economic changes around the world, focusing on transforming the operations of carbon-intensive industries to those that produce low-carbon or zero-emissions. The Institute of Development Studies conducted a study to find evidence of whether changes in the use of renewable energy around the world will be able to deliver a cheaper, lower-carbon change.

South Korea's increasing interest in the development of green ships, there is a great desire and intention from the Chinese shipbuilding industry to improve technological innovation and design capabilities, as well as an increase in strategic partnerships between European and Asian companies in the development of green ship innovations [7], shows that efforts to decarbonize the shipping sector are getting clearer. Proving that countries in the world increasingly want zero emissions in shipping sector operations.

Table 1 Potential renewable energy technology application [7]

Renewable energy type		Retrofit (RF)/ New build (NB)	Vessel category, application and potential			
			<400 tonnes e.g. recreation, arti-sanal/ small fishery tourism, passenger, break, landing craft, barges research coastal patrol and security	400 < 10,000 tonnes e.g. large landing craft, small-medium fishery, domestic Ro-Ro break bulk, bulk, container, tanker tramp	10,000 < 50,000 tonnes e.g. Ro-Ro, deep sea fishery bulk, container, tanker car carrier, cruise liner	>50,000 tonnes e.g. Very Large Crude Carrier (VLCC), Panamax, Aframax, large container ships
Wind	Soft sails	RF	...	...	...	.
		NB	...	...	...	.
	Fixed wings	RF	.	.	.	
		NB	.	...	...	.
	Rotors	RF	.	.	.	.
		NB	...	...	...	.
	Kites	RF/NB	.	.	.	
	Turbines	RF/NB				
Solar photovoltaics	Main propulsion	RF	N/A	N/A	N/A	N/A
		NB		N/A	N/A	N/A
	Auxiliary propulsion	RF	.	N/A		N/A
		NB	.	N/A		N/A
	Ancillary power	RF/NB	.	N/A		N/A
BIOFUELS	1stGeneration	RF	.	.	.	
		NB	.	.	.	
	2nd Generation	RF	N/A	N/A	N/A	N/A
		NB	...	...	...	.
	3rd Generation	RF	N/A	N/A	N/A	N/A
		NR	...	...	...	.
Table description						
In commercial use			...	High potential (scores well or all three metrics economic, environmental, and social)		
Proven						

(continued)

Table 1 (continued)

Renewable energy type	Retrofit (RF)/ New build (NB)	Vessel category, application and potential			
		<400 tonnes e.g. recreation, arti-sanal/ small fishery tourism, passenger, break, landing craft, barges research coastal patrol and security	400 < 10,000 tonnes e.g. large landing craft, small-medium fishery, domestic Ro-Ro break bulk, bulk, container, tanker tramp	10,000 < 50,000 tonnes e.g. Ro-Ro, deep sea fishery bulk, container, tanker car carrier, cruise liner	>50,000 tonnes e.g. Very Large Crude Carrier (VLCC), Panamax, Aframax, large container ships
Proof-of-concept		•	Medium potential (Scores on two of the three metrics)		
Design			Limited (scores or only one of the three metrics)		
Concept					
Uncertain		N/A	Not available		

1.3 Performance and Cost

Talking about the application of technology to new and renewable energy is not just about costs and return on investment. But there are a few other reasons that need to be explained: first, many types of applications and designs are still in the development stage; second, data on costs and benefits are often insufficient and not collected completely; and third, there is insufficient comparative data on the operating costs of other vessels, which makes it difficult to generate relevant data to support a comprehensive analysis of overall costs and profits.

The shipping sector uses different types of renewable energy, depending on the scale, function, and operational location of the ship. However, technology providers are confident that research and innovation on the use of renewable energy, along with efficient design, have resulted in significant direct energy savings. Enercon announced in 2013 that the E-Ship 1 prototype resulted in a 25% savings, and OCIUS Technology Ltd announced solar panel applications could result in savings of between 5 and 100%. In addition, it also claims that the cost of renewable energy solutions can reach between 10 and 15% of the ship's capital cost and return on investment (ROI) between 2 and 4 years. B9 Shipping and Fair Transport BV says that there are additional costs for construction and maintenance of 10 and 15% of the overall cost of the asset, and the resulting fuel savings can be up to 60% [7].

Many shipping companies continue to use fossil fuels to generate energy for ship operations. However, in recent years, many shipping companies have begun to consider switching to renewable energy technologies. However, it is very difficult to obtain information on the actual costs of the various renewable energy solutions that have been implemented in the shipping industry. There has been no implementation that describes a viable commercial solution for the shipping industry that can lower operational costs. Ultimately, the market forces in the regulations implemented will determine the absorption and application of renewable energy technologies in the shipping industry.

2 Renewable Energy in the Shipping Industry

To maximize the use of renewable energy sources such as solar, wind, and ocean waves, the ship's operational paradigm must change. This is necessary for effective performance. Many new designs that use new technologies can result in significant fuel savings depending on the type of boat, route, and speed. These designs also aim to reduce the need for fossil fuels.

Implementing renewable energy solutions for the shipping industry can be used in 2 ways as follows [7]:

1. Applying renewable energy as a retrofit (adding new technology or features to systems that are already in operation).
2. Incorporating renewable energy into new ship designs.

Zero-emission technology in the shipping industry, utilizing renewable energy as the main propulsion of ships (Greenheart, Ecoliner, B9, OCIUS), as additional propulsion (such as UT Wind Challenger, E-Ship 1, SkySails) [7]. In addition, renewable energy can be used in coastal infrastructure as an offshore supply, especially for alternative power plants.

2.1 Renewable Energy Applications—Wind

Sails used the wind to move the ship before the advent of steam engines. Wind is a renewable energy source that is easily accessible and used. One of its main drawbacks is fluctuations in wind speed and difficulty harnessing its full propulsion potential while sailing. Different types of ships, from small-scale ships to large cargo ships, are targeted as key and additional drivers in today's innovations, which include the use of various renewable energy technologies. Wind propulsion technologies include soft-sail, fixed-sail, rotors, kites, and turbines.

1. **Soft-Sails Technology**

Conventional mast-mounted soft sails offer technology capable of directly harnessing the force of wind propulsion. They can be used as a primary or auxiliary mover, and they can be installed on existing vessels or incorporated into new designs. The 75 dwt Greenheart cargo ship, the 3,000 dwt bulker B9 Shipping, and the 7,000 dwt Dykstra/Fair Transport Ecoliner use this technology (Fig. 3).

2. **Fixed-Sails Technology**

Basically, a fixed—sails is a "wing" assigned to a rotating pole. It was first used on large vessels such as the UT Wind Challenger and the EffShip5. However, the tests did not show significant savings, and the design needed to solve some technological problems. In 2015, the British company Oceanfoil reviewed and proposed new patents that were updated, improved, and used. To power the ferry, the Australian company OCIUS Technology Ltd. uses fixed—sails and photovoltaic panels. OCIUS patented a fixed-sail shape that can be folded to adapt to various wind conditions (Fig. 4).

3. **Rotors Technology**

The Magnus effect is an aerodynamic phenomenon that occurs when an object rotates through a fluid, such as air. The Magnus effect causes the relative motion between a rotating body and an air-fluid, which produces an aerodynamic force. This feature is used by Feltner Rotors. It occurs when the wind flows through a cylinder that has rotated to provide propulsion (Fig. 5).

Fig. 3 Examples of ship designs deploying soft-sail technology [7]. **a** Greenheart,**b** B9; **c** Ecoliner; **d** Seagate delta wing sail

Fig. 4 Examples fixed wing sail designs [7]. **a** UT Wind Challenge,**b** Oceanfoil

Fig. 5 Ships using rotor sail (Magnus effect) [7]. **a** Alcyone,**b** E-Ship 1

4. **Kite Sails Technology**

To maximize wind speed, kite sails are mounted on the bow of the ship at a certain altitude. For more than ten years this technology has been used. In 2008, the world's first commercial container cargo ship partially driven by a 160 m^2 kite sail was called MS Beluga Skysails (Fig. 6).

5. **Wind Turbines**

For many years, wind turbines for ship propulsion have been the subject of debate. To date, no prototype has been used. During its development, wind turbines were more widely used for shore-side power plants—OPS (ONSHORE POWER SUPPLY) [5]. For port operations, especially for loading and unloading operations. This strategy aims to improve air quality in ports and cities through the reduction of ship exhaust emissions, especially CO_2, SO_2, and NOx.

Due to public pressure and strong environmental laws on pollutant emissions, ports that have OPS installations on docks are currently located on the East coast of North America and Northern Europe [1, 11, 12]. Ports in Northern Europe can rely on wind power as a major source of renewable energy. In contrast, ports in the Mediterranean can rely on solar power as a major source of renewable energy. This

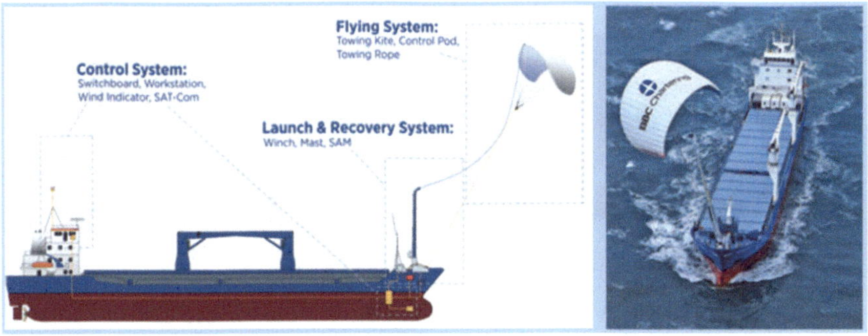

Fig. 6 Kite-assisted sail [7]

concept emerged in the Cartagena basin in Southern Spain. It is a Spanish government port in the Region of Murcia. Autoridad Portuaria de Cartagena is the name of this port. Two basins comprise this port: Cartagena and Escombreras. The area has a high level of solar irradiation energy, with an average of 5,582 kWh/m^2 per day [1, 13, 14]. On the other hand, the mountains that surround the two basins are an ideal place to build wind turbines. Accessible area of 51,395 m^2. For panels with a 30° angle and an azimuth angle of 0°, the average solar radiation is 5,582 kWh/m^2. With 280 WP modules, the solar power plant can generate 31.05 MWh per day [1]. In addition, the average wind speed in the region is about 6 m/s. The installation of a wind turbine with a unit power of 850 kW makes it possible to generate 2.6 MWh every day. There are 30 wind turbines that can generate 76.9 MWh every day. With an average monthly production of 108 MWh, which is enough to meet the average monthly energy requirements for port operations of 45% of the total power required.

OPS technology at all ports of Cartagena port is equivalent to removing emissions from around 178,000 cars from the city's streets [7].

2.2 *Renewable Energy Applications—Solar Photovoltaics and Hybrid Systems*

Electrical energy is generated by photovoltaic (PV) cells. This technology has come a long way in marine transportation for several operations. Its main limitation is the need for a large area for the installation of photovoltaic panels and the necessary energy storage. To provide additional propulsion for its main sail rig, Greenheart's design for a 220-tonage cargo ship uses lead-acid solar-powered batteries. As long as recharging does not increase other harmful emissions, the battery can serve as a hybrid solution for powering small to medium-sized boats (Fig. 7).

OCIUS Technology designed the SolarSailor12 with a hybrid fixed sails that uses solar energy. These sails can be mounted on the deck of the ship or the sails. It is now used on competitive port ferries in Australia, Hong Kong, and Shanghai, and shows strong potential for use on larger vessels. The large solar-sailed Aquarius MRE

Fig. 7 Fixed wing and solar PV hybrids [7]. **a** Eco marine power's aquarius MRE system; **b** OCIUS technology's SolarSailor; **c** NYK's Auriga leader

Fig. 8 Tûranor PlanetSolar [15]

(Marine Renewable Energy) system is being developed by Eco Marine Power13, based in Japan.

In the Auriga Leader15 project carried out by NYK and Nippon Oil Corporation between 2008 and 2009, 328 solar panels were installed on a car carrier weighing 60,000 tons. These solar panels produce 40 kW, which is about 10% of the ship's power while stationary in the dock. In addition, the ship is the first to synchronize solar power to the ship's main power grid (Fig. 8).

The largest solar-powered ship to date is the Tûranor PlanetSolar. The ship weighs 85 tons, with a length of 31 m and a width of 15 m. The vessel produces a maximum power of 93.5 kW as it is lined with 537 m^2 of solar panels and is one of the best solar-powered vessels in the world [15].

2.3 *Renewable Energy Applications—BioFuels*

In the transportation sector, biofuels are the best alternative to replace fossil fuels. However, it is still very rarely used in the shipping industry. The International Energy Agency is still conducting a thorough evaluation of alternative fuels for maritime applications, including biofuels (IEA-AMF 2013). In addition, research has been conducted to evaluate the possible use of biofuels in the shipping industry [7] (Fig. 9).

Hydrogenated vegetable oils (HVOs), pyrolysis oils, dimethyl ethers (DME), bioethanol, biodiesel, biomethane, pure vegetable oils (SVOs), or other derivatives are some examples of biofuels. The image shows how this fuel is made from raw materials.

In this next chapter, we will discuss mainly the use of hydrogen fuel cells.

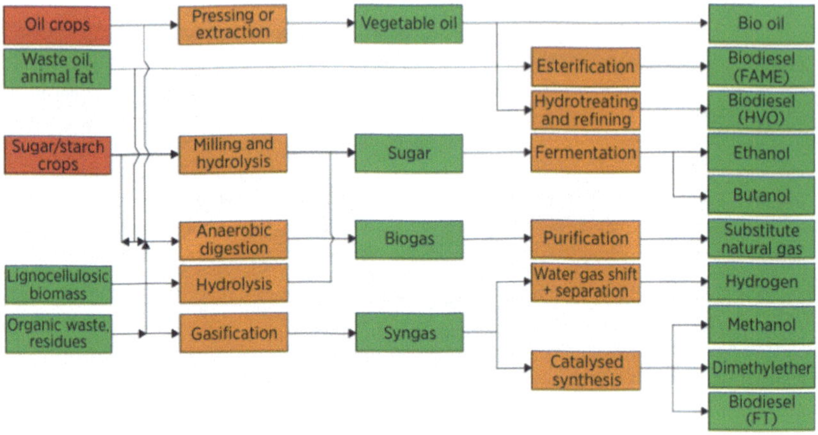

Fig. 9 Biofuels production [7]

1. **Hydrogen Fuel Cell**

Hydrogen fuel cells offer an efficient and environmentally friendly energy solution by converting hydrogen into electricity, producing only water vapor as a byproduct. Hydrogen, which is abundant in the universe, can be produced from various sources, including water through electrolysis, especially when using renewable energy [16]. Although this technology continues to evolve, with innovations in materials and designs that enhance efficiency, challenges such as refueling infrastructure, costs, and storage still need to be addressed [17]. With appropriate policy support and investment, hydrogen fuel cells hold significant potential in the transition toward clean and sustainable energy [18].

2. **Operating Principles of Hydrogen Fuel Cells and Their Applications in Ships**

Hydrogen fuel cells convert the chemical energy from hydrogen into electrical energy through an electrochemical process. Hydrogen entering the anode undergoes oxidation, producing hydrogen ions and electrons. These electrons flow through an external circuit to generate electricity, while hydrogen ions move toward the cathode, where they react with oxygen to produce water vapor as the only emission, making it a highly environmentally friendly technology [19].

Below is a flowchart illustrating the operating principles of hydrogen fuel cells and their applications in ships (Fig. 10).

The operation of hydrogen fuel cells begins with the filling of hydrogen and oxygen. At the anode, hydrogen undergoes oxidation, producing hydrogen ions (H^+) and electrons (e^-). The hydrogen ions then move through the electrolyte membrane toward the cathode, where they combine with oxygen to form water. This process generates electrical energy and water vapor as byproducts, with the electricity produced powering ships or other applications [19]. The main advantage of hydrogen

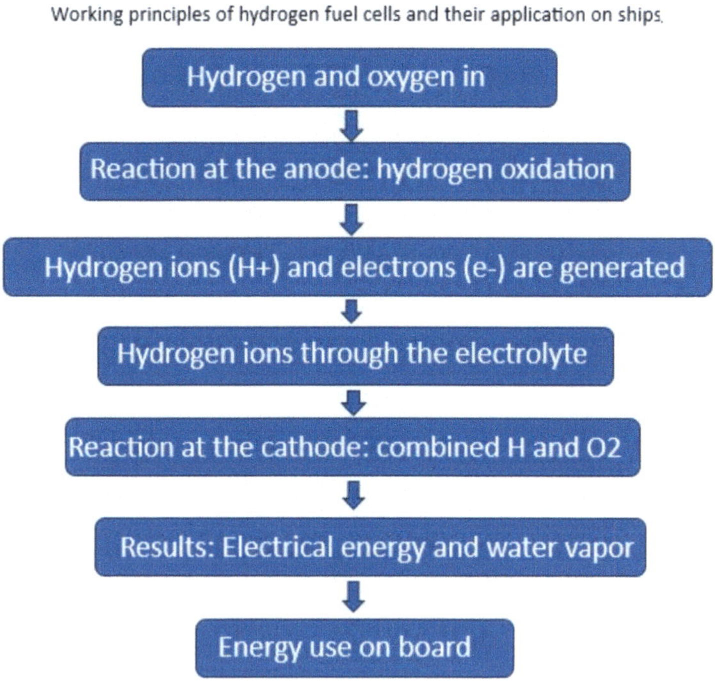

Fig. 10 Working principle of hydrogen fuel cell

fuel cells is their efficiency, which can exceed 60%, significantly higher than that of internal combustion engines [20]. This technology also offers rapid refueling times, similar to conventional fuel vehicles, making it suitable for transportation applications, including the maritime sector [18]. Ships using hydrogen fuel cells can operate without producing greenhouse gas emissions, helping to reduce air pollution in waterways and ports.

However, the main challenges in adopting this technology are the still-limited hydrogen refueling infrastructure and the relatively high production costs. With government support and private investment, the development of refueling stations at major ports is expected to encourage the use of this technology [18]. Hydrogen fuel cells offer a sustainable solution for the maritime industry, potentially reducing emissions and supporting a cleaner energy transition.

3. **Benefits of Hydrogen Fuel Cells Compared to Conventional Fossil Fuels**

Hydrogen fuel cells have a significantly lower environmental impact than fossil fuels, as they produce only water vapor as emissions. This technology has great potential to reduce air pollution, carbon footprints, and climate change. With an energy efficiency of over 60%, fuel cells are much more efficient than traditional combustion engines, which only achieve 20–30%. Hydrogen, the most abundant element in nature, can be produced from sources such as water or natural gas, providing a strategic advantage for regions that struggle to access fossil fuels. Additionally, this technology fosters innovation and new industries, creating jobs and promoting clean energy.

Although challenges related to production costs and infrastructure remain, policy support and technological advancements can accelerate the adoption of hydrogen fuel cells across various sectors. Benefits such as reduced emissions, high efficiency, and diversified energy sources make this technology promising for creating a cleaner and more sustainable energy system.

4. **Hydrogen Fuel Cell Ship Pilot Project and Technical Challenges**

The H2Ship project in Norway aims to build the world's first hydrogen-powered passenger ship to reduce greenhouse gas emissions in the shipping sector. Using hydrogen fuel cells, this ship can cut CO_2 emissions by up to 95% compared to fossil fuel-powered ships while maintaining efficiency [21]. Major challenges include safe hydrogen storage and the development of refueling infrastructure at ports [20].

Despite high production costs, the development of corrosion-resistant materials and crew training is crucial to ensure the ship can operate in harsh sea conditions. If successful, H2Ship could become a global model for the transition to clean energy and help reduce emissions in the maritime transportation sector [18].

5. **Long-Term Operational Savings Through Fuel Consumption Reduction**

The use of renewable energy technology in maritime operations, such as hydrogen, can reduce operational costs by up to 50%, depending on the type of vessel and route [22]. In addition to significant savings, shipping companies can save up to $5 million over 20 years and reduce greenhouse gas emissions to comply with environmental regulations and improve their reputation [23]. The use of clean energy also lowers the risk of accidents associated with hazardous fuels, decreases insurance costs, and provides cost stability by reducing dependence on fossil fuels, enabling companies to plan their budgets more sustainably [18].

6. **Funding Sources and Government Support for Renewable Energy Technology in the Maritime Sector**

Funding renewable energy projects in the maritime sector often poses challenges, especially for small to medium-sized shipping companies. To address this, governments in various countries have introduced initiatives such as public financial institutions, grant programs, and low-interest loans to support investments in environmentally friendly technology [24]. Additionally, tax incentives such as tax credits are offered to reduce the initial investment costs, making renewable technologies more accessible (DNV [23]. Government-supported research and development (R&D) programs can drive technological innovation, while international organizations like the International Maritime Organization (IMO) and the World Bank provide technical and financial support for sustainable projects [25]. However, a lack of awareness regarding available funding opportunities remains an issue, making it essential to improve access to information and training on how to leverage this support. Collaboration between the public and private sectors is also crucial to creating effective policies and a conducive investment environment, thereby supporting the transition to renewable energy and the sustainability of the shipping industry as a whole.

7. **Subsidy Programs and Incentives for Renewable Energy Adoption in the Maritime Sector**

Government subsidies and incentives are critical to promoting the adoption of renewable energy in the maritime sector. These supports include direct subsidies to cover technology costs such as hydrogen fuel cells, tax incentives that make investments more affordable, and low-interest loans from financial institutions [18]. Additionally, technical assistance and staff training help companies implement this technology effectively, ensuring that they meet operational requirements (DNV [23]. Collaboration among governments, industry, and research communities is essential, with long-term support needed to ensure a successful transition to renewable energy and provide positive impacts on both the environment and the economy [24].

8. **Partnerships with Private Companies and Government in Financing Green Energy Projects**

Partnerships between the private sector and government are key to financing green energy projects amid the global energy transition. This collaboration overcomes government funding limitations, with private companies bringing innovation and technology, while governments provide policy support, incentives, and guarantees to attract investors [18]. Public–private partnership (PPP) models allow for risk-sharing and more efficient resource utilization, encouraging the development of renewable energy projects [24]. Consistent policy support also creates a conducive investment environment, accelerating the transition to clean energy and contributing to economic growth and job creation (DNV [23].

3 Economic Analysis of the Integration of Renewable Energy in Maritime Operations

The renewable energy potential that Indonesia has is very extraordinary. As much as 18 GW of marine energy to 208 GW of solar energy, has not been utilized and implemented properly.

Indonesia still needs investment from around the world to encourage the development of renewable energy, especially from developed countries that have developed it. Indonesia has been actively encouraging foreign investment by implementing various favorable policies and regulations to attract international investors. One of them is Law No.11/2020 concerning Job Creation, which was later amended into a Government Regulation instead of Law No.2/2022 concerning Job Creation. It is hoped that this Government Regulation will improve the investment environment in Indonesia [26] (Fig. 11).

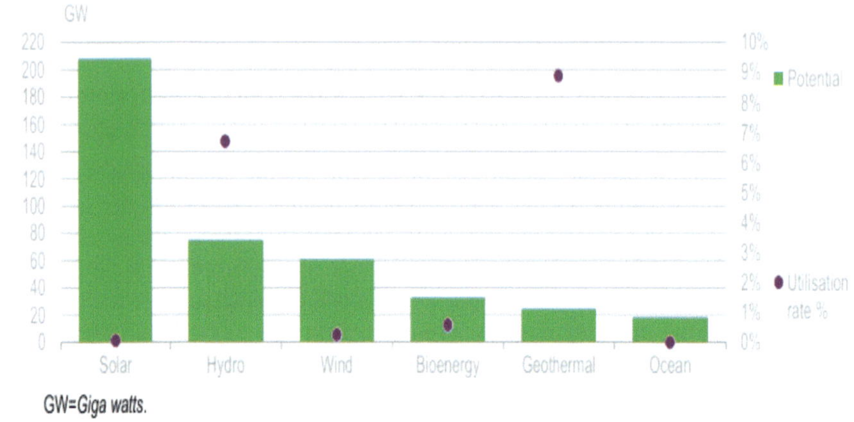

Fig. 11 Potential of renewable energy and realization of utilization in Indonesia [27]. *Source* OECD Based on Statistical Data from The Ministry of Energy and Mineral Resources (ESDM Ministry) and PLN

The General Plan for National Energy sets Indonesia's energy development goals from 2025 to 2050, including renewable energy. "**Others" includes diesel energy as well as biofuels (biofuels) and marine energy (tidal and heat) (Fig. 12).

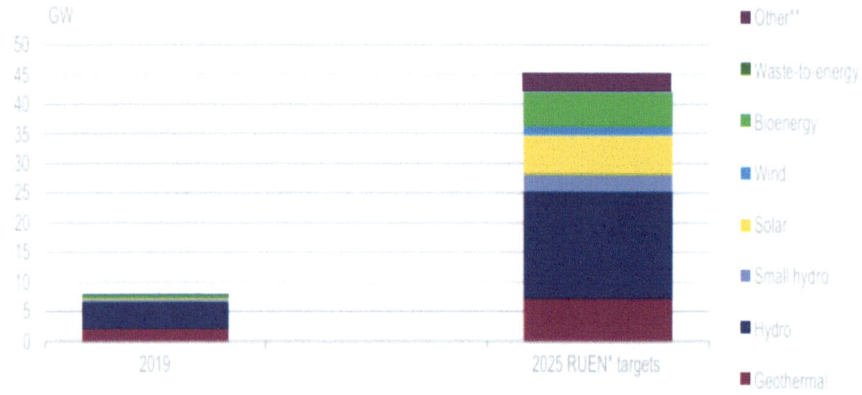

Fig. 12 Targets and Installed Capacity of Renewable Energy Power Plants [27]. *Source* OECD Based on Presidential Regulation No. 22/2017 on The National Energy General Plan and Statistical Data of The Ministry of Energy and Mineral Resources and PLN

3.1 Utilization of Renewable Energy in Green Ports

Many authorities and port operators around the world are committed to achieving zero emissions targets by incorporating renewable energy into maritime operations. Many ports have implemented renewable energy such as photovoltaic panels, wind turbines, and more. However, the application of renewable energy in environmentally friendly ports faces several problems, such as high initial costs, limited space, and technical complexity [6]. The following table shows the countries that have implemented renewable energy use programs in their green ports. To look at the economic analysis of the application of renewable energy to maritime integration, the concept of Levelized Cost of Energy (LCOE) is used. LCOE is a lifetime cost measure, which is the total cost of building and operating a power plant over its estimated lifetime, divided by energy production [28]. Table 2 shows a list of green port applications spread throughout the world along with the energy used to support their activities. Meanwhile, Fig. 13 illustrates the Levelized Cost of Energy (LCOE) Concept proposed by [28]. In this illustration there are 5 (five) main components that need to be taken into account, such as: Initial Costs, Annual Expenses, Annual Cost Per Year, Site Characteristics, and Annual Energy Production.

3.2 Economic Analysis of Offshore Wind Power Plants

"Offshore wind power is the most expensive energy generation technology for large-scale deployments in ports," according to the U.S. Energy Information Administration (EIA) in 2010. Labor costs range from €2.5 million/MW to €3 million/MW. By the end of 2011, 53 wind farms off the coast of Europe in Belgium, Denmark, Finland, Germany, Ireland, the Netherlands, Norway, Sweden, and the United Kingdom had been operating with a capacity of 3,813 and 5,603 MW installed. In line with its development, the cost of offshore wind energy is falling faster than expected. At a lower and more competitive price than gas energy, several contracts were made in the UK in September 2017 at 57.50 pounds/MWh (Fig. 14).

The contract was signed for Vineyard Wind in Massachusetts in September 2018 at a cost of between $65 and $74/MWh. According to a report by the National Renewable Energy Laboratory (NREL), the Levelized Cost of Energy (LCOE) for fixed and floating offshore wind energy in 2021 was $78/MWh and $133/MWh, respectively. According to the Report, the average LCOE for commercial-scale offshore wind projects in the United States fell by 13% to $84/MWh, which covers the price range of $61/MWh to $116/MWh. Is projected to drop again to an average of $60/MWh by 2030 [6].

Table 2 Programs of utilizing renewable energy in green ports worldwide [6]

Port name	Country	Implemented/Planned RET
Port of North Sea	Belgium and The Netherlands	Photovoltaic
Port Vienna	Austria	Hydrogen
Port of Solomon Islands	Salomon Islands	Photovoltaic
Port of London	United Kingdom	Hydrogen
Port of Amsterdam	The Netherlands	Hydrogen
Port of Fiji	Fiji	Photovoltaic
Port of Colombo	Sri Lanka	Photovoltaic
Port of Gothenburg	Sweden	Hydrogen biogas, hydrogenated vegetable oils, wind
Port of Yokohama	Japan	Hydrogen
Port of Los Angeles	United States	Hydrogen, fuel cell, photovoltaic
Port of Rotterdam	The Netherlands	Wind energy
Port of Auckland	New Zealand	Biofuel, hydrogen
Port of Marseille	France	Hydrogen fuel cell, photovoltaic
Port of Helsinki	Finland	Biofuel, photovoltaic
Port of Long Beach	United States	Hydrogen fuel cell, photovoltaic
Port of Anewerp	Belgium	Hydro turbine, photovolaic
Ports of Niederscheen	Germany	Hydrogen
Port of Batangas	Philippines	Photovoltaic
Ports of Associated British	United Kingdom	Photovoltaic, wind turbine
Port of Valencia	Spain	Hydrogen fuel cells, photovoltaic
Ports of Tenerifo	Spain	Photovoltaic

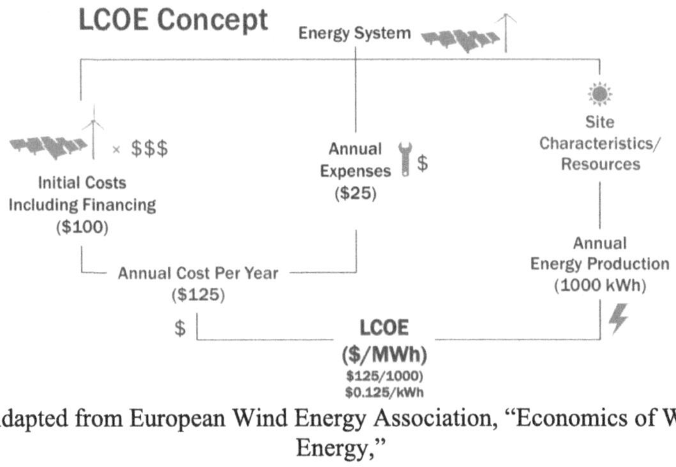

Fig. 13 Levelized Cost of Energy (LCOE) concept [28]. Adapted from European Wind Energy Association, "Economics of Wind Energy,"

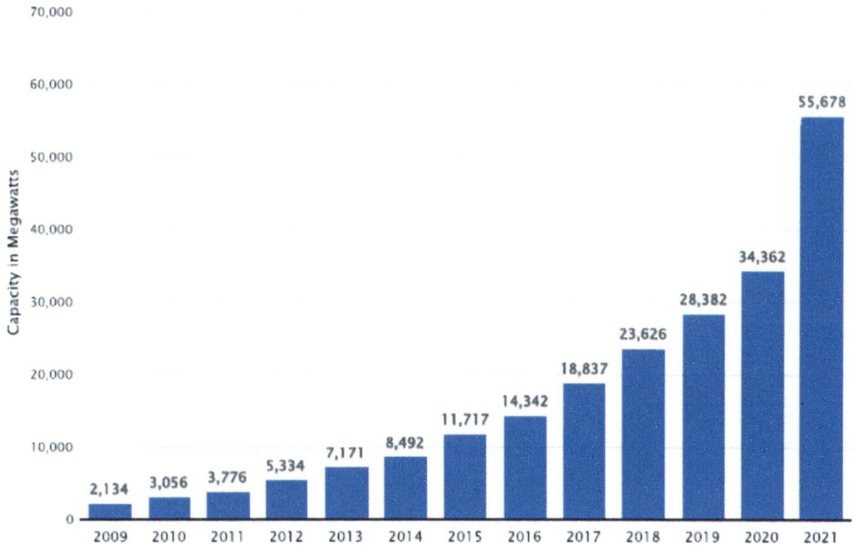

Fig. 14 Cumulative offshore wind energy capacity worldwide [6]

3.3 Economic Analysis of Floating Solar Power Plants

The various advantages of photovoltaic (PV) systems include reduced greenhouse gas emissions, unlimited energy sources, ease of use, and low maintenance requirements [6]. Although photovoltaic (PV) systems require large areas of land, most regions of the world have large water areas, which allows the installation of photovoltaic (PV) systems to reduce land costs. From 2007 to 2013, the first twenty plants with a capacity of tens of kilowatts were built. After that, in 2015, there were only 10 MW. However, since 2016, floating solar technology has increased. In 2020, the utilization of floating solar panels worldwide reached 3 GW. This is in stark contrast to land-based solar power systems that reach more than 700 GW (Fig. 15).

About 4,000 GW of solar capacity could be generated if 10% of hydropower reservoirs worldwide were covered with floating solar panels. This is equivalent to the power generation capacity of all fossil fuel power plants operating around the world. The project in Batam, Indonesia, will use solar panels over 16 km^2 of water to generate 2.2 GW and is planned to be completed in 2024 [6].

An on-ground solar power plant costs employees $40 per hour, while a floating solar power plant costs $60 per hour. In 2018, the LCOE for floating solar power plants was €53/MWh, much higher than the LCOE for land-based power plants, which ranged between €35/MWh and €40/MWh [6]. According to the NREL (National Renewable Energy Laboratory) report, the LCOE for floating solar power plants is $57/MWh without an Investment Tax Credit (ITC) and $38/MWh with an ITC. An on-site solar power plant costs $47/MWh without ITC and $32/MWh with ITC [6].

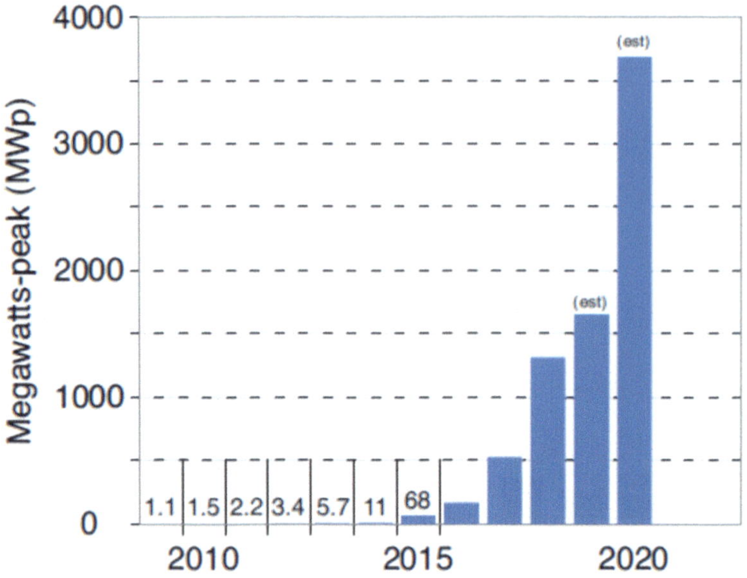

Fig. 15 Cumulative installed capacity of floating solar plants worldwide [6]

3.4 Economic Analysis of Biofuels Plants—Hydrogen Fuel Cells

Integrating renewable energy sources such as wind, solar, and hydrogen into maritime operations can significantly reduce long-term operational costs, with potential fuel savings of 30–50% [22]. Although the initial costs for renewable energy infrastructure are quite high, policy support and financial incentives can encourage adoption [18]. Additionally, renewable energy-based ships offer a competitive advantage in terms of environmental reputation and compliance with stringent emission regulations (DNV [23]. These investments not only provide financial benefits but also support public health and overall environmental sustainability [24].

1. **Initial Costs and Long-Term Savings**

Implementing renewable energy technology in maritime operations requires significant initial costs, such as for the procurement and installation of hydrogen systems, solar panels, or wind turbines. However, these investments can lead to long-term savings through reduced fuel costs of up to 50% and lower maintenance expenses [22]. With government support in the form of subsidies and financial incentives, shipping companies can more easily transition to this technology [18]. Overall, despite the high initial costs, the long-term financial and environmental benefits make this transition a favorable choice [23].

2. **Initial Investment Analysis in Renewable Energy Technology**

Initial investments in renewable energy technologies in the maritime sector, such as hydrogen fuel cells, solar panels, and wind turbines, require high costs, particularly for small companies that may need financial support or subsidies [18]. Although the upfront spending is significant, this investment can reduce fuel costs by up to 50% and lower maintenance costs, accelerating capital recovery in the long run [22]. Additionally, companies that invest in environmentally friendly energy can attract sustainability-conscious customers, enhancing their market position (DNV [23]. However, they must also consider the risks of energy price fluctuations and technological developments, which could impact their financial planning [24].

Suppose we have a ship that wants to switch from a conventional diesel engine to a hydrogen fuel cell system. Here are some data to be used in the analysis:

- Initial Cost for the Hydrogen Fuel Cell System: $1,500,000
- Annual Diesel Fuel Cost (before transition): $500,000
- Fuel Cost Reduction After Transition: 50% (after using hydrogen)
- Annual Maintenance Cost for Diesel Engine: $100,000
- Annual Maintenance Cost for Hydrogen Fuel Cell System: $50,000
- The lifespan of the Hydrogen Fuel Cell System: 20 years

The hydrogen fuel cell system requires an initial investment of $1,500,000, while the diesel engine does not incur additional upfront costs since it is already installed. Switching to hydrogen fuel cells reduces annual fuel costs from $500,000 to $250,000, saving $250,000 per year. Additionally, the annual maintenance cost for diesel decreases from $100,000 to $50,000, yielding an additional savings of $50,000. The total annual operating cost for the diesel engine reaches $600,000, while the hydrogen system only costs $300,000, resulting in a total annual savings of $300,000. Over 20 years, the total cost for the diesel engine amounts to $12,000,000, whereas the hydrogen fuel cell system costs only $7,500,000, meaning total savings over that period is $4,500,000 (Table 3).

3. **Economic Feasibility Study**

An economic feasibility study is a crucial step in planning renewable energy projects, aiming to analyze potential profitability and economic impact. Beyond financial aspects, this study also considers the environmental and social benefits of the proposed projects, such as solar power generation in remote areas of Indonesia [29]. Another example, a wind power project on the coast, highlights the importance of

Table 3 Comparison of diesel engine usage and hydrogen fuel usage

Component	Diesel engine ($)	Hydrogen fuel cell ($)	Savings ($)
Initial cost	–	1,500,000	–
Fuel costs/Year	500,000	250,000	250,000
Maintenance costs/Year	100,000	50,000	50,000
Total annual cost	600,000	300,000	300,000
Total cost for 20 years	12,000,000	7,500,000	4,500,000

a thorough evaluation of initial costs and revenue projections to ensure investment feasibility [30]. In the maritime context, a cost–benefit analysis of replacing fossil fuels with hydrogen fuel cells in Europe shows that despite high initial costs, long-term savings and carbon emission reductions support its economic viability (DNV [23]. Besides financial analysis, it is also essential to assess project risks and social and environmental impacts, such as job creation and energy access for local communities [24]. Overall, economic feasibility studies help ensure that renewable energy projects are not only financially profitable but also provide broad benefits for society and the environment.

4. Economic Feasibility of Integrating Wind, Solar, and Hydrogen Energy

In this renewable energy integration project, a study was conducted in the coastal region of Europe to evaluate the economic feasibility of developing an integrated energy system utilizing wind, solar, and hydrogen energy. The project aims to generate clean energy by harnessing available natural resources and reducing dependence on fossil fuels.

Based on the table, this integrated energy project involves three main components: Wind Power Plant, Solar Power Plant, and Hydrogen Storage System. The total initial cost amounts to $3,300,000, with an annual operational cost of $125,000. The wind power plant requires an initial investment of $1,500,000, while the solar power plant and hydrogen storage require $1,000,000 and $800,000, respectively. The annual operational costs for these three components range from $20,000 to $75,000. The project is estimated to generate an annual revenue of $650,000 from the sale of energy produced by the three systems (Table 4).

Based on the data above, we can conduct an economic feasibility analysis of this project. With a total annual revenue of $650,000 and total annual operational costs of $125,000, the annual profit of the project is calculated as follows:

$$\begin{aligned}\text{Annual Profit} &= \text{Annual Revenue} - \text{Annual Operational Costs}\\ &= 650,000 - 125,000 = 525,000\text{USD}\end{aligned}$$

with an Annual Profit of $525,000.

Table 4 Energy integration project cost and revenue breakdown

Component	Initial cost (USD)	Annual operating cost (USD)	Annual revenue (USD)	Description
Wind power plant	1,500,000	75,000	300,000	1 MW capacity, project life 20 years
Solar power	1,000,000	30,000	200,000	500 kW capacity, 25-year project life
Hydrogen storage system	800,000	20,000	150,000	200 kWh storage capacity, 20 years project life
Total	**3,300,000**	**125,000**	**650,000**	

Table 5 Breakdown of ship costs and revenues

Components	Conventional ship	Hydrogen-fueled ship	Description
Initial investment cost (USD)	2,500,000	4,000,000	Initial cost for ship construction
Annual operating costs (USD)	300,000	150,000	Fuel, maintenance, and other costs
Annual revenue (USD)	600,000	700,000	Revenue from transportation or other services
Annual profit (USD)	300,000	550,000	Revenue—Operating costs
Payback period (years)	8.33	7.27	Investment payback time

5. **Comparison of Economic Models: Conventional Ship vs. Hydrogen-Fueled Ship**

This project compares two economic models for ships: conventional ships using fossil fuels and hydrogen-fueled ships. The analysis includes initial investment costs, annual operational costs, and projected annual revenues for each model. This comparison aims to evaluate the economic feasibility and sustainability of both types of ships.

This table compares two ship models: Conventional Ship and Hydrogen-Fueled Ship. The initial investment cost for the conventional ship is $2,500,000, while the hydrogen ship amounts to $4,000,000 due to its more complex technology. However, the annual operational costs for the conventional ship reach $300,000, whereas the hydrogen ship only incurs $150,000, thanks to fuel savings and lower maintenance costs. The conventional ship is estimated to generate an annual revenue of $600,000, while the hydrogen ship can achieve $700,000 due to government incentives. The annual profit for the conventional ship is $300,000, while the hydrogen ship reaches $550,000. With a payback period of 8.33 years for the conventional ship and 7.27 years for the hydrogen ship, the investment in the hydrogen ship proves to be recouped more quickly and is more profitable in the long term (Table 5).

4 Conclusion

The conclusions that can be shown related to the integration of new and renewable energy in maritime operations are as follows:

1. One of the main pillars of globalization and the pillars of world trade is maritime transportation. About 80% of global trade takes place in all ports in the world. More than 2,000 ports around the world, economic growth, and pollution emissions are the main problems.

2. Friedlingstein et al. in 2022 showed an increasing trend in CO_2 emissions. One of the main greenhouse gases contributing to global warming and the greenhouse effect is CO_2, which is one of the big problems.
3. The use of renewable energy in maritime operations can help reduce carbon dioxide emissions and overcome global warming. Government Regulation No. 79/2014 concerning Indonesia's National Energy Policy sets the level of new and renewable energy use in Indonesia to be at least 23% by 2025 and 31% by 2050.
4. The Nordic region, is an area of the European Union that always pays attention to the integration of the use of new and renewable energy. The Norwegian government, one of the governments in the European Union region, has made a policy to conserve natural resources so that they can be used sustainably since the 1990s in response to the problem of global warming. Norway aims to reduce greenhouse gas emissions by thirty percent in its energy policy by 2020. Therefore, the Norwegian Ministry of Oil and Gas and Energy uses renewable energy from water, solar, and wind to generate electricity.
5. Integration of renewable energy *(Wind Energy)* in maritime: using *soft-sails*, applied to Greenheart's 75 dwt freighter, B9 Shipping's 3,000 dwt bulker and Dykstra/Fair Transport's 7,000 dwt Ecoliner; *fixed-sails*, applied to UT Wind Challenger and the EffShip's project; *Flettner Rotors,* applied to the Alcyone and Enercon's 12,800 dwt E-Ship 1; *kite sails* are applied to MS Beluga Skysails; *Wind turbines* (no successful prototypes until now).
6. Integration of renewable energy *(solar photovoltaics)* in maritime, hybrid models with other energy sources applied to small boats, such as NYK's Auriga Leader and SolarSailor by OCIUS Technology.
7. The largest solar-powered ship to date is the Tûranor PlanetSolar. The ship weighs 85 tons, with a length of 31 m and a width of 15 m. The vessel produces a maximum power of 93.5 kW as it is lined with 537 m^2 of solar panels and is one of the best solar-powered vessels in the world.
8. This analysis shows that although the initial investment in a hydrogen fuel cell system is quite high, the savings generated from fuel and maintenance costs can provide significant financial benefits in the long run. By considering total costs over 20 years, using renewable energy in maritime operations is not only environmentally friendly but also economically advantageous.
9. To look at the economic analysis of the application of renewable energy to maritime integration, the concept of Levelized Cost of Energy (LCOE) is used. LCOE is a lifetime cost measure, which is the total cost of building and operating a power plant over its estimated lifetime, divided by energy production.
10. This economic feasibility study demonstrates that the integrated energy project utilizing wind, solar, and hydrogen can be a profitable investment. With a relatively short payback period (less than 7 years), this project has the potential to make a significant contribution to carbon emissions reduction and the transition towards clean energy sources. The implementation of this integrated system will not only provide economic benefits but also enhance energy resilience and environmental sustainability.

11. This comparative analysis shows that although the initial investment cost for hydrogen-fueled ships is higher, the long-term economic model is more advantageous compared to conventional ships. Lower operational costs and higher revenue potential make hydrogen-fueled ships a more sustainable and economical choice. Considering environmental and social aspects, hydrogen-fueled ships can significantly contribute to energy transition efforts and sustainability in the maritime industry.
12. Many authorities and port operators around the world are committed to achieving zero emissions targets by incorporating renewable energy into maritime operations. Many ports have implemented renewable energy such as photovoltaic panels, wind turbines, and more.
13. According to the Report, the average LCOE for commercial-scale offshore wind projects in the United States fell by 13% to $84/MWh, which covers the price range of $61/MWh to $116/MWh. Is projected to drop again to an average of $60/MWh by 2030.
14. In 2018, the LCOE for floating solar power plants was €53/MWh, much higher than the LCOE for land-based power plants, which ranged between €35/MWh and €40/MWh. According to the NREL (National Renewable Energy Laboratory) report, the LCOE for floating solar power plants is $57/MWh without an Investment Tax Credit (ITC) and $38/MWh with an ITC. An on-site solar power plant costs $47/MWh without ITC and $32/MWh with ITC.

References

1. Esteve Pérez J, Gutiérrez Romero JE (2015) Renewable energy supply to ships at port. UPC Universitat Politècnica de Catalunya, Cartagena
2. Lomax OC, Wilkinson G, Martinez V (2021) Bahan bakar pelayaran rendah karbon untuk sektor pelayaran Indonesia. Global Maritime Forum, Indonesia
3. Petrychenko O, Levinskyi M, Goolak S, Lukoševičius V (2024) Solar energy: revolutionizing shipping industry towards sustainability and environmental stewardship. MDPI
4. Sarante J (2024) Energi Baru dan Terbarukan (EBT) Sebagai Teknologi Alternatif Dimasa Depan Dalam Mendukung Pertahanan Negara, Jakarta: s.n.
5. Apriliyanti K, Rizki D (2023) Kebijakan Energi Terbarukan: Studi Kasus Indonesia dan Norwegia dalam Pengelolaan Sumber Energi Berkenlanjutan. Jurnal Ilmu Pemerintahan Widya Praja 49:186–209
6. Parhamfar M, Sadeghkhani I, MohammadAdeli A (2023) Towards the application of renewable energy technologies in green ports: Technical and economic perspectives. IET Renew Power Gener
7. Mofor L, Nuttall P, Newell A (2015) Renewable energy options for shipping technology brief, s.l. IRENA
8. DNV (2014) Alternative fuels for shipping. *DNV*
9. Greig AR (2013) Future ship powering options: exploring alternative methods of ship propulsion. Royal Academy of Engineering, London
10. Rojon I, Dieperink C (2014) Blowin' in the wind? Drivers and barriers for the uptake of wind propulsion in international shipping. Energy Policy 67:394–402
11. Bailey D, Solomon G (2004) Pollution prevention at ports: clearing the air. Environ Impact Assess Rev 24:749–774

12. Tzannatos E (2010) Ship emissions and their externalities for the Port of Piraeus - Greece. Atmos Environ 44:400–407
13. Chang CC, Wang CM (2012) Evaluating the Effects of Green Port Policy: Case Study of Kaohsiung Harbor in Taiwan. Transp Res Part D 17:185–189
14. United Nations Conference on Trade and Development (2012) Review of maritime transport. UNCTAD, Geneva, Switzerland
15. McAllister L, Wang H (2024) Techno-economic and environmental analysis of the integration of PV systems into hybrid vessels. MDPI
16. Staffell I, Scamman D, Abad AV, Balcombe P, Dodds PE, Ekins P, Shah N, Ward KR (2019) The role of hydrogen and fuel cells in the global energy system. Energy Environ Sci 12(2):463–491
17. Zhang X, Zhao T (2020) Advances in hydrogen storage and transportation: enabling fuel cell deployment in remote regions. Renew Sustain Energy Rev 130:109947
18. International Energy Agency (IEA) (2021) Global hydrogen review 2021. IEA
19. Shiva Kumar S, Himabindu V (2019) Hydrogen production by PEM water electrolysis – a review. Mater Sci Energy Technol 2(3):442–454
20. Liu Y, Zhang D, Zhou J, Cai W, Wang J (2020) Progress in fuel cell applications in the maritime industry: a review. J Power Sources 448:227590
21. NCE Maritime CleanTech (2020) H2Ship project: hydrogen-powered shipping for a sustainable future. NCE Maritime CleanTech
22. Bouman EA, Lindstad E, Rialland AI, Strømman AH (2017) State-of-the-art technologies, measures, and potential for reducing GHG emissions from shipping – a review. Transp Res Part D: Transp Environ 52:408–421
23. DNV GL (2020) Maritime forecast to 2050: energy transition outlook 2020. DNV GL
24. World Bank (2020) The World Bank's role in fostering sustainable maritime transportation. World Bank
25. International Maritime Organization (IMO) (2021) IMO's efforts towards sustainable shipping. IMO
26. Permana SH (2023) TANTANGAN INVESTASI ENERGI BARU DAN ENERGI TERBARUKAN MENUJU INDONESIA NET ZERO EMISSION. INFO Singkat, Juni XV:16–20
27. Cormann M (2021) Clean energy finance and investment policy review of Indonesia. OECD, Paris
28. Krohn S, Morthorst PE, Awerbuch S (2009) The economics of wind energy, s.l. European Wind Energy Association
29. International Renewable Energy Agency (IRENA) (2020) Renewable power generation costs in 2019. IRENA
30. Lemaire E, Grégoire M, Goeury C, Roudier S (2019) The costs of offshore wind energy in Europe: a comparison of the costs of three offshore wind farms. Renew Energy 135:290–299

Chapter 8
Case Studies of Successful Energy Management Initiatives

Anggara Trisna Nugraha, Rikky Leonard, Imaniah Sriwijayasih, Eky Novianarenti, and Am Maisarah Disrinama

Abstract The implementation of digital technology in the maritime and port industries has significantly enhanced energy efficiency and reduced carbon emissions. Technologies like digital twins and predictive analytics enable real-time energy monitoring, allowing ports to identify inefficiencies and optimize operations, resulting in up to 25% energy savings in specific cases. Additionally, these technologies support compliance with international environmental regulations, such as IMO 2020, which sets strict limits on sulfur and carbon emissions. Beyond operational and environmental benefits, investments in digital technologies deliver long-term financial gains by reducing operational costs and enhancing efficiency. However, challenges remain, including high initial implementation costs, infrastructure requirements, and cybersecurity risks. Integrating renewable energy sources like solar and wind power strengthens sustainability strategies, making port operations more environmentally friendly and energy-efficient.

A. T. Nugraha (✉)
Department of Marine Electrical Engineering, Shipbuilding Institute of Polytechnic Surabaya, Surabaya, Indonesia
e-mail: anggaranugraha@ppns.ac.id

R. Leonard · I. Sriwijayasih
Department of Shipbuilding Engineering, Shipbuilding Institute of Polytechnic Surabaya, Surabaya, Indonesia
e-mail: rikkyleonard@ppns.ac.id

I. Sriwijayasih
e-mail: imaniahsriwijayasih@ppns.ac.id

E. Novianarenti · A. M. Disrinama
Department of Marine Engineering, Shipbuilding Institute of Polytechnic Surabaya, Surabaya, Indonesia
e-mail: ekynovianarenti@ppns.ac.id

A. M. Disrinama
e-mail: dokteram@ppns.ac.id

M. Elsisi et al. (eds.), *Maritime Infrastructure for Energy Management and Emission Reduction Using Digital Transformation*, Studies in Infrastructure and Control,
https://doi.org/10.1007/978-981-96-4438-4_8

Keywords Digital twin · Predictive analytics · Energy efficiency · Carbon emission reduction · Maritime industry · Renewable energy integration · Cybersecurity

1 Introduction

The maritime sector is grappling with urgent and significant energy challenges, primarily from heavy reliance on fossil fuels for propulsion and operations. This dependence leads to exorbitant fuel costs, environmental issues, and regulatory pressures. It is essential to implement effective energy management strategies to address these challenges. By optimizing fuel consumption, minimizing energy loss, and embracing sustainable practices, the maritime industry not only improves operational efficiency and reduces costs but also plays a critical role in shaping a greener future [1].

Digital transformation holds immense potential as a solution to overcome energy challenges in the maritime sector. By harnessing advanced technologies such as IoT, AI, and data analytics, maritime operators can gain invaluable insights into energy consumption patterns, identify inefficiencies, and optimize operations. Real-time monitoring, predictive maintenance, and intelligent route planning can significantly reduce fuel consumption, minimize emissions, and improve energy efficiency. Moreover, digital solutions can facilitate the integration of renewable energy sources and support the transition to cleaner fuels, marking a more sustainable and promising future for the maritime sector [2].

Digital transformation in the maritime sector promises more efficient energy management and emission reduction. Advanced technologies such as IoT, AI, and data analytics are being leveraged to improve ship operations, monitor fuel consumption, and determine areas for improvement. IoT sensors collect real-time data on the ship's performance, allowing for early detection of anomalies and proactive maintenance [3]. AI algorithms analyze this data to optimize routes, reduce fuel consumption, and predict equipment failures. Notably, digital platforms play a pivotal role in integrating renewable energy sources and supporting the transition to cleaner fuels, instilling hope for a more sustainable maritime industry [4].

This chapter explores real-world case studies on implementing digital strategies for successful energy management. The following section contains three case studies from around the world that can inspire you about the maritime sector.

1.1 Overview of Digital Technologies in Maritime Energy Management

The digital transformation in the maritime sector has introduced various technologies aimed at making energy management more effective and efficient. These technologies seek to optimize ship and port operations by reducing energy consumption, enhancing predictive maintenance, and enabling real-time monitoring capabilities. Key technologies in maritime energy management include the Internet of Things (IoT), Artificial Intelligence (AI), and digital twins. Together, these technologies play a crucial role in improving operational efficiency while reducing the environmental impact of this energy-intensive industry.

1.2 Internet of Things (IoT) and Its Applications

IoT technology enables the continuous collection of data through sensors installed on ships and port infrastructure. These sensors measure essential parameters, such as fuel consumption, engine temperature, pressure, and vessel speed. Through seamless connectivity, this data is integrated into energy monitoring systems to provide in-depth, real-time insights into energy use. For instance, IoT sensors allow ship captains and port operators to track fuel usage accurately, identify energy inefficiencies, and optimize navigation routes and vessel speed to lower energy consumption. Studies from reputable sources, like the *Journal of Maritime Energy Systems*, have demonstrated that IoT in energy management can reduce fuel consumption by up to 15% on large vessels [5].

1.3 Artificial Intelligence (AI) for Predictive Maintenance

In addition to energy monitoring, Artificial Intelligence (AI) has become an invaluable tool for accurately predicting equipment maintenance needs. In the context of predictive maintenance, AI algorithms analyze sensor data to identify patterns or anomalies that might go unnoticed by human operators. For example, AI can predict when a ship's main engine or generator might require maintenance based on detected wear patterns and operating conditions. By anticipating these maintenance needs, shipping companies can prevent unnecessary downtime, maintain operational efficiency, and control energy costs. Research presented in the *Proceedings of the Maritime Technology Conference* supports this, showing that implementing AI for predictive maintenance can enhance fuel efficiency by up to 10% [6].

1.4 Digital Twins for Real-Time Monitoring

The concept of digital twins creates a detailed virtual representation of a vessel, providing a digital replica that enables real-time monitoring. In energy management, digital twins deliver deeper insights into a ship's operational status, from engine conditions to energy usage patterns. By integrating data from IoT and AI, digital twins allow for highly accurate simulations, enabling more efficient energy management through predictive scenarios. For instance, before altering a ship's route or conducting specific maintenance, digital simulations can predict the potential impact on overall energy consumption. According to an article in *Marine Engineering Review*, using digital twins can optimize energy consumption by 12% compared to conventional monitoring methods [7].

The adoption of digital technologies such as IoT, AI, and digital twins in maritime energy management is transforming this sector. These technologies not only optimize energy consumption but also bolster operational resilience through predictive maintenance and real-time monitoring. International research and publications have confirmed the reliability of these technologies in enhancing energy efficiency, reducing operational costs, and minimizing the environmental impact of the maritime industry. As these technologies advance, shipping companies are expected to continue innovating in their applications to create a more sustainable maritime industry.

1.4.1 Case Study 1: Digital Energy Management in Global Shipping Companies Company Background

Maersk Line, a Danish shipping company and global leader in container shipping, operates a vast fleet of over 700 vessels, comprising container ships, tankers, and bulk carriers. The company faces numerous energy challenges, including fluctuating fuel prices, stringent environmental regulations, fuel consumption optimization, and managing vast amounts of data. However, these challenges also present opportunities for innovative solutions that can significantly enhance energy efficiency, reduce emissions, and improve operational performance, ultimately shaping a brighter future for the global shipping industry [1].

- **The Crucial Role of IoT and AI in Energy Management**

The implementation of IoT in Maersk Line's operations includes ship monitoring, remote conditions, and smart containers. IoT sensors play a pivotal role in collecting real-time data on various parameters, which is then transmitted to a central platform for analysis and optimization. This technological advancement in data collection not only informs but also empowers industry professionals, shipping company executives, and environmental regulators about the current state of the shipping industry. Artificial Intelligence proactive approach to predictive maintenance, route optimization, and energy efficiency [8] (Fig.1).

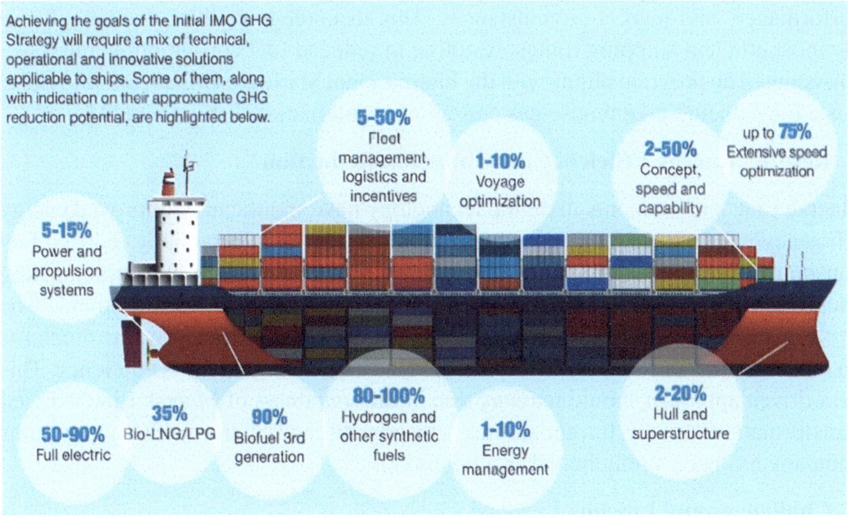

Fig. 1 Case study: a wide variety of design, operational, and economical solutions

Digital Technologies Assist in Fulfilling Regulatory Needs Digital technology is essential for the maritime sector to adhere to stringent regulatory requirements. For example, IoT sensors and data analytics platforms make real-time monitoring of vessel performance, fuel consumption, and emission levels possible. This real-time data facilitates prompt remedial action by identifying departures from regulatory restrictions. Additionally, AI-powered systems may optimize delivery routes by analyzing weather data, vessel performance, and market conditions, which lowers emissions and fuel usage [6, 9].

AI is also used in predictive maintenance, enabling businesses such as Maersk Line to plan maintenance in advance and reduce unscheduled repairs' associated downtime and fuel usage. Artificial intelligence (AI) algorithms and Internet of Things (IoT) sensors are examples of digital tools that may analyze energy usage trends and spot areas for improvement [8].

Maersk Line optimizes its shipping routes using AI-powered systems that analyze weather conditions, vessel performance, and market circumstances. This technology enables them to choose the most efficient shipping routes, resulting in reduced fuel consumption and lower emissions. The program aligns with the International Maritime Organization's (IMO) goals for reducing greenhouse gas emissions in international shipping [10].

Additionally, Maersk Line employs AI to identify equipment faults and schedule preventive maintenance. This proactive approach minimizes downtime and reduces fuel consumption caused by unscheduled maintenance, further lowering emissions. The digital systems also allow Maersk Line to accurately assess its carbon footprint and monitor progress toward emission reduction targets. Maersk Line optimizes its shipping routes using AI-powered systems that analyze weather conditions, vessel performance, and market circumstances. This technology enables them to choose the most efficient shipping routes, resulting in reduced fuel consumption and lower emissions. The program aligns with the International Maritime Organization's (IMO) goals for reducing greenhouse gas emissions in international shipping [11].

- **Results: Energy Efficiency and Emission Reduction**

Maersk Line's investments in digital technology have significantly improved energy efficiency and reduced CO2 emissions. Maersk Line has substantially reduced fuel consumption by optimizing routes, enhancing vessel performance, and implementing energy-saving measures, directly contributing to a healthier environment. The streamlining of operations, reduced downtime, and, most importantly, the emphasis on data-driven decision-making have all played a crucial role in this efficiency. This data-driven approach should reassure you of the soundness of Maersk Line's digital transformation strategy. Its reputation as a responsible and environmentally conscious company has been strengthened as a bonus [1].

- **Challenges and Lessons Learned**

Like many large organizations, several challenges and lessons have been learned during their digital transformation journey. Data management and integration must be prioritized when integrating data from various sources. Security in protecting sensitive data and systems from cyber threats is essential—adaptive implementation strategy regarding digitalization acceptance in the workforce. Investing in data management is never a loss. Prioritize implementing robust measures to protect sensitive data and systems. In the process of digital transformation, it is essential to address problems and build support. Additionally, track the progress and results of digital initiatives to demonstrate their value and identify areas for improvement. By carefully considering these challenges and applying the lessons learned, other organizations can successfully implement digital transformation and reap the benefits of increased efficiency, sustainability, and competitiveness [12].

1.4.2 Case Study 2: Use of Renewable Energy Sources in Maritime Operations

Indonesia is located on the equator, which is the area of movement. Geographically, Indonesia is an archipelagic country with two-thirds of the ocean area larger than the land area. This can be seen from the coastlines on almost every island in Indonesia (±81,000 km). This advantage makes Indonesia rank second after Canada as the country with the longest coastline in the world. Apart from providing additional natural wealth, the sea also provides certain functions to Indonesia, in particular influencing welfare/socio-economics and the level of progress in the field of science and technology (Science and Technology). This strength has great potential to advance the Indonesian economy (Setyawati et al. 2021) [13]. Ships are one of the means of transportation at sea. Ships have advantages in transportation, namely they can carry more cargo compared to other means of transportation, such as cars or trains. Using

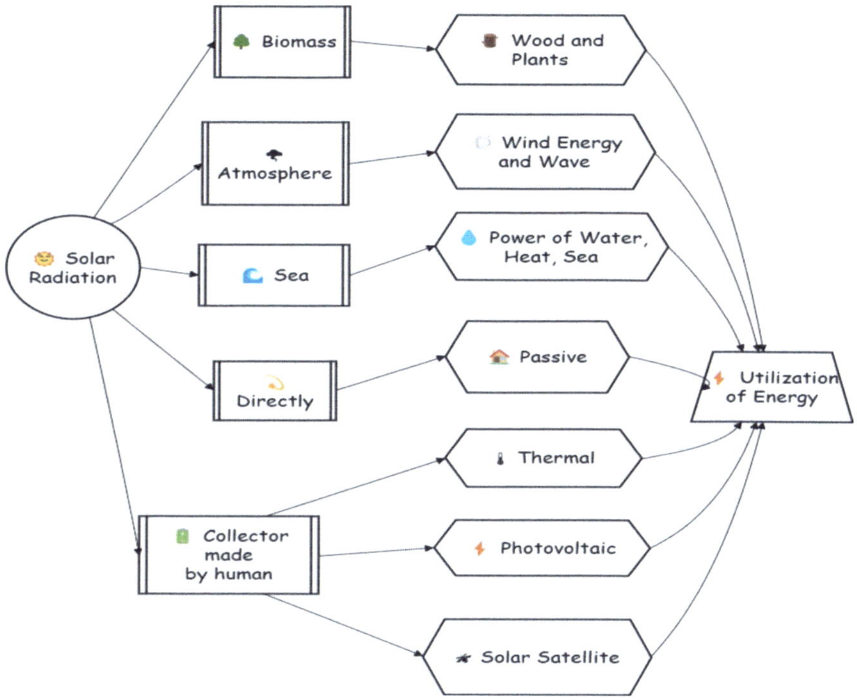

Fig. 2 Schematization of solar energy utilization

ships as a means of transportation has a relatively lower accident rate because, in the construction of ships, facilities are equipped to prevent and overcome them [14] (Nugraha et al. 2021). Apart from using diesel engines, power plants on ships can also utilize energy from sunlight as a source of electrical energy [15, 16]. This can be used as an alternative source of electrical energy because the need for electrical energy continues to increase every year (Fig. 2).

Utilization of solar energy as electrical energy can be done using Solar PV installed on the ship. The electrical energy produced can be used to power electrical equipment and ship lights at night and can be used as a fishing aid. The use of solar energy as a source of electrical energy is expected to improve the health (sanitation in the engine room) of fishermen, as well as reduce the use of fossil energy which is not good for the environment [15] (Nugraha et al. 2018, 2019, 2021).

- **How Digital Technologies Help Meet Regulatory Requirements**

Digital technology can help maritime companies comply with emissions and carbon regulations in various ways, namely monitoring and reporting emissions by utilizing a monitoring, reporting, and verification (MRV) system using sensors and IoT devices to monitor greenhouse gas emissions in real time and report them by regulations as stipulated by IMO. Route Optimization by utilizing algorithms and data analysis optimizes shipping routes, reducing travel time to reduce fuel consumption and carbon emissions. Energy management involves optimizing the energy management system, monitoring and managing energy consumption on ships, identifying inefficiencies, and providing recommendations for energy savings. By adopting digital technology, maritime companies can monitor, manage, and optimize their vessel operations more effectively, ensure compliance with emission regulations, and contribute to environmental sustainability.

- **Technological and Operational Changes**

Indonesia has quite a large potential for new and renewable energy (EBT), including mini/micro hydro of 450 MW, biomass of 50 GW, solar energy of 4.80 kWh/m2/day, wind energy of 3–6 m/s, and nuclear energy of 3 GW. The latest EBT potential data was presented by the Director of New, Renewable Energy, and Energy Conservation in a Focus Group Discussion on Supply–Demand. Policies to reduce greenhouse gases have been implemented in various parts of the world, such as the Paris Agreement policy. The Paris Agreement is an international treaty or agreement that initiates mitigation and adaptation related to climate change which was agreed in 2015 (Mukhtarov et al. 2023). This policy is designed to guide countries in efforts to reduce emissions of carbon dioxide and other greenhouse gases. Indonesia implements it in the National Energy Policy contained in Government Regulation No. 79 of 2014 which was later updated through Presidential Decree No. 11 of 2023. Implementing the national energy policy is expected to improve renewable energy infrastructure to solve the problem of providing and distributing electricity. The use of microgrid and electric vehicle technology can also be used as an alternative option to increase the renewable energy mix [17].

Along with increasing urbanization, the need for energy in cities is also increasing. Most of Indonesia's energy needs still use non-renewable energy which has a negative impact on the environment and also disrupts the ecosystem. Carbon reduction is a discourse that continues to attract the attention of the global community, especially regarding policies to reduce carbon emissions from industrial activities to boost a country's productivity. There are three key elements in energy policy, or energy policy to maintain a country's energy security. First is rationing in the form of allocating existing energy resources by limiting their use. The aim is to reduce demand for certain energy sources (such as petroleum, natural gas, and coal), with the hope that these restrictions will provide additional time to address energy limitations. Second, stockpiling is carried out by storing reserves of the most widely used energy sources, such as petroleum. This aims to reduce dependence on oil-exporting countries, maintain energy security, and avoid uncontrolled fluctuations in world oil prices. Third, diversification is an effort to ensure the supply of energy resources by conducting research, exploration, and finding new energy sources (alternative energy) to replace existing energy sources. The country is trying to develop alternative energy resources such as biofuels, hydropower, solar cells, and nuclear power. International cooperation can also be part of this diversification strategy (Bollino & [18]. Deforestation is a crucial issue when talking about environmental degradation and survival. Concerns

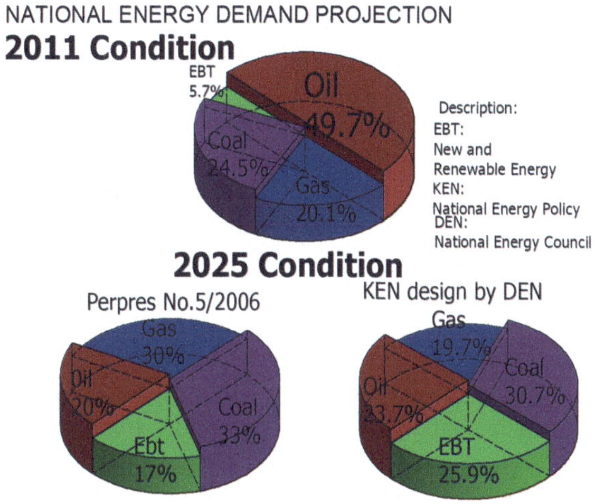

Fig. 3 National energy demand projection

about uncontrolled deforestation are causing ongoing ecological damage. Countries implementing forest utilization programs are now under international pressure to take steps to slow the rate of deforestation. Deforestation is a form of land loss that can be a threat to the survival of living things; forest areas are decreasing due to forest engineering for human interests, then causing serious impacts in the form of global warming (Wahyuni & [19]). Various activities caused by deforestation will have a direct impact and be felt by humans, so efforts are needed to encourage environmental awareness among the public to preserve Indonesia's nature (Fig. 3).

- **Results and Environmental Impact**

Environmental damage and low public awareness of the environment encourage the need to implement sustainable development by utilizing renewable energy. Human activities that use fossil energy, for example coal, oil, and flammable gas, cause higher carbon dioxide emissions, such as deforestation, and can also increase temperatures throughout the earth, resulting in global warming which is not environmentally friendly. Utilization of petroleum energy is still the main energy source used by many countries. The use of energy, for example, coal, oil, and gasoline, has become the main use in every country. Fossil fuels have become a global primary energy demand and have become dependent on consuming fossil energy to support a country's economy. This is supported by World Bank data which states that as much as 80% of global demand is for fossil energy. Coal, oil, and combustible gas are some of the non-renewable energy sources that are the biggest contributors to environmental change around the world. Thus, the use of energy from petroleum products is a fundamental part of what drives CO2 emissions.

1.4.3 Case Study 3: Real-Time Monitoring and Predictive Analytics for Energy Efficiency

The digital twin technology has brought significant changes to port operations, especially in real-time energy consumption monitoring. This technology enables ports to create virtual replicas of physical assets like ships, cranes, and other infrastructure, allowing for direct performance monitoring. By leveraging data from connected sensors, digital twins can identify areas of energy inefficiency, enabling rapid corrections. Research by [20] found that the use of digital twins in European ports reduced energy consumption by up to 15%, indicating that this technology can significantly improve port energy efficiency.

Digital technologies also assist maritime companies in meeting strict emission regulations and carbon reduction targets. Integrated digital systems allow for automatic emission monitoring, ensuring that ports and maritime operators comply with regulations like IMO 2020, which limits sulfur content in marine fuels. Additionally, these technologies generate auditable reports, making it easier for companies to adapt to regulatory changes. Reports from the [21] indicate that digital technology accelerates compliance with carbon emission targets more efficiently than manual methods (Table 1).

One successful example of digital technology application is Maersk Line, which met IMO sulfur emission regulations. In 2020, Maersk implemented digital twin systems to monitor fuel consumption and emissions in real time. This system allowed them to monitor every ship in their fleet and adjust operations to meet regulatory requirements. As reported in the [22], Maersk reduced sulfur emissions by 10% in the first year without significant operational disruption, showcasing digital technology's potential to help companies meet global environmental standards (Table 2).

Predictive analytics also plays a key role in optimizing energy consumption. By analyzing historical and real-time data, these predictive systems allow port companies to forecast future energy needs and identify inefficient energy consumption patterns. According to the Proceedings of the Maritime Energy Efficiency Symposium [23], predictive analytics reduced port energy consumption by up to 20%. With such analyses, operators can take preventive steps to avoid unnecessary energy peaks, optimize resource usage, and prevent energy waste.

The results of implementing digital technology in port operations demonstrate significant energy savings and carbon emission reductions. For example, the Port of Rotterdam reduced energy consumption by up to 25% after adopting digital technology, as reported in [24]. Additionally, the port's carbon emissions decreased substantially, aiding compliance with strict EU environmental targets. This illustrates that digital energy management technologies not only support economic goals but also contribute to broader environmental sustainability.

The long-term impact of digital technology initiatives in ports is substantial. The energy savings achieved are sustainable, and this technology can be implemented at larger scales in other ports. In the long run, these technologies will enhance overall operational efficiency and enable ports to respond more effectively to evolving environmental regulations. According to the [25], the scalability of these technologies

Table 1 Implementation of digital technology in case study on real-time monitoring and predictive analytics for energy efficiency

No	Digital technology aspect	Case study implementation	Key outcomes
1	Digital twin for port operations	Implementation of digital twin technology for real-time energy consumption monitoring on port assets	15% energy savings at ports in Europe
2	Emission regulation compliance	Digital systems automatically monitor and report emissions to meet IMO standards and carbon reduction targets	Simplifies IMO compliance and process efficiency
3	Maersk line case study	Maersk uses digital twin technology to monitor fuel consumption and emissions, complying with IMO 2020	10% sulfur emission reduction in the first year
4	Predictive analytics	Predicts energy consumption and identifies inefficiency patterns, allowing for resource optimization in ports	Energy consumption reduced by up to 20%
5	Energy savings and emission reduction	Digital technology helps ports achieve significant energy savings and reduce carbon emissions	25% energy consumption reduction at Port of Rotterdam
6	Long-term impact and scalability	Technology can be applied to other ports, supporting long-term efficiency initiatives in the port sector	Sustainable and eco-friendly efficiency
7	Renewable energy integration	Use of solar and wind energy combined with predictive technology for energy source optimization	Operational efficiency increased by up to 30%
8	Long-term financial benefits	ROI obtained from energy savings and lower operational costs	Significant cost reduction
9	Challenges and cybersecurity	Challenges include initial costs and cybersecurity threats	Cybersecurity is a top priority
10	Human resources development	Workforce training in operating digital twin and predictive systems	Improved operational staff efficiency
11	Big data for energy efficiency	Analysis of big data from IoT sensors to enhance energy management decisions	Optimized resource usage
12	5G connectivity	Enhanced connectivity for IoT devices in ports, accelerating response and decision-making	Higher operational efficiency
13	Research and innovation	Development of AI and research collaborations for energy optimization	Support for further energy efficiency

(continued)

Table 1 (continued)

No	Digital technology aspect	Case study implementation	Key outcomes
14	Future regulatory challenges	Possible stricter IMO regulations require ports to keep innovating	Drives innovation and sustainability
15	Maritime industry sustainability	Digital twin and predictive analytics as key elements in achieving sustainability	A more sustainable maritime industry

presents substantial opportunities for ports worldwide to adopt similar solutions and achieve comparable energy efficiency.

In addition to digital technology, many ports are integrating renewable energy into their operations for greater efficiency. The use of solar and wind energy for port power needs is increasingly common, especially with predictive analytics technology ensuring optimal use of these energy sources. The [26] reports that ports that integrate renewable energy with digital technology improve operational efficiency by up to 30%, reducing dependence on fossil fuels and supporting carbon emission reduction targets more effectively.

Beyond environmental benefits, investment in digital technology has proven financially beneficial in the long term. While initial implementation costs are high, the return on investment (ROI) through energy savings and lower operational costs can be achieved within two to three years. According to the [27], ports that adopted digital technology reported significant cost reductions due to increased efficiency, decreased unscheduled maintenance, and optimized energy resource use.

However, implementing digital technology is not without challenges. Ports face issues such as high initial costs and the need for adequate infrastructure. Additionally, cybersecurity remains a major concern, given the increasing number of cyber-attacks on complex digital systems. The report [28] notes a 15% increase in cyber-attacks in the port sector over the past five years, highlighting cybersecurity as a critical component in digital technology implementation.

Human resource development is also essential to support digital technology adoption. Port employees must be trained to understand and effectively use these new technologies. Proper training can boost productivity and operational efficiency. A study in the [29] found that ports investing in workforce training observed significant efficiency improvements as staff could fully leverage digital technology capabilities.

Big data also plays a crucial role in port energy management. By collecting data from various IoT-connected sensors, big data enables deeper analysis of energy consumption patterns, which is then used to optimize operations and prevent energy waste. The article [30] notes that big data integration assists ports in making better decisions regarding energy resource allocation, ultimately leading to increased efficiency.

Table 2 IMO emission reduction regulations and technology implementation in case study

No	IMO regulation	Description	Case study implementation	Results and impact
1	IMO 2020 (sulfur emission cap)	Limits sulfur content in ship fuel to a maximum of 0.5%	Digital twin used to monitor fuel consumption at Maersk Line	10% sulfur emission reduction in the first year
2	IMO greenhouse gas reduction strategy	Aims for a 50% reduction in GHG emissions from ships by 2050 compared to 2008 levels	Predictive technology for optimizing energy and reducing carbon emissions	Significant carbon emission reduction at Port of Rotterdam
3	Ship Energy Efficiency Management Plan (SEEMP)	Requires ship operators to implement effective energy management systems	Digital twin for real-time monitoring and predictive analytics at ports	Improved energy efficiency, cost savings, regulatory compliance
4	Energy Efficiency Design Index (EEDI)	Sets minimum emission limits for new ships built after a certain date	Port innovation with renewable energy technology and big data	Enhanced energy efficiency and IMO compliance
5	IMO long-term carbon emission target	Ensures gradual carbon emission reductions through technological innovation and renewable energy development	Solar and wind energy usage combined with digital integration for energy source optimization	30% energy efficiency and carbon emission reduction
6	IMO DCS emission monitoring and reporting framework	Requires CO_2 emission monitoring, reporting, and verification for ships over 5,000 GT	Digital twin with automatic monitoring at Maersk line	IMO DCS compliance, automated reporting
7	Energy Efficiency Existing Ship Index (EEXI)	Sets energy efficiency standards for ships, requiring structured reporting and management	Digital monitoring for energy optimization and efficiency at ports	Sustainable port energy efficiency
8	Innovative technology for emission reduction	Encourages the maritime industry to adopt digital and innovative technologies as a long-term strategy for emission reduction and efficiency enhancement	Use of predictive analytics, AI, and digital twin technology	Sustainable carbon emission reduction and long-term compliance

5G technology has also been adopted by some modern ports to support faster and more reliable device connectivity. With low latency and high data transfer speeds, 5G allows energy and operational monitoring systems to work more efficiently. A study published in [31] shows that ports adopting 5G technology experienced significant improvements in response times to operational issues and energy optimization, ultimately enhancing efficiency and energy savings.

The adoption of digital technology also opens opportunities for further research and innovation in the port and maritime sectors. Collaboration between academia, industry, and government can yield more innovative solutions, such as using artificial intelligence (AI) to monitor and manage energy automatically. Research from the [32] suggests that with the support of these technological innovations, the port industry can achieve higher efficiency and reduce carbon emissions on a larger scale.

However, future regulatory challenges continue to loom over the industry. Stricter regulations from IMO and other organizations are anticipated to evolve, pushing companies to innovate continuously to comply with standards. According to the Global Maritime Regulations [33], the maritime industry that adopts digital technology early on will be better prepared to meet future regulations and ensure sustainable long-term operations.

In conclusion, the application of digital technologies in the maritime and port industries, such as digital twins and predictive analytics, has brought significant benefits in terms of energy savings and emission reduction. Case studies like Maersk Line demonstrate that these technologies not only help companies comply with international emission regulations but also provide substantial economic benefits. With wide-scale implementation and a high potential for growth, digital technology will continue to be a key element in supporting the future sustainability of the maritime industry.

2 Cybersecurity Considerations in Digital Maritime Systems

The introduction of technology is critical in the development of Indonesian Maritime Affairs. Computer-based design and simulation have provided significant benefits to port development, shipbuilding, and maritime transportation. Cybersecurity is essentially an issue in security studies that is still very new. This issue arises when all aspects of political, military, economic, social, and cultural life are connected to cyberspace. Cyber threats that have the potential to become threats are cyber terrorism, cybercrime, and cyber war. Southeast Asia, as one of the important regions in the world with a fairly high level of economic growth, cannot be separated from this threat. Information security has become an expensive matter nowadays, so ethical hacking is needed to ensure that a company's information system is reliable enough. This way you can maintain your reputation in the eyes of your partner [34]. Based on a survey conducted by the Indonesian Internet Service Providers Association (APJII) in 2022, more than 77.02% of Indonesians use the Internet. This means that there are 210,026,769 people out of a total population of 272,682,600 Indonesians who are connected to the Internet.

The increase in Internet users occurred during the COVID-19 pandemic due to the use of Internet technology to replace face-to-face activities. This is also in line with data released by the Organisation for Economic Co-operation and Development, where there is an increase in Internet access which continues to increase year over year by both households and individuals throughout the world (OECD 2021). From the technological aspect of GCI, Indonesia is currently in the process of creating national standards and evaluation tools for how information security should be carried out by government agencies and industry. One of the interesting points in this strategy is how Indonesia implements a form of Internet filtering through the Trust+ and Nawala filtering programs. The debate regarding Indonesian filtering is also an important topic in the Indonesian cyber landscape [35] and should be part of future national cybersecurity discussions in Indonesia.

3 Future Trends and Innovations in Maritime Energy Management

Digital transformation is an effective solution to improve the efficiency and effectiveness of maritime law enforcement in Indonesia. By adopting advanced technologies such as satellite monitoring systems, big data analysis, and artificial intelligence, law enforcement can be faster and more accurate in detecting and handling illegal activities at sea. New regulations, including Government Regulation No. 13 of 2022, support the integration of information systems that enable real-time data exchange between agencies, improving coordination, transparency, and accountability in maritime law enforcement [36]. AI and monitoring automation can be an effective solution for marine ecosystem conservation by reducing monitoring bottlenecks and long-term costs. Although implementation challenges remain, research and emphasis on automated tools and AI across disciplines suggest that the future of monitoring and management in marine science can be facilitated and enhanced by the implementation of automation. There are several things that need to be considered in the use of deep learning, big data processing, and data-driven modeling [37].

A thorough examination of efforts to enhance energy efficiency in ports through operational strategies, innovative technologies, and energy management systems. Amid rising energy prices and a global focus on reducing greenhouse gas (GHG) emissions, many ports have implemented measures to transition from a carbon-based industry to a low-emission port model. This study highlights the significant potential for achieving better energy efficiency in ports, as well as the high-impact research opportunities to support the sustainability goals of this industry. The things that can be done to support the sustainability of the industry are operational strategies for energy efficiency, technologies for energy efficiency, intelligent energy management systems, opportunities for renewable energy and clean fuels, and challenges and future research directions [38].

4 Conclusion and Recommendations

4.1 Conclusion

The integration of digital technology in the maritime and port industries has proven highly effective in enhancing energy efficiency and reducing carbon emissions. Through digital twins, predictive analytics, and real-time monitoring systems, ports can optimize resource use, identify inefficiencies, and meet stringent environmental regulations. Notable successes, such as energy consumption reductions of up to 25% in European ports and 10% sulfur emission reductions at Maersk Line, demonstrate the practical value of these technologies. Beyond environmental gains, the adoption of renewable energy sources and advanced digital systems also leads to long-term operational and financial benefits. However, initial costs, infrastructure needs, and cybersecurity remain pressing challenges that require careful attention and strategic investment.

4.2 Recommendations

1. Investment in Digital Infrastructure: Ports and maritime operators should prioritize investments in digital infrastructure, including digital twins and predictive analytics, to drive energy efficiency and support emission compliance.
2. Cybersecurity Measures: With rising cybersecurity threats, it is essential for ports to implement strong security protocols to protect digital systems and safeguard critical data.
3. Training and Workforce Development: Effective adoption of digital technology requires a skilled workforce. Investing in training for port staff to manage and maintain new systems will improve operational efficiency and drive greater returns on technology investments.
4. Expansion of Renewable Energy Use: Integrating renewable energy sources such as solar and wind, coupled with predictive analytics, can further improve energy efficiency and support sustainability goals.
5. Collaboration for Future Innovation: Partnerships between industry, academia, and government are vital for ongoing research and innovation, particularly in developing new digital tools and strategies to improve energy management and regulatory compliance.

6. Implementing these recommendations will support the maritime industry's transition toward more sustainable, efficient, and resilient operations in response to evolving environmental regulations.

References

1. Kumar et al (2021) A review of energy efficiency in the maritime sector. J Mar Sci Technol 29(1):2021
2. Zhang et al (2020) Digital twin technology for energy efficiency in maritime. J Transp Syst Eng Manag 10(4):2020
3. Gupta A, Kumar R, Singh S (2021) Artificial intelligence for maritime energy optimization: a systematic review. J Mar Sci Eng 9(1)
4. Park et al (2022) Blockchain technology for maritime energy management: a review. J Mar Sci Technol 30(2):2022
5. Smith D, Wang X (2020) IoT-based fuel optimization in maritime transport. J Marit Energy Syst
6. Lee J, Kim H (2021) Predictive maintenance using AI in maritime operations. In: Proceedings of the maritime technology conference
7. Nguyen P, Larson T (2022) Digital twins for energy efficiency in maritime vessels. Mar Eng Rev
8. Sharma M, Gupta R (2022) The role of digital technology in achieving sustainable shipping: a case study of Maersk line. J Sustain Transp 10(2)
9. Lee J, Kim S (2021) IoT and AI in maritime logistics: a case study of Maersk line. J Ind Inf Syst 15(4)
10. Calado G, Castro R (2021) Hydrogen production from offshore wind parks: current situation and future perspectives. Appl Sci 11(12). https://doi.org/10.3390/app11125561
11. Park J, Lee K (2023) Blockchain technology for supply chain transparency and sustainability: a case study of Maersk line. J Sustain 15(4)
12. Wang J, Chen H (2023) AI-powered predictive maintenance in maritime shipping: a case study of maersk line. J Intell Manuf 12(3):2023
13. Sukamto (2017) Management of Indonesian marine potential in the spirit of Islamic economics. J Islam Econ 35–61
14. Gultom ER (2017) Refunctioning Indonesian sea transportation through the sea highway for the economic development of eastern Indonesia. ITS Sci Arts J 1–12
15. Dwicaksana MP (2021) Review and analysis of PLTS development transportation means. J Resist 105–118
16. I Made Aditya Nugraha (2022) Study of the use and availability of PLTS as a source of energy. Natl Mar J 123–130
17. Utami (2022) Implementation of Indonesia's electric vehicle policy to support national energy security. Energy Secur J 49–65
18. Galkin CA (2021) Energy security and portfolio diversification: conventional. J Energ 1–24
19. Suranto HW (2021) The impact of large scale deforestation on global warming. JIIP: Sci J Gov Sci 148–162.
20. Cordeiro PR, Nunes MC, Lopes JA (2020) Digital twins for energy efficiency in European ports. Int J Marit Technol 8(3):58–66
21. International Journal of Maritime Technology (2021) Emission compliance through digital solutions in the maritime industry. IJMT 9(4):245–259
22. Journal of Cleaner Production (2022) Sulfur emissions management through digital monitoring in the shipping industry: The Maersk case study. JCP 325:129035

23. Proceedings of the Maritime Energy Efficiency Symposium (2021) Advances in predictive analytics for energy optimization in port operations. MEES 7(2):112–120
24. Energy Efficiency in Maritime Studies (2023) Digital technology and carbon reduction in the Port of Rotterdam. EEMS 14(1):31–44
25. Journal of Maritime Logistics and Transport (2022) Scalability of digital solutions for global port energy management. JMLT 6(1):97–105
26. Renewable Energy Journal (2021) Integration of renewable energy in modern port operations. REJ 15(4):183–198
27. International Journal of Port Management (2022) Financial returns of digital technology adoption in port management. IJPM 11(2):59–74
28. Cybersecurity in Maritime Operations (2022) Cybersecurity risks in digital port systems. CMJ 3(4):45–62
29. Journal of Maritime Workforce Development (2023) Impact of workforce training on digital technology adoption in ports. JMWD 5(1):77–85
30. Big Data in Maritime Operations (2022) Using big data to optimize port energy management. BDMO 12(3):101–115
31. Telecommunications for Maritime (2023) 5G integration in port operations for energy efficiency. TFM 9(2):203–220
32. Journal of Port Innovation (2023) AI and digital solutions in energy-efficient port management. JPI 4(3):11–28
33. Review GMR (2022) Emerging regulations and the role of digital technology in compliance. GMRR 7(4):201–218
34. Muzaki AR (2019) The effect of ethical hacking on company data security. CyberSecurity dan Forensik Digit 77–81
35. Walsh R (2021) A guide to internet censorship in Indonesia - And how to. https://proprivacy.com/guides/indonesia-privacy
36. Maruf IR (2024) The importance of digital transformation in maritime law enforcement at sea. Fox Justi: Jurnal Ilmu Hukum 14(2):118–127. https://doi.org/10.58471/justi.v14i02
37. Ditria EM, Buelow CA, Gonzalez-Rivero M, Connolly RM (2022) Artificial intelligence and automated monitoring for assisting conservation of marine ecosystems: a perspective. Front Mar Sci 9:1–14. https://doi.org/10.3389/fmars.2022.918104
38. Iris Ç, Lam JSL (2019) A review of energy efficiency in ports: operational strategies, technologies and energy management systems. Renew Sustain Energy Rev 112:170–182. https://doi.org/10.1016/j.rser.2019.04.069

Chapter 9
The Application of Artificial Intelligence of Things (AIoT) to Smart Port Emission

Azlina Idris, Wan Norsyafizan W. Muhamad, Aidatul Julia Abd Jabar, Mohammad Basuki Rahmat, and Idris Taib

Abstract This chapter explores the transformative integration of Artificial Intelligence of Things (AIoT) in the development of smart ports, emphasizing its role in emission monitoring and reduction. AIoT, combining artificial intelligence and the Internet of Things, enhances port efficiency, sustainability, and decision-making through real-time data analysis and automation. The discussion highlights key technologies, including predictive analytics, autonomous systems, and IoT-enabled sensors, which optimize operations, reduce emissions, and improve resource allocation. Case studies of ports in Singapore, Busan, and Hong Kong demonstrate successful implementations, showcasing reduced environmental impact, cost savings, and operational enhancements. Addressing challenges such as regulatory compliance, infrastructure integration, and high implementation costs, the chapter underscores the importance of stakeholder collaboration and phased strategies. Future directions advocate for advanced AIoT applications like high-resolution sensor networks, edge computing, and predictive emission controls to meet the increasing demands for sustainable port operations. This work concludes that AIoT is pivotal in redefining maritime logistics and achieving environmental objectives.

Keywords Artificial Intelligence of Things (AIoT) · Smart ports · Emission reduction · Sustainability · Predictive analytics

---

A. Idris (✉) · W. N. W. Muhamad · A. J. A. Jabar
Wireless Communication Technology Group (WiCOT), School of Electrical Engineering, Universiti Teknologi MARA, College of Engineering, Shah Alam, Selangor, Malaysia
e-mail: azlina831@uitm.edu.my

W. N. W. Muhamad
e-mail: syafizan@uitm.edu.my

M. B. Rahmat
Department of Marine Electrical Engineering, Shipbuilding Institute of Polytechnic Surabaya, Surabaya, Indonesia
e-mail: mbasuki.rahmat@ppns.ac.id

I. Taib
Infrastructure University Kuala Lumpur, Kajang, Selangor, Malaysia
e-mail: idristaib@iukl.edu.my

M. Elsisi et al. (eds.), *Maritime Infrastructure for Energy Management and Emission Reduction Using Digital Transformation*, Studies in Infrastructure and Control,
https://doi.org/10.1007/978-981-96-4438-4_9

1 Introduction to AIoT and Smart Ports

Artificial Intelligence of Things (AIoT) refers to the integration of Artificial Intelligence (AI) with the Internet of Things (IoT). IoT encompasses a network of interconnected devices like sensors and cameras that exchange data. The incorporation of AI enhances these devices' capabilities, enabling them not only to gather information but also to analyze, make decisions, and execute tasks independently, adapting to real-time conditions and addressing complex issues. This synergy is pivotal in boosting efficiencies, optimizing processes, and refining decision-making across various sectors, especially in the development of smart industries, like port operations (Fig. 1).

1.1 Smart Port Overview

Smart Ports are ports that leverage advanced digital technologies, including AI, IoT, and data analytics, to improve their operations and services. Key aspects of smart ports include automation, where systems handle cargo, vessel traffic, and logistics with minimal human intervention, significantly speeding up processes [1]. Digital transformation allows for real-time monitoring through AI and IoT sensors, while virtual replicas of port assets enable predictive management. Finally, sustainability efforts focus on minimizing environmental impact through better energy management, waste reduction, and compliance with environmental regulations, thus promoting greener operations within the port industry.

Fig. 1 Port of Hong Kong, Basenton

1.2 Port Emissions: A Critical Environmental Issue

Port emissions, including carbon dioxide (CO2), nitrogen oxides (NOx), sulfur oxides (SOx), and particulate matter (PM), pose a significant environmental challenge by contributing to air pollution and climate change. These emissions primarily arise from marine vessels, diesel-powered port equipment, and land transport like trucks. Ships burn bunker fuel, leading to substantial CO2, NOx, and SOx emissions, while port machinery and trucks further exacerbate the situation. Given their critical role in global trade, ports account for a considerable share of emissions, prompting increased calls for reductions from governments and environmental advocates. Regulatory initiatives, such as the International Maritime Organization's greenhouse gas (GHG) reduction targets and the European Union's Green Deal, aim to decarbonize the shipping sector [2].

1.3 AIoT Technologies in Reducing Port Emissions

AIoT technologies are becoming essential for reducing port emissions by optimizing operations through IoT sensors and enhancing energy efficiency. The application of AIoT technological advances in smart port facilities improves operations and decision-making by efficiently collecting and analyzing massive quantities of information from sensors, cameras, and self-driving vehicles as mentioned in Fig. 2. Key areas include the following:

1. Optimized Port Operation: AI-driven predictive maintenance enhances machinery performance, minimizing emissions.
2. Emission Tracking and Analytics: Continuous tracking by AIoT sensors allows for identifying inefficiencies, enabling operational adjustments like vessel rerouting, which decreases emissions.

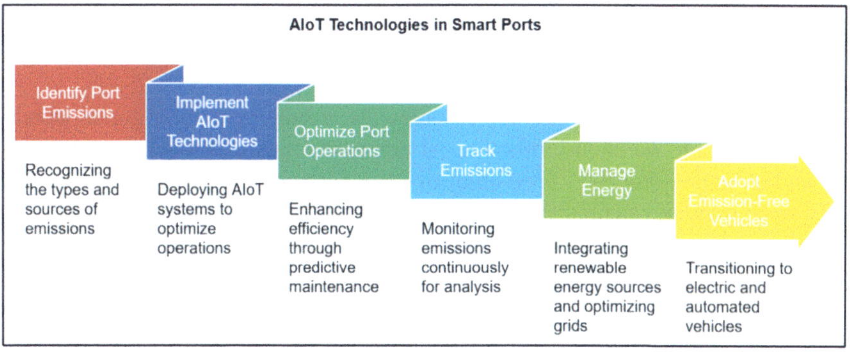

Fig. 2 AIoT technologies in smart port applications

3. Energy Management: Integrating renewable energy sources enables AI to optimize energy grids, improve distribution, and reduce fossil fuel dependence by analyzing consumption patterns for real-time adjustments.
4. Emission-Free Port Vehicles: The adoption of electrified vehicles and automated systems, coordinated by AIoT, streamlines operations to further cut down emissions.
5. AI-Driven Traffic Flow Optimization: AI mitigates congestion at ports, optimizing vessel and cargo movement to reduce idle time and fuel consumption.
6. Sustainability Reporting: AI analytics provide real-time environmental impact data, fostering transparency and aiding in tracking emission reduction efforts.

2 AIoT Technologies and Their Application in Ports

The integration of the Internet of Things (IoT) and Artificial Intelligence (AI) is transforming transportation and logistics, presenting significant advantages for port operations. This synergy enhances logistics networks, cargo management, and vehicle operations through IoT's data-gathering capabilities and AI's analytical skills [3].

2.1 Autonomous Vehicles and Fleet Management

AIoT applications in logistics encompass fleet management and the use of autonomous vehicles. Self-driving trucks equipped with IoT sensors gather data on location, speed, and cargo, which AI utilizes for real-time decision-making such as route optimization. Fleet management systems leverage IoT and AI to monitor vehicle health, predict maintenance needs, and reduce breakdowns. In the port context, autonomous vehicles can handle container movement independently.

2.2 Smart Warehousing and Inventory Management

Smart warehouses, utilizing IoT devices, regulate inventory, monitor goods' conditions, and manage warehouse robotics. AI algorithms facilitate stock level estimation and automate ordering, preventing overstocking and enhancing efficiency. In ports, smart systems can monitor container dynamics, automate sorting, and optimize storage.

2.3 Route Optimization and Traffic Management

AIoT enhances logistics through improved fuel efficiency and congestion avoidance via real-time traffic data. Employing GPS and sensors, AI algorithms optimize vehicle paths. In ports, similar technologies enable real-time tracking and management of shipping activities, minimizing delays and congestion, which streamlines operations and maximizes dock efficiency.

2.4 Predictive Maintenance for Fleet and Equipment

AIoT technologies are transforming predictive maintenance in the transportation sector, employing IoT sensors to monitor crucial performance indicators, such as engine temperature, oil pressure, and tire health. This data aids logistics organizations in minimizing downtime, enhancing equipment longevity, and optimizing fleet efficiency. AI-driven systems can forecast maintenance needs based on wear patterns, effectively cutting operational costs and boosting vehicle uptime. Particularly in ports reliant on heavy machinery, AIoT-based predictive maintenance enhances operational efficiency, lowers energy consumption, reduces emissions, and accelerates port processes.

2.5 Environmental Monitoring and Emission Control

Industries including ports are increasingly embracing AIoT for environmental monitoring. Sensors measure air quality, temperature, and vehicle emissions, feeding data to AI systems that assess emission footprints and predict pollution levels. This technology supports green logistics initiatives by optimizing clean fuel usage and employing electric vehicles, enabling real-time emission tracking and pollution management in port operations.

2.6 Fleet Tracking and Logistics Visibility

AIoT enhances fleet tracking, allowing logistics firms to monitor vehicle movements and improve scheduling, resource distribution, and disruption handling. It facilitates end-to-end visibility for customer shipments, which boosts satisfaction through precise delivery estimates and cargo handling efficiency.

2.7 Demand Forecasting and Supply Chain Optimization

AIoT is critical in forecasting demand and optimizing supply chains by analyzing extensive IoT data. This capability enables logistics firms to balance supply and demand effectively, ensure timely deliveries, and avert inventory issues, including congestion in ports through better container and traffic management.

The integration of Artificial Intelligence of Things (AIoT) in transportation and logistics paves the way for enhancing operations in ports, focusing on efficiency, sustainability, and emission reduction [4]. Many advancements in logistics—such as autonomous vehicles, predictive maintenance, and real-time environmental monitoring—can be adapted for port applications. With increasing pressure on ports to lessen their environmental impact and manage higher trade volumes, AIoT presents innovative solutions that improve operational sustainability and profitability as illustrated in Fig. 3.

Key AIoT applications for port operations include

1. Smart Sensors for Real-Time Emission Monitoring: These sensors track pollutants like Nitrogen Oxides (NOx), Sulfur Oxides (SOx), and Carbon Dioxide

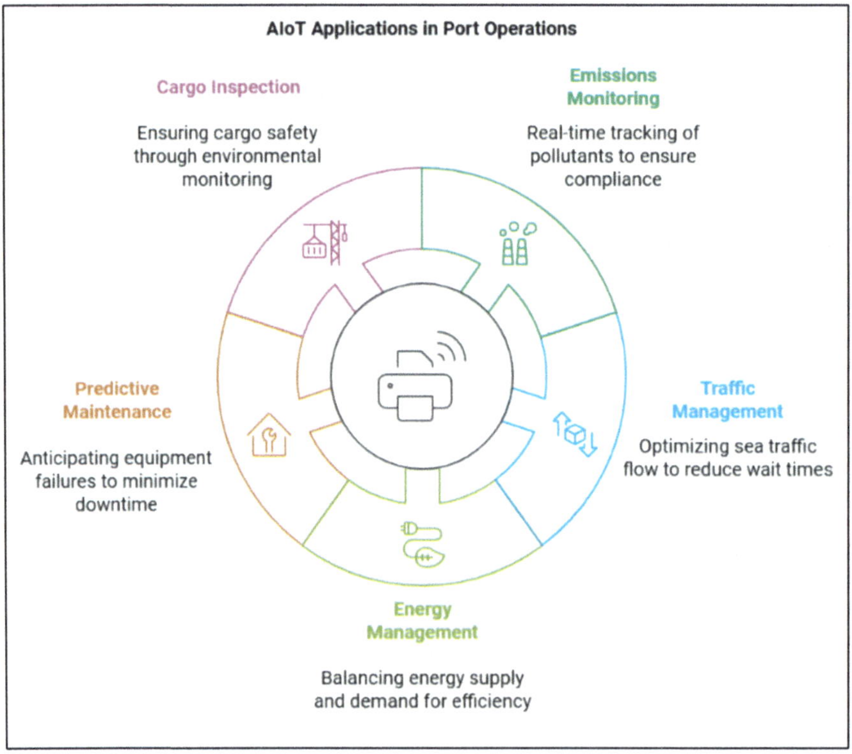

Fig. 3 Summary of AIoT technologies in port operations

(CO2) [2]. AI algorithms can analyze the data to predict pollution patterns, facilitating proactive measures in emission reduction and regulatory compliance.

2. Predictive Analytics for Traffic Management: AIoT systems enhance scheduling by accurately forecasting sea traffic, thereby reducing wait times and optimizing dock utilization. Real-time data from IoT sensors regarding cranes and vehicles can streamline traffic flow for ships, trains, and trucks.
3. Energy Management: AI algorithms can optimize energy distribution by predicting peak demand and regulating power supply to minimize outages. This capability allows for better management of renewable energy use, cost savings, and enhanced operational efficiency at ports.
4. Predictive Maintenance of Port Equipment: It leverages IoT sensors to enhance system efficiency by tracking equipment health, foreseeing potential breakdowns, and streamlining inventory and asset management. These sensors detect deviations from normal operational parameters, enabling operators to proactively address issues such as misplacements or unexpected delays, ultimately minimizing downtime and reducing expensive repair costs.
5. Automated Cargo Inspection and Safety Management: It utilizes AIoT technology to ensure optimal cargo conditions by monitoring factors like temperature, humidity, and vibrations in shipping containers. Real-time detection of anomalies facilitates prompt problem resolution, thus enhancing the overall safety and integrity of the cargo. Furthermore, by analyzing data from diverse sources, such as cameras, motion sensors, and sensor networks situated in high-risk areas, AIoT can identify potential hazards, contributing to a safer operational environment.

3 Emission Challenges in Ports

Ports encounter considerable emission challenges due to the high activity of ships, cargo-handling equipment, trucks, and trains, which release pollutants such as nitrogen oxides (NOx), sulfur oxides (SOx), particulate matter (PM), and greenhouse gases (GHGs). These emissions contribute to air pollution, impacting local communities and ecosystems, while also driving climate change, necessitating urgent mitigation. Reducing emissions is complex and involves adopting cleaner technologies, implementing stricter regulations, and innovating operational practices. As global trade grows, addressing emissions in ports is essential for sustainable development and environmental health (Fig. 4).

3.1 Types of Port Emissions

Ports are major sources of multiple emissions, stemming from the complexity of their operations, which includes shipping, cargo handling, and transportation activities. Key pollutants include NOx, a group of gases like nitrogen monoxide (NO) and

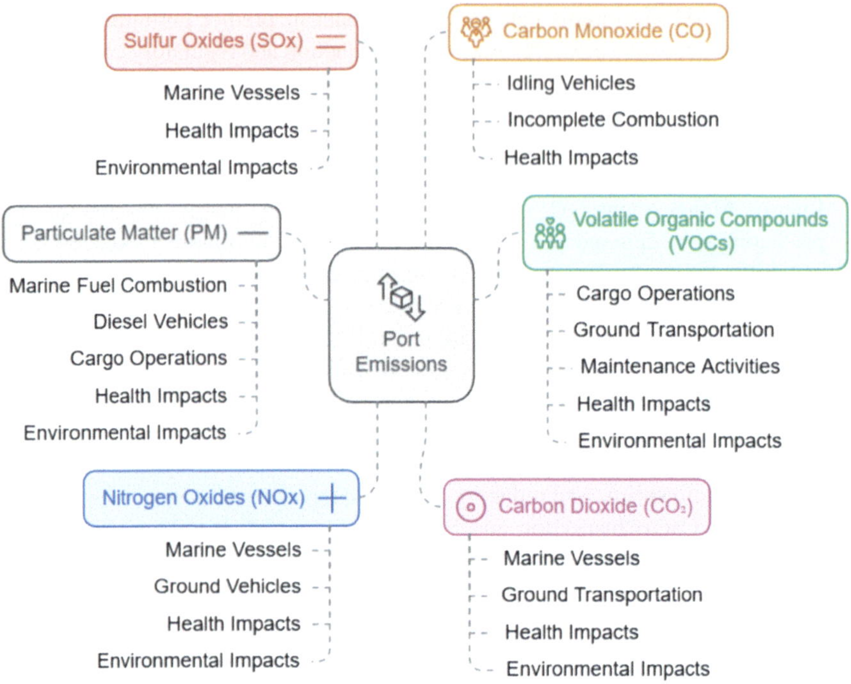

Fig. 4 Emission challenges in ports

nitrogen dioxide (NO_2), generated primarily by high-temperature fossil fuel combustion. Ships, especially those using high-sulfur fuels, and ground vehicles like diesel trucks and rail systems contribute to NOx emissions. These emissions are associated with respiratory issues, ground-level ozone, and acid rain. Another significant pollutant, SOx, mainly comprises sulfur dioxide (SO_2), produced from burning high-sulfur fossil fuels. Marine vessels using heavy fuel oil (HFO) are primary sources, contributing to acid rain and respiratory issues. CO_2, a greenhouse gas, is another critical emission linked to fossil fuel combustion by ships, ground transportation, and port facilities, driving climate change and its effects, such as rising sea levels. PM emissions from marine fuel combustion, diesel vehicles, and dust from cargo operations are also concerning due to health impacts, including respiratory issues and ecosystem damage. VOCs, which contribute to smog, arise from cargo operations, ground transportation, and maintenance activities. CO, mainly emitted by idling vehicles and incomplete combustion in port equipment, poses health risks and impairs oxygen transport in the bloodstream. In sum, emissions from port operations pose challenges to air quality and public health, requiring targeted emission reduction strategies through cleaner technologies, operational efficiencies, and alternative fuels.

3.2 Current Methods of Emission Control

To mitigate environmental impact, ports employ traditional emission control technologies. Ships use Exhaust Gas Cleaning Systems (EGCS), or scrubbers, to remove SOx from exhaust gases, meeting IMO sulfur standards. Ports are increasingly shifting to low-sulfur fuels like marine gas oil (MGO) and liquefied natural gas (LNG), which reduce both SOx and PM emissions during ship operation. Shore power systems, which allow ships to connect to local electrical grids while docked, are also widely implemented, reducing CO_2, NOx, and PM emissions from auxiliary engines. Ports have optimized ground transportation logistics by establishing designated truck lanes to minimize congestion and idle time, and promoting hybrid or electric cargo vehicles. Alternative energy sources, including solar panels and wind turbines, have been adopted to further reduce fossil fuel reliance. Regulatory frameworks play a vital role, as ports work closely with government agencies to enforce standards, including Hong Kong's Air Pollution Control Regulation and other policies across Asia. Additionally, many ports participate in voluntary programs, adopting Environmental Management Systems (EMS) and initiatives like the Green Marine program, which encourages best practices for environmental performance.

3.3 Regulatory Framework: International and Local Regulations

A mix of international and local regulations shapes the regulatory framework for port emissions. Internationally, the IMO's MARPOL Convention, particularly Annex VI, limits SOx and NOx emissions from ships, with a global sulfur cap for fuel oil reduced to 0.5% m/m in 2020. This cap encourages the adoption of cleaner fuels to reduce SOx emissions. Many Asian countries also enforce national regulations that complement IMO standards. Singapore's Maritime Singapore Green Initiative incentivizes emission reductions in port operations, while South Korea has established air quality control areas around major ports like Busan to mandate low-sulfur fuel usage. Hong Kong's Air Pollution Control Regulation and Jakarta's port authority initiatives align with local air quality standards, aiming to address port activity emissions. Some ports also adopt voluntary agreements like the Clean Ports Program, encouraging low-emission technologies and collaboration among stakeholders to mitigate port emissions. These international and local regulations collectively promote sustainable practices in port operations, driving the use of cleaner technologies and fuels.

3.4 Limitations of Traditional Systems

Despite the effectiveness of traditional emission control methods, these systems have notable limitations that highlight the need for smarter solutions. Traditional systems are often reactive, targeting emission reduction only at specific points, such as scrubbers and low-sulfur fuel switching, without fully covering emissions from ground transportation and cargo-handling equipment. Lacking real-time monitoring and data analysis capabilities, traditional systems struggle to track emissions continuously and adapt to operational changes, often resulting in inefficiencies and higher operational costs due to manual monitoring. These systems also involve high installation and maintenance expenses, particularly challenging for smaller ports. As port operations grow increasingly complex and regulatory standards tighten, traditional systems may fall short of achieving emission reduction goals. Conventional solutions, like scrubbers, primarily address SOx emissions but lack the scope to tackle other pollutants, such as NOx and PM [5, 6]. Additionally, most traditional systems lack integration with digital solutions that provide advanced analytics and predictive maintenance capabilities. These limitations underscore the need for more comprehensive approaches, such as AIoT technologies, which offer real-time data collection, predictive maintenance, and automated emission management. AIoT-enabled solutions are poised to address the shortcomings of traditional systems, allowing for cleaner, more efficient port operations that align with modern environmental standards.

4 AIoT for Emission Monitoring and Reduction

This section explores how Artificial Intelligence of Things (AIoT) technology transforms emission monitoring and reduction in ports. As ports strive to comply with strict environmental regulations and reduce carbon emissions, AIoT offers a dynamic approach through real-time data, advanced analytics, and automation. Key AIoT components include IoT-enabled sensors for tracking pollutants and AI-driven analytics for predictive insights and optimization. By enabling continuous monitoring across various emission sources, AIoT supports proactive adjustments and optimizes maintenance, fostering more efficient and sustainable port operations. Through real-world examples, this chapter demonstrates how AIoT addresses traditional emission control limitations, helping ports meet regulatory standards, enhance efficiency, and minimize environmental impact.

4.1 AIoT-Based Emission Monitoring

AIoT-based monitoring combines IoT technology and AI to provide real-time, precise tracking of emissions across port operations. IoT sensors positioned throughout port facilities continuously measure pollutants like NOx, SOx, CO_2, and PM from sources such as ships, cargo equipment, and vehicles. Drones and satellites are also used to capture emission data in hard-to-reach areas, offering a comprehensive view of air quality across the port and surrounding areas. AI processes this data, identifying patterns and predicting peak emissions, which allows port operators to make timely adjustments to reduce emissions. This integrated data from sensors, drones, and satellites creates a detailed emission picture, supporting ports in managing air quality and adhering to environmental regulations.

4.2 Data Collection and Analytics

In an AIoT framework, AI analyzes vast data generated by IoT devices, transforming raw data into actionable insights that improve emission management. Machine learning algorithms identify patterns and correlations, helping predict high-emission events based on historical data and real-time conditions. For instance, AI can anticipate emission spikes during certain operational activities, allowing ports to proactively adjust equipment usage or scheduling. AI-powered analytics provide real-time visualizations and automated reports, enhancing compliance monitoring and supporting data-driven strategies to reduce environmental impact.

4.3 Predictive Emission Control

Predictive emission control uses AI to forecast emissions by analyzing real-time and historical data, enabling proactive management. Machine learning algorithms identify emission trends, such as those occurring during peak port activity, and provide early warnings. This allows port operators to adjust operations, such as rescheduling equipment or using lower emission machinery, to prevent emissions from exceeding limits. Predictive control also optimizes resource allocation, improving efficiency and ensuring compliance with air quality standards.

4.4 Success Stories/Case Studies

AIoT technology has been successfully implemented in several major Asian ports, setting a new standard for sustainable operations.

In Singapore, one of the world's busiest ports, the Maritime Singapore Green Initiative, employs AIoT systems to track and reduce emissions [7]. Through a network of IoT sensors, the port monitors air quality, fuel consumption, and pollutant levels across its operations. AI-driven analytics then process this data, enabling port authorities to optimize vessel traffic flow, reduce fuel usage, and lower greenhouse gas emissions. The initiative also incentivizes lower emission vessels by offering discounts on port fees, further encouraging sustainable practices within the port community. This holistic approach highlights how AIoT not only enhances operational efficiency but also actively contributes to Singapore's environmental sustainability goals.

The Port of Busan in South Korea, the country's largest and most active port, integrates AIoT technology as part of its Smart Port initiative, aligning with South Korea's Green New Deal. AIoT sensors monitor pollutant levels, including sulfur and nitrogen oxides, and provide real-time air quality data across port areas. This continuous monitoring enables Busan to make on-the-spot adjustments, such as providing shore-to-ship power (which allows ships to switch off engines while docked) and adjusting port traffic flow during high-emission periods [8]. These strategies not only improve air quality but also showcase how smart technology can make port operations more eco-friendly while supporting South Korea's national goal of a greener economy.

The Port of Hong Kong has adopted an AIoT-driven Air Quality Monitoring Network that combines data from sensors across the port with advanced AI analytics to detect, predict, and mitigate high-emission events. By identifying potential emission spikes in real time, the system enables immediate operational adjustments, like rescheduling heavy machinery or restricting high-emission activities during certain hours. Additionally, Hong Kong's port enforces mandatory fuel switching for vessels and provides shore power, which further reduces emissions and enhances the port's environmental performance. These initiatives align with Hong Kong's objective to protect local air quality and minimize the port's environmental footprint.

In Indonesia, the Port of Tanjung Priok is leveraging AIoT technology to support its sustainable development goals. The port has installed a network of sensors that track emissions, and AI systems analyze this data to identify inefficiencies and optimize energy usage. By implementing electric-powered port equipment and promoting the use of cleaner fuels, Tanjung Priok has taken significant steps to reduce emissions and improve its overall energy efficiency. These efforts contribute to Indonesia's national goal of reducing greenhouse gases, particularly within key industrial sectors, and highlight the role of AIoT in helping developing nations meet their sustainability objectives.

5 Implementation Strategies for AIoT in Ports

Implementing Artificial Intelligence and Internet of Things (AIoT) in ports requires a structured approach to ensure efficient integration and minimal disruption to existing systems. The process begins with a thorough assessment of the port's operational needs and a detailed plan, identifying specific areas where AIoT can optimize functions, such as real-time cargo tracking, energy management, predictive maintenance, and traffic flow control. This phase should define measurable objectives and expected

ROI, providing a clear vision and metrics for success. Once a plan is established, the port's infrastructure must be prepared to handle AIoT requirements, including enhancing network capabilities, ensuring stable Internet, and deploying necessary sensors, cameras, and edge devices at strategic locations. A centralized data platform, integrated with existing databases and applications, then enables seamless data flow for real-time analytics and AI-driven insights.

Integration with existing systems is critical and requires careful coordination between new AIoT technologies and current port software and hardware. AIoT systems must seamlessly interact with Terminal Operating Systems (TOS), ERP systems, and equipment used in daily port operations. This is often achieved through APIs and middleware, which bridge AIoT solutions with traditional systems. Hardware compatibility is also essential; retrofitting legacy equipment may be necessary to enable IoT connectivity. Once these foundational steps are in place, a pilot project is advisable to test AIoT effectiveness on a small scale. This pilot allows ports to refine the setup, adjust workflows, and troubleshoot any issues before scaling the solution to the full operation.

Challenges in AIoT deployment in ports are significant, but proactive strategies can address them. Cost is a common concern, as AIoT infrastructure, devices, and training require substantial investment. Phased implementation can help manage expenses, focusing on high-impact areas initially, while partnerships with private firms or seeking government grants can provide financial support. Technical expertise is another hurdle, as ports often lack in-house skills in AI and IoT. Collaboration with technology providers and training for port staff can fill this gap, allowing the port to build technical capacity over time. Data security and privacy are critical issues as well; safeguarding sensitive information requires implementing strict cybersecurity protocols, regular audits, and robust access controls. Interoperability between AIoT systems and existing technologies can be challenging, too, but relying on open standards and fostering close collaboration with technology providers can help smooth out integration.

Collaboration and stakeholder engagement are essential throughout the AIoT adoption process. Port authorities play a central role in planning and allocating resources, while government agencies can offer regulatory guidance, funding, and infrastructure investment, supporting the port's strategic goals. Partnerships with private sector technology providers bring specialized knowledge and advanced solutions, offering ongoing support for the port's AIoT systems. Including port employees and union representatives in the conversation ensures that workers' concerns are addressed, especially regarding job impacts from automation; offering reskilling opportunities and emphasizing a human–machine collaboration approach fosters workforce buy-in. Engaging end users, such as shipping lines and logistics companies, is also crucial, as their feedback can shape AIoT solutions to meet specific service needs. By aligning all stakeholders, port authorities can maximize the benefits of AIoT, fostering a more efficient, secure, and sustainable port environment (Fig. 5).

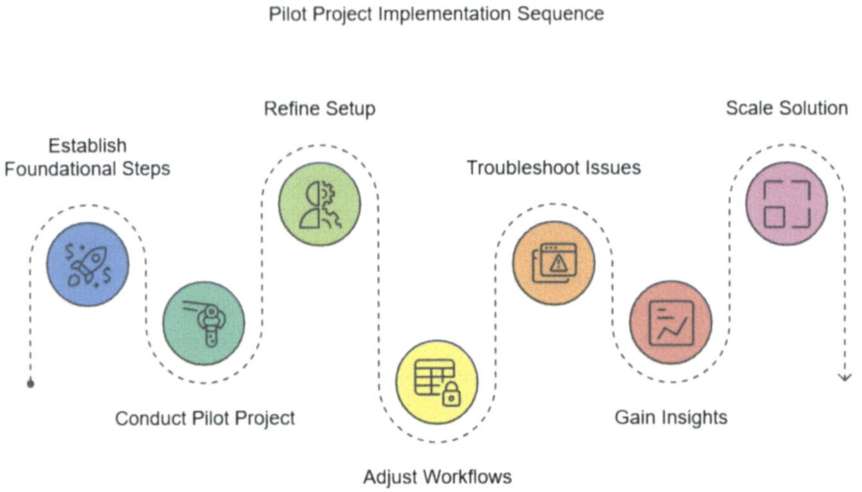

Fig. 5 Pilot project implementation using AIoT for smart port emission

6 Economic and Environmental Impact

Ports play a crucial role in global trade by connecting continents and facilitating the movement of goods. However, port operations face challenges, including environmental pollution and inefficiencies, leading to costly delays. Artificial Intelligence of Things (AIoT)—combining AI with IoT technology—provides a solution by enabling ports to optimize processes, make data-driven decisions, and reduce environmental impacts. This section examines the transformative effects of AIoT in ports such as Singapore, Busan, and Hong Kong, focusing on economic benefits and environmental sustainability.

6.1 Economic Benefits of AIoT in Port Operations

(1) Cost Savings
AIoT technology helps ports cut costs by improving resource allocation, minimizing bottlenecks, and enabling predictive maintenance. The Port of Singapore uses AIoT-driven predictive maintenance systems that have reduced equipment downtime by 20%, lowering unexpected repair costs [7, 9]. Optimized vessel scheduling and cargo management also boost fuel efficiency, making AIoT-enabled ports more attractive to logistics companies that prioritize reliability and cost-effectiveness.
(2) Enhanced Efficiency
Effective resource management is critical in port operations. AIoT enables real-time adjustments and precise management of equipment, as seen at the

Fig. 6 Singapore Port: the world's largest transshipment hub

Port of Busan, where dynamic scheduling has lowered congestion and reduced container processing times. Similarly, the Port of Hong Kong leverages AI-driven systems to optimize crane and truck usage, handling higher cargo volumes without additional costs.

(3) Competitiveness and Market Growth
AIoT provides a competitive edge by increasing throughput and reducing labor costs. For example, Singapore's Pasir Panjang Terminal uses full automation, powered by AI and robotics, reducing container handling times by 30% [7, 9]. This has solidified Singapore's position as a top choice for international shipping lines, enhancing its global standing as a logistics hub.

(4) New Revenue from Data
Beyond operational savings, AIoT generates new revenue streams by providing data-based services. Singapore's Port Operations & Trade Intelligence Platform (PORT) offers stakeholders real-time data on container movements, delivering valuable insights that enhance PSA's role as an innovator in digital port management [7].

See Fig. 6.

6.2 *Environmental Benefits of AIoT in Port Operations*

(1) Reducing Air Pollution
AIoT assists ports in minimizing emissions by reducing ship idle times and optimizing fuel use. For instance, Singapore has lowered CO_2 emissions by 10% using sensors to monitor air quality, while Hong Kong's AIoT energy

management systems enhance energy efficiency by reducing reliance on fossil fuels.

(2) Noise and Water Pollution Control
AIoT systems aid in reducing noise pollution and protecting marine life. The Port of Busan uses noise data to adjust operations and minimize disruptions, while Japan's Port of Yokohama employs AIoT to detect water contaminants, triggering automatic containment [10].

(3) Energy and Waste Management
AIoT systems improve energy efficiency and waste management, as seen in Singapore and Kaohsiung, where smart systems optimize power use and waste disposal schedules, reducing environmental impact.

7 Future Directions and Opportunities

As regulations on the environment tightened and the global sustainability movement gets prominence, smart ports are increasingly implementing Artificial Intelligence of Things (AIoT) technologies to more efficiently manage and reduce emissions. AIoT offer significant prospects to improve port environmental friendliness along with operational efficiency, hence reducing carbon footprints.

7.1 Emerging AIoT Technologies

As AIoT offers significant possibilities in emission management, particularly real-time, precise emission monitoring, the next advances are probably going to include high-resolution sensor networks that can identify a wider range of pollutants, giving precise understanding into the sources of emissions and how they affect air quality. Artificial intelligence (AI)-powered systems can automate compliance reporting, monitoring pollution levels over time, and producing reports for stakeholders [11]. This reduces the administrative load for port operators while also ensuring environmental compliance. Furthermore, AIoT can predict when emissions could exceed allowed levels, setting off warnings and preventative actions that allow ports to regulate equipment usage, reroute big vehicles, or temporarily stop operations.

As the maritime sector progresses toward sustainability, the integration of Artificial Intelligence (AI) with the Internet of Things (IoT), termed AIoT, is vital for enhancing smart port operations and emission control. Key emerging AIoT technologies include the following:

1. AI-Driven Predictive Analytics: AIoT may identify high-emission periods and propose proactive pollution-reduction actions. It can predict peak port traffic hours and adapt vessel or cargo-handling activities accordingly. AI-driven predictive algorithms are able to recognize periods of high emissions and modify the operation of equipment effectively. By controlling unnecessary emissions, such

idling machinery or improper cargo handling can be helped by behavioral observations [12]. Ports can assess previous and present environmental data, predicting emission and pollution for optimal management and energy consumption by using AI with IoT sensor data.
2. Edge Computing: This technology enables local data processing in busy ports, reducing delays and facilitating prompt decision-making critical for real-time emission monitoring.
3. Autonomous Ships and Drones: These AIoT-integrated vehicles can autonomously gather vital data on air quality and fuel consumption, enhancing emission tracking and operational efficiency.
4. Advanced IoT Sensors: Future AIoT advancements are expected to involve more sensitive and widespread sensor networks that can detect a wider range of pollutants, including carbon dioxide (CO2), sulfur oxides (SOx), nitrogen oxides (NOx), and particulate matter [2]. Through the provision of higher resolution data throughout a port region, these sensors will allow for more accurate visibility into the sources of emissions and how it affects the quality of the air.
5. AI for Optimization: AI can streamline ship docking, loading/unloading, and departure schedules, optimizing resource use and reducing emissions from idling vessels and equipment.

7.2 Additional Research is Needed in Areas Such as Real-Time Emission Tracking and Upgraded IoT Devices for Better Data Collection

Additionally, unlocking AIoT's full potential requires significant research and development in these areas to advance emission control in the maritime industry. The enhancement of AI algorithms for real-time emission tracking is crucial, as existing models typically depend on historical data and static variables. There is a growing need for advanced models that can interpret and adapt to real-time data, including variations in weather and fuel quality [13]. Furthermore, IoT device technology must advance to improve sensor accuracy and durability in maritime environments, with research focused on developing multi-modal sensors for diverse pollutant tracking. Real-time data transfer demands energy-efficient communication systems, especially for AIoT applications in remote areas like ports, necessitating new low-power communication protocols. Additionally, the transparency of AI decision-making is vital for emission control, prompting a need for explainable AI (XAI) to assure human operators of its reliability.

7.3 AIoT Principles Can Be Applied to Various Maritime Logistics Areas

With the integration of AIoT into port operations, robust cybersecurity measures are essential to safeguard sensitive emission control data from cyber threats. Beyond emission control, AIoT principles can significantly enhance other maritime logistics areas, such as cargo management. These systems enable real-time tracking and condition monitoring of cargo, optimize loading/unloading processes, and predict supply chain disruptions, ultimately promoting efficient and sustainable logistics management.

8 Conclusion

AIoT is revolutionizing port operations by enhancing efficiency, safety, and reduced environmental impact. Ports may lower emissions, enhance air quality, and promote sustainable growth by incorporating IoT and AIoT technology. AIoT-powered solutions can be profitable and ecologically conscious, serving as a model for other areas. In order to integrate AIoT with current systems, a methodical strategy is required, which includes network capabilities, dependable Internet, and the deployment of sensors, cameras, and edge devices. A pilot project is suggested to test AIoT effectiveness on a modest scale. AIoT can transform the maritime industry by boosting traffic management, energy efficiency, security, and fleet management. To realize its full potential in smart operations and emission management, it is critical to continue developing AI algorithms, IoT sensor technologies, and real-time data processing infrastructure.

References

1. Paraskevas A, Madas M, Zeimpekis V, Fouskas K (2024) Smart ports in industry 4.0: a systematic literature review. Logist MDPI 8(1):1–29. https://doi.org/10.3390/logistics8010028
2. Ducruet V et al (2024) Ports and their influence on local air pollution and public health: a global analysis. Sci Total Environ 915:1–18. https://doi.org/10.1016/j.scitotenv.2024.170099
3. Postolache O et al (2019) Modelling IoT technology for a Cargo Port. J Phys: Conf Ser IOP Publishing 1–7. https://doi.org/10.1088/1742-6596/1379/1/012072.
4. Min H (2022) Developing a smart port architecture and essential elements in the era of industry 4.0. Marit Econ Logist 24(2):189–207. https://doi.org/10.1057/s41278-022-00211-3
5. Busan Port for a sustainable world - observatory of public sector innovation. https://oecd-opsi.org/about/. Accessed 12 Nov 2024
6. Roh S, Thai VV, Jang H, Yeo GT (2023) The Best practices of port sustainable development: a case study in Korea. Marit Policy Manag Flagship J Int Ship Port Res 50(2):254–280. https://doi.org/10.1080/03088839.2021.1979266
7. Edgar M (2023) Sustainability 2023 at PSA Singapore, green horizons: going the distance. https://doi.org/10.1007/978-1-4842-9054-5_3

8. Hong H et al (2022) An air pollutants prediction method integrating numerical models and artificial intelligence models targeting the area around Busan port in Korea. Atmos. MDPI 13(1462):1–29. https://doi.org/10.3390/atmos13091462
9. PSA Singapore, “PSA Singapore,” eco-cities and green transport. https://www.singaporepsa.com/-/media/Feature/PDF/News-And-Media/PSA_Singapore.pdf
10. Camarero Orive A, Parra Santiago JI, Esteban-Infantes Corral MM, González-Cancelas N (2020) Strategic analysis of the automation of container port terminals through BOT (business observation tool). Logist MDPI 4(3):1–14. https://www.mdpi.com/2305-6290/4/1/3
11. Alanhdi A et al (2023) Artificial intelligence-based sensors for next generation IoT applications: a review. IEEE Internet Things J 21(4):346–353. https://doi.org/10.1109/JIOT.2020.2993411
12. Valero CI, Pla EI, Vaño R, Garro E, Boronat F, Palau CE (2021) Design and development of an aiot architecture for introducing a vessel eta cognitive service in a legacy port management solution. Sensors MDPI 21(8133):1–15. https://doi.org/10.3390/s21238133
13. Kuo YH, Wu EHK (2023) Advanced, innovative AIoT and edge computing for unmanned vehicle systems in factories. Electronics MDPI (Switzerland) 12(1843):1–17. https://doi.org/10.3390/electronics12081843

Chapter 10
Regulatory Framework and Policy Implications

Dewi Kurniasih, Desi Tri Cahyaningati, Hayy Nur Abdillah, Nanda Dwi Wuryaningrum, and Arfiana Dewi

Abstract Indonesia, as one of the largest archipelagic countries in the world, has great potential in maritime energy management. The maritime sector plays an important role in the national economy, but it also becomes a significant source of emissions. Therefore, Indonesia needs to address these issues through regulation. The regulation of maritime energy management and emissions reduction in Indonesia are based on the international regulations such as the Paris Agreement and the United Nations Framework Convention on Climate Change (UNFCCC). This chapter discussed the regulatory landscape governing maritime energy management and emissions reduction in Indonesia including international conventions, regional regulations, policy initiatives driving industry compliance and innovation and the implementation of these regulations with the real condition in Indonesia maritime sector. These regulations are formulated to achieve net zero emissions for global environment.

Keywords Maritime energy management · Regulations · Net zero emissions

D. Kurniasih (✉)
Department of Marine Engineering, Shipbuilding Institute of Polytechnic Surabaya, Surabaya, Indonesia
e-mail: dewi.kurniasih@ppns.ac.id

D. T. Cahyaningati · H. N. Abdillah · N. D. Wuryaningrum · A. Dewi
Department of Shipbuilding Engineering, Shipbuilding Institute of Polytechnic Surabaya, Surabaya, Indonesia
e-mail: dtricahyaningati@ppns.ac.id

H. N. Abdillah
e-mail: hayyabdillah@ppns.ac.id

N. D. Wuryaningrum
e-mail: nanda.dwi@ppns.ac.id

A. Dewi
e-mail: arfianadewi@ppns.ac.id

M. Elsisi et al. (eds.), *Maritime Infrastructure for Energy Management and Emission Reduction Using Digital Transformation*, Studies in Infrastructure and Control,
https://doi.org/10.1007/978-981-96-4438-4_10

1 International Convention

The International Convention on Energy Maritime Management encompasses various initiatives and agreements aimed at enhancing energy efficiency and sustainability within the maritime sector. This area of focus is critical due to the significant environmental impacts associated with maritime activities, particularly concerning greenhouse gas (GHG) emissions.

In this case, Indonesia plays a significant role in international conventions related to maritime energy management. As a maritime country, Indonesia is a member of several international conventions related to maritime safety, such as The International Ship and Port Facility Security Code (ISPS Code) and International Safety Management Code (ISM Code). These regulations ensure safety at sea, prevent human injury, loss of life, and avoid marine environmental pollution. The international policies implemented in Indonesia are as follows:

1. **Paris Agreement**
   Indonesia is committed to reducing GHG emissions by 29% with its efforts and up to 41% with international support by 2030. This commitment encourages the development of renewable energy and more aggressive climate change mitigation policies.
2. **United Nations Framework Convention on Climate Change (UNFCCC)**
   Indonesia is following UNFCCC because as one of the countries participating in this convention. Indonesia participates in various global initiatives to reduce emissions and enhance resilience to the impacts of climate change.

2 Regulations Governing Maritime Energy Management

Indonesia, as one of the largest archipelagic countries in the world, has great potential in maritime energy management. The maritime sector plays an important role in the national economy, but it also becomes a significant source of emissions. According to the Ministry of Energy and Mineral Resources, the maritime transportation sector contributes about 15% of the total Greenhouse Gas (GHG) emissions in Indonesia [1]. Therefore, effective management is essential to achieve emission reduction targets. Maritime energy management and emission reduction in Indonesia is an increasingly urgent issue in line with global commitments to address climate change. In this context, Indonesia has established various regulations and strategies to manage energy resources sustainably and reduce GHG emissions [2].

Maritime energy management in Indonesia is regulated by various regulations that cover aspects of new renewable energy, energy conservation, as well as occupational safety and health [3, 4]. These regulations aim to ensure that maritime energy management is carried out sustainably, efficiently, and safely, taking into account environmental aspects and public health. The following are some important relevant regulations:

1. **New Renewable Energy Regulation**
   - Minister of Energy and Mineral Resources Regulation Number 25 of 2013 and Number 20 of 2014: Regulates the provision and use of biofuel as an alternative to other fuels.
   - Presidential Regulation Number 11 of 2023: Regulates additional concurrent government affairs in the field of new renewable energy, including national and regional energy management.
2. **Energy Conservation**
   - Minister of Energy and Mineral Resources Regulation Number 4 of 2011 and Number 14 of 2012: Establishing energy management and awards for energy efficiency.
   - Minister of Energy and Mineral Resources Regulation Number 39 of 2017: Regulates the implementation of physical activities for the utilization of new renewable energy and energy conservation.
3. **Occupational Safety and Health**
   - Minister of Energy and Mineral Resources Regulation Number 33 of 2021: Regulates occupational safety and health as well as environmental protection in the utilization of energy resources, including technical aspects for the utilization of geothermal energy.
4. **Supervision and Evaluation**
   - Minister Regulation Number 48 of 2017: Provides a framework for supervision of enterprises in the energy and mineral resources sector, including evaluation of the implementation of maritime energy management.

Legislation Regarding Maritime Energy Management in Indonesia considers the following:

1. **The Vision of the World Maritime Axis**
   Indonesia has a vision to become a world maritime axis country, which means becoming a sovereign, advanced, independent, and strong maritime nation, capable of providing positive contributions to the security and peace of the region and the world in accordance with national interests. This vision is reflected in Indonesia's Marine Policy aimed at accelerating the implementation of the World Maritime Axis [4, 5].
2. **Law No. 32 of 2014**
   Implementing the law made, namely Law No. 32 of 2014 concerning the Sea, is a legal umbrella that regulates the utilization of Indonesia's seas comprehensively and integratively. This law emphasizes the need for optimal and sustainable management of marine resources to provide the greatest benefits for the prosperity of the Indonesian people.

3. **Maritime Zones in Indonesia**

   Indonesia has maritime zones that are determined in accordance with the provisions of international maritime law, namely:

   - Territorial Sea, sea area with a maximum distance of 12 nautical miles from the baseline.
   - Additional Zone, sea area with a distance of more than 12 nautical miles to the EEZ boundary.
   - Exclusive Economic Zone (EEZ), sea area bordering the continental shelf and a maximum distance of 200 nautical miles from the baseline, unless Indonesia can scientifically prove the existence of a wider seabed.
   - Continental Shelf: Seabed that is an integral part of the territory of the Republic of Indonesia.

4. **Sovereign Rights and Authorities**

   Indonesia has sovereign rights to manage the resources of the water column in the EEZ, as well as the authority to regulate and manage the seabed and its resources in the Continental Shelf. These sovereign rights must be in accordance with international law and pay attention to the strategic interests of Indonesia, such as maintaining the integrity of the state's territory and the welfare interests of the entire nation. In the EEZ, Indonesia has several authorities, including:

   - **Exploration and Exploitation**: Conducting activities to search for and utilize natural resources in the waters and seabed.
   - **Conservation of Natural Resources**: Managing and preserving natural resources to remain sustainable.
   - **Scientific Research**: Conducting research related to the ocean to enhance understanding of marine ecosystems.

5. **Infrastructure and Human Resources**

   The management of maritime energy in Indonesia also requires infrastructure improvements, human resource development, and technology modernization. This is to provide economic benefits for the country and society, as well as to build a strong maritime power. Infrastructure improvements in Indonesia's maritime sector are very important to enhance operational efficiency and security. Several aspects of infrastructure that need to be improved include:

   - **Ports**: Development and renovation of major ports to increase cargo capacity and turnaround time for ships. This can be done by updating unloading and loading facilities, as well as improving accessibility to the ports.
   - **Transportation Network**: Development of a more complete and efficient sea transportation network, including sea toll routes to enhance inter-island connectivity. This can help reduce transit time and logistics costs.
   - **Security Facilities**: Construction of security facilities such as radar stations, monitoring centers, and foreign ship detection systems to improve surveillance and response to maritime security threats.

3 Regulations Governing Emission Reductions

Regulations governing emission reductions in Indonesia include various regulations aimed at addressing climate change and reducing greenhouse gas emissions (GHG). Here are some key relevant regulations, both domestic and adopted from abroad:

1. Law No. 32 of 2009 on Environmental Protection and Management. This regulation establishes a legal framework to protect the environment and manage natural resources, including efforts to reduce GHG emissions.
2. Presidential Regulation No. 98 of 2021 on the Implementation of Carbon Economic Value. This regulation governs the carbon economic value mechanism to achieve GHG emission control targets, focusing on the imposition of carbon taxes and the establishment of carbon markets. This regulation also serves as a basis for implementing the Nationally Determined Contribution (NDC) target and controlling greenhouse gas emissions. Moreover, it also provides a guideline for reducing greenhouse gas emissions through policies, measures, and activities [6–8].
3. Financial Services Authority's (OJK) Regulation No. 14 of 2023 on Carbon Trade. This regulation governs the carbon trading mechanism through carbon exchanges as a way to reduce GHG emissions by buying and selling carbon units.
4. Regulation of the Minister of Environment and Forestry No. 7 of 2023. This regulation governs general provisions regarding emissions trading and offset emissions of GHG in the forestry sector, as well as non-tax state revenues from carbon trading [9].
5. Minister of Energy and Mineral Resources Regulation No. 2 of 2023 on GHG Emission Reduction. Regulates certification of GHG emission reductions through carbon injection and storage, as well as monitoring and reporting procedures [10].

The regulations governing emission reduction and economic policy in emission reduction are closely linked, as effective environmental policies often require economic incentives to drive meaningful change. Regulations set legally binding limits on emissions, establish standards, and mandate reporting requirements, compelling industries to adopt cleaner technologies and reduce their carbon footprint. Economic policies, such as carbon pricing (taxes or cap-and-trade systems), subsidies for renewable energy, or tax credits for green innovations, support these regulatory goals by making emission reduction financially viable and attractive for businesses. Together, these mechanisms create a comprehensive approach: regulations enforce minimum environmental standards, while economic policies promote voluntary reductions and innovation by balancing costs and benefits for businesses. This synergy encourages sustainable practices and investment in green technology, aligning environmental targets with economic growth.

Economic policy has an impact on emission reduction. The existence of regulations governing financial emission reductions will make society more cautious in using environmentally unfriendly equipment. For example, the addition of costs for those who damage the environment will support a sustainable environment, and conversely, those who protect the environment will be given incentives for their responsibility towards their surroundings. Economic policies aimed at sustainable development will better preserve the environment, which can provide greater welfare in the future.

3.1 Carbon Tax

To support the implementation of emission reductions, there needs to be encouragement for society to comply with existing regulations. One such policy is the imposition of taxes on those who operate businesses and emit emissions from their activities. The carbon tax policy does not yet have international regulations due to differing conditions in each country. The Paris Agreement of 2015 is a regulation that encourages countries to implement carbon taxes; this agreement commits to keeping the global temperature increase below 2 °C and pursuing efforts to limit the increase to around 1.5 °C, which can reduce the impacts and risks of climate change globally [11]. There is no obligation to implement a carbon tax in the Paris Agreement; however, this agreement encourages each country to take significant steps in reducing carbon emissions.

Finland is one of the first countries to implement a carbon tax; research shows that the implementation of the carbon tax has a positive influence, with carbon emissions reduced by 1.69% compared to without the imposition of a carbon tax. Meanwhile, Sweden has the highest tax rate up to 2021 at US$137 per ton of CO_2 equivalent. Countries in Asia also implement carbon taxes, including Japan and China. Japan imposes a very low carbon tax rate of $2.61 for each ton of CO_2 equivalent, so Japan's GDP is not affected by the imposition of the carbon tax, and the impact on emission reduction is minimal. The implementation of the carbon tax in China is considered quite successful as it achieves the target for carbon emission reduction. The implementation of the carbon tax can be a source of revenue for a country, as taxes have a budgetary function in the country's financial system. Indonesia has not officially implemented a carbon tax; however, the government is committed to climate change with the enactment of the Tax Regulation Harmonization Law (UU HPP) on October 7, 2021 [12].

The existence of a carbon tax will increase industry awareness to have cleaner and more efficient technological innovations that can reduce the tax burden. The use of environmentally friendly technology initially will increase investment costs and ultimately will be passed on to consumers; however, due to the additional carbon tax burden, companies will shift to this investment. In 2024, the maritime industry contributes 2–3% of global greenhouse gas emissions [13], which initiates business actors to develop environmentally friendly technology. One of them is by utilizing renewable energy (EBT) in the form of solar panels as a source of lighting energy and various work equipment energy sources. However, the use of EBT incurs high maintenance costs, so it is recommended for maintenance management and operational management, as it will significantly contribute to the reduction of maritime carbon emissions. The International Maritime Organization (IMO) has conducted various studies on ways to reduce carbon emissions from ships operating in international waters. This organization has also taken concrete steps to create binding international rules to regulate the level of carbon emissions from these ships. The policies considered include environmental indices, carbon taxes, emission standards, and emission trading schemes.

In addition to providing carbon taxes for companies that produce carbon emissions, the government also provides subsidies for companies that implement sustainable management. Subsidies can encourage companies to create larger job opportunities because consumers receive more affordable prices, thus increasing demand. Germany is one of the countries that has succeeded in implementing these subsidies. Germany has an ambitious program called "***Energiewende***," aimed at a full transition to renewable energy. This program is supported by various types of subsidies, such as feed-in tariffs and investment taxes. Germany is gradually reducing the role of emission-free energy generation, namely nuclear energy. To replace the nuclear electricity supply, Germany must rebuild fossil fuel and gas power plants because the natural fluctuations of renewable energy production still require reserves from fossil power plants that produce greenhouse gases (GHG) (CO_2). This situation makes the GHG reduction targets increasingly difficult to achieve [14]. Electricity prices and government subsidies continue to show a tendency to rise, and household users must bear these increases. High electricity prices are caused by high electricity production costs plus expenses for the expansion and modernization of networks to accommodate the electrical system with renewable energy. Although electricity prices are high and CO_2 reductions do not meet targets, the ***Energiewende*** program still receives good acceptance from the public and has strong political support [15].

3.2 Carbon Market

The carbon market has existed in Europe since 2005, named the EU ETS or European Union Emission Trading System, where industries in European countries that joined the European Union (EU) are strictly given a limited quota for emissions (allowances). The EU ETS became the first emission trading system in the world and is considered successfully implemented. This carbon market contributed more than 75% of international carbon trading in 2021 and reduced emissions by up to 24% in 2019. According to the Carbon Pricing for Climate Action Report in the Indonesian Carbon Trading Handbook [16], the existence of the carbon market has several benefits, including:

1. Reducing pollution levels, thus improving air quality.
2. Increasing the availability of clean water due to reduced water pollution.

3. Improving soil quality due to reduced soil contamination and the use of hazardous chemicals.
4. Reducing traffic congestion and transportation accidents.
5. Minimizing forms of tax evasion and reducing administrative costs in fiscal policy.
6. Increasing macroeconomic stability due to reduced uncertainty risks.
7. Encouraging technological shifts that lead to low-carbon technological innovations.

In Indonesia, the carbon market is outlined in Presidential Regulation No. 98 Article 1 Paragraph 17 of 2021, which explains that the carbon market is a market-based trading mechanism useful for reducing the presence of greenhouse gas (GHG) emissions through the buying and selling of carbon units. Carbon units are proof of carbon ownership in the form of certificates or technical approvals stated in one ton of carbon dioxide equivalent (CO_2e). Carbon emission units generated from human and economic activities have been regulated through the Economic Value of Carbon (NEK). The Economic Value of Carbon is an instrument in realizing contributions to the reduction of greenhouse gas emissions that have been established [6–8].

3.3 Carbon Emission Disclosure (CED)

Carbon Emission Disclosure (CED) is based on the recording of factors that have been identified in the information request sheet developed by the Carbon Disclosure Project (CDP). CDP is an international non-profit organization established as a global carbon disclosure project in 2000. This organization was formed to encourage companies to disclose information related to corporate activities concerning energy use and its impact on the environment (Carbon Disclosure Project 2023). Disclosure of information is divided into the following categories and items:

Carbon emission disclosure in Indonesia is not yet mandatory but has begun to be implemented by companies listed on the Indonesia Stock Exchange. This is because it affects the company's value in the capital market, which is related to the company's reputation. The main motivation for companies to disclose information is to increase market value and maintain legitimacy. Signaling theory and voluntary disclosure theory support this view. Signaling theory emphasizes the importance of transparent information for investors in making investment decisions, while voluntary disclosure theory highlights the tendency of companies to disclose information that is beneficial to them. Issues regarding sustainable finance have become an interesting topic in recent research, as companies committed to reducing carbon emissions are driven not only by profitability but also by ethical awareness of social responsibility. They understand that their actions can contribute to saving human lives threatened by the adverse effects of climate change caused by carbon emissions (Table 1) [17].

Table 1 Disclosure of information item

Category	Disclosure items
Climate change risks and opportunities (CC)	1. Climate risk or climate change 2. Opportunity or business or climate change
Greenhouse gas (GHG) emissions	1. GHG protocol or ISO or Panel on Climate Change (IPCC) 2. Verification or third party 3. Total emissions or CO2 4. Greenhouse gases or emissions or Carbon 5. Coal or electricity or mining 6. Machine or vehicle 7. Previous year's emissions
Energy consumption (EC/energy consumption)	1. The amount of energy or energy consumption or consumption or amount 2. Biodiesel or biofuel or vegetable or renewable 3. Energy or Electricity
Greenhouse gas reduction and costs (RC)	1. Reducing emissions or strategies 2. Emission reduction targets or levels 3. Environmental costs or environmental budget 4. Environmental capital expenditure or capital expenditure
Accountability of emission carbon (AEC)	1. Climate or climate change (seen from the board of directors' report) 2. Company climate or development (seen from the board of directors' report)

Source [18, 17]

4 Regulations Governing Biofuels Usage

Minister of Energy and Mineral Resources Regulation Number 25 of 2013 does not directly affect maritime energy management. This regulation governs amendments to Minister of Energy and Mineral Resources Regulation Number 32 of 2008 concerning the provision, utilization, and trade of biofuels as alternative fuels. However, this regulation plays an important role in energy management in general, especially in the context of the use of biofuels as an alternative to other fuels. This includes the implementation of a minimum policy for the utilization of biodiesel (B 100) in the mixture of fuel oil, starting from 10% in 2014, increasing to 20% in 2020, and 25% in 2025. The use of regulated biofuels can have broad implications in overall energy management. Here are some ways these regulations can indirectly affect maritime energy management:

1. Energy Source Diversification

   The use of biofuels can help diversify energy sources, which in turn can reduce dependence on fossil energy sources. This diversification can enhance energy security and reduce the environmental impact of fossil energy exploitation.

2. Emission Reduction
   Biofuels generally have lower greenhouse gas emissions compared to fossil fuels. With the increasing use of biofuels, these regulations can assist in efforts to reduce greenhouse gas emissions, which is an important aspect of sustainable energy management.
3. Technology Development
   These regulations encourage the development of technology in the production and utilization of biofuels. This technology can be applied in various sectors, including the maritime industry, to improve the efficiency and sustainability of energy use.
4. Community Involvement
   By expanding the use of biofuels, these regulations can enhance community involvement in energy management. A more engaged community can better understand the importance of sustainable energy management and participate in these efforts.

Regulations governing emission reductions in Indonesia include various regulations aimed at addressing climate change and reducing greenhouse gas emissions (GHG). Here are some key relevant regulations, both domestic and adopted from abroad:

1. Law No. 32 of 2009 on Environmental Protection and Management. This regulation establishes a legal framework to protect the environment and manage natural resources, including efforts to reduce GHG emissions.
2. Presidential Regulation No. 98 of 2021 on the Implementation of Carbon Economic Value. This regulation governs the carbon economic value mechanism to achieve GHG emission control targets, focusing on the imposition of carbon taxes and the establishment of carbon markets [6–8].
3. Financial Services Authority's (OJK) Regulation No. 14 of 2023 on Carbon Trade. This regulation governs the carbon trading mechanism through carbon exchanges as a way to reduce GHG emissions by buying and selling carbon units.
4. Regulation of the Minister of Environment and Forestry No. 7 of 2023. This regulation governs general provisions regarding emissions trading and offset emissions of GHG in the forestry sector, as well as non-tax state revenues from carbon trading.
5. Minister of Energy and Mineral Resources Regulation No. 2 of 2023 on GHG Emission Reduction. Regulates certification of GHG emission reductions through carbon injection and storage, as well as monitoring and reporting procedures.

The implementation of the Minister of Energy and Mineral Resources Regulation Number 25 of 2013 and Number 20 of 2014 at the local level in Indonesia faces several challenges. Here are some of the main challenges identified:

1. Fossil fuel dominance creates a dependency on fuel oil. Although there are efforts to increase the use of biofuels, there is still a dominance of fuel oil in power generation and the transportation sector. This complicates the transition to broader biofuel use.
2. The Geographical condition imposes challenges in island regions. Indonesia, which consists of thousands of islands, has diverse geographical conditions that can affect the distribution and accessibility of biofuels. Inadequate infrastructure in remote areas can hinder the effective implementation of regulations.
3. Human resources such as the lack of experts and trained human resources in the field of renewable energy is a constraint in policy implementation. This includes technical skills in the production and use of biofuels as well as understanding existing regulations.
4. Challenges in supervision include the implementation of oversight of biofuel utilization often being suboptimal, either due to a lack of resources or a monitoring system that is not yet fully effective. This can lead to non-compliance with established provisions.
5. The development of infrastructure for the production and distribution of biofuels is still limited, thus hindering the ability to meet biofuel usage targets in the mixture of fuel oil.
6. Community Awareness, the obstacle is Lack of Awareness and Education. The community and industry players often lack understanding of the benefits of using biofuel and its potential positive impacts on the environment, thus reducing support for the transition to renewable energy.
7. Regulatory and Other Policy Constraints, the obstacle is Policy Changes. The existence of changes in authority in the management of the energy and natural resources sector can cause confusion in the implementation of regulations, thus requiring regulatory adjustments to ensure alignment between central and regional policies.

5 The Implementations of Maritime Energy Management and Emissions Reduction Regulation in Indonesia

The implementation of maritime energy management and emissions reduction regulations in Indonesia represents a critical step towards sustainable maritime practices and environmental stewardship. As one of the world's largest archipelagic nations with a rapidly growing shipping sector, Indonesia faces both opportunities and challenges in reducing its maritime emissions and improving energy efficiency. In line with global standards such as those set by the International Maritime Organization (IMO), Indonesia has adopted policies aimed at minimizing greenhouse gas emissions from shipping activities, optimizing energy use, and promoting cleaner, low-carbon technologies in the maritime industry. These regulations not only help Indonesia meet its international environmental commitments but also align with national economic goals by fostering a greener maritime industry that can support long-term economic resilience. The success of these efforts depends on effective coordination among government agencies, maritime operators, and other stakeholders, as well as the ongoing development of supportive policies and infrastructure to facilitate compliance and innovation in energy management.

5.1 Emission Reduction Policy

The Indonesian government has set emission reduction targets through the Nationally Determined Contribution (NDC) adapted from the Paris Agreement. In 2022, Indonesia updated its NDC target to a 32% unconditional emission reduction and 43% conditional by 2030 [19]. This revised target demonstrates a strong commitment to contributing to global efforts in addressing climate change. This update reflects Indonesia's increasing ambition in facing climate change challenges, where the previous target was a 29% unconditional and 41% conditional emission reduction [20, 21]. In the Enhanced NDC document submitted to the UNFCCC, the government also emphasizes the importance of integrating long-term strategies to achieve net zero emissions by 2060 or sooner [22]. With these steps, Indonesia not only strives to fulfill its international obligations but also commits to creating a better environment for its society. This action is in line with the vision of sustainable development that prioritizes the balance between economic growth and environmental protection [21].

Indonesia's decarbonization roadmap to 2060 envisions a gradual transition towards a carbon–neutral society through multiple phases (Fig. 1). Starting from 2030, the country plans to intensify the adoption of renewable energy sources and implement policies like deforestation prevention, carbon taxation, and other regulatory measures to reduce CO_2 emissions. Moving into the 2040s, Indonesia aims to shift to non-carbon-based fuels such as hydrogen and ammonia, supporting the transition in hard-to-abate sectors. By 2050, the roadmap includes advancing synthetic carbon fuels derived from carbon–neutral resources, aided by affordable CO_2 separation technologies. These efforts collectively target net-zero carbon emissions by 2060, aligning with global climate goals and ensuring a sustainable, low-carbon future for Indonesia [23].

Indonesia's decarbonization roadmap aligns closely with the development of CO_2 capture and utilization (CCU) technologies as the country strives to reduce its greenhouse gas emissions and reach net-zero targets. CCU plays a critical role in Indonesia's strategy to lower carbon emissions, especially in sectors where emissions are hard to abate, such as cement, steel, and energy production. By capturing CO_2 from maritime sectors and converting it into valuable products like fuels, chemicals, or materials, CCU can help Indonesia balance economic growth with environmental sustainability. Integrating CCU into the decarbonization roadmap supports Indonesia's commitment to global climate goals, creating pathways for carbon reduction that complement the expansion of renewable energy and energy efficiency initiatives.

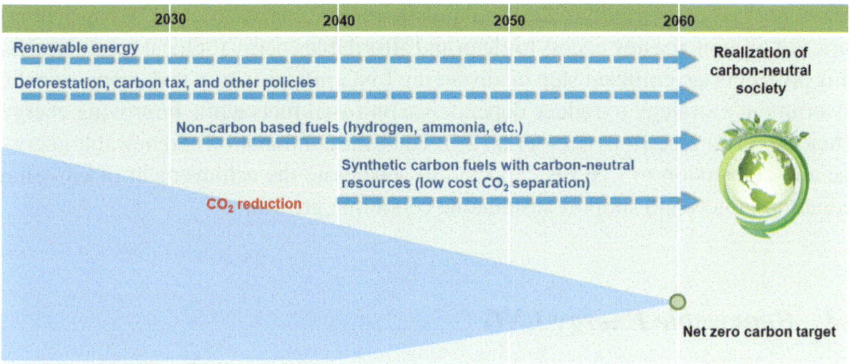

Fig. 1 General roadmap for decarbonization in Indonesia (*Source* [24])

CO_2 capture and utilization (CCU) has promising applications in the maritime sector, particularly in reducing emissions from ships, which are major contributors to global CO_2 levels. In this sector, CO_2 capture technology can be integrated into ship engines and exhaust systems to capture emissions directly from the combustion process. The captured CO_2 can then be stored onboard in pressurized tanks for later utilization or offloading at ports equipped for CO_2 handling. Utilization options for the captured CO_2 include converting it into synthetic fuels like methanol, which can be reused as marine fuel, effectively creating a closed-loop system. Additionally, the CO_2 can be mineralized to produce stable carbonates or other materials used in construction, potentially offsetting emissions in other industries. Implementing CCU in the maritime sector not only helps reduce direct ship emissions but also supports the sector's transition towards carbon neutrality by enabling the use of captured CO_2 in innovative and sustainable ways.

5.2 Renewable Energy Strategy

One of the main strategies in emission reduction is the transition to renewable energy. The government encourages the use of New and Renewable Energy (EBT) such as solar, wind, and biomass in the maritime sector. For example, the use of solar panels in ports to meet energy needs [25, 26]. In addition, the conversion of fossil fuels to Liquefied Natural Gas (LNG) is also a primary focus to reduce dependence on more polluting fuels such as coal, as well as to produce lower carbon dioxide emissions during combustion, making it a cleaner alternative in the global effort to reduce GHG emissions (Indonesia Safety Center 2024) [27].

This transformation is not only limited to the use of LNG as an alternative fuel but also includes innovations in infrastructure and technology that support this transition. For example, PT PLN Energi Primer Indonesia (PLN EPI) has partnered with the

Indokorea Gas consortium to develop midstream LNG infrastructure in Nusa Tenggara, aimed at increasing access to clean and affordable energy for local communities. This project is a significant step in advancing LNG infrastructure and supporting the government's strategy to reduce dependence on fossil fuels while improving energy efficiency in the electricity sector [28, 29]. Thus, the combination of renewable energy use and conversion to LNG is expected to accelerate the achievement of emission reduction targets and support sustainable economic growth.

5.3 Renewable Energy LNG

The management of maritime energy is increasingly becoming a main concern amid the challenges of climate change and the need to reduce carbon emissions. The shipping sector, which has long relied on fossil fuels, contributes a significant percentage to global emissions. Therefore, the development and implementation of alternative energy sources, such as LNG, become a very important solution. LNG offers various advantages, including lower carbon dioxide emissions and significant reductions in harmful pollutants such as sulfur dioxide and nitrogen oxides. The utilization of LNG in the maritime industry not only helps reduce the carbon footprint but also promotes the operational efficiency of ships. With cleaner combustion technology, ships using LNG can operate more efficiently, reducing fuel consumption and operational costs.

In addition, the use of LNG as fuel supports the development of technological innovations, such as dual fuel engines that can operate with both LNG and conventional fuels. This provides flexibility for the fleet to adapt to various changing conditions and environmental policies. However, to maximize the benefits of LNG, there needs to be investment in supporting infrastructure, such as refueling terminals and distribution systems at ports. The government and the private sector must work together to create an ecosystem that supports the use of LNG, including adequate regulations and incentives for shipping companies to switch to cleaner energy. With these steps, the management of maritime energy can be carried out sustainably, leading to significant emission reductions and creating a more environmentally friendly future for the shipping industry. Several regulations in Indonesia that govern the utilization of renewable energy LNG are as follows:

1. Decree of the Minister of Energy and Mineral Resources No. 13 K/13/MEM/2020 revised to Decree of the Minister of Energy and Mineral Resources No. 2 K/TL.01/MEM.L.2022. The regulations are related to the Development of Supply Chains System to Distribute LNG to 52 Converted Power Plants.
2. Decree of the Minister of Energy and Mineral Resources No. 08 of 2017. The regulations are related to Domestic Market Obligation (DMO) set at 25% of Gas Production Results.
3. Decree of the Minister of Energy and Mineral Resources No. 45 of 2017 revised to Decree of the Minister of Energy and Mineral Resources No. 10 of 2020. Regulation related to the Utilization of Natural Gas for Power Generation and

the Establishment of the Upper Price Limit of Natural Gas at the Plant Gate, which is 14.5% of the Indonesia Crude Price (ICP) Index.

4. Presidential Regulation No. 79 of 2014 was revised to Presidential Regulation No. 22 of 2017. Presidential Regulation No. 22 of 2017 establishes national energy policy and the national energy general plan that is cross-sectoral. This includes long-term strategies to achieve energy resilience and sustainability. This presidential regulation serves as a reference for various ministries and agencies in formulating harmonious and integrated energy policies [30]. Its implementation involves the development of new and renewable energy sources as well as improvements of energy efficiency across all sectors, although there are still challenges in terms of inter-agency coordination and budget allocation. The regulations are related to the National Energy Policy and the National General Energy Plan which is Cross-Sectoral [31].

   - Decree of the Minister of Energy and Mineral Resources No. 2 K/TL.01/MEM.L.2022 regulates the development of the supply chain system to distribute LNG to 52 converted power plants. The main objective of this policy is to ensure the availability and efficient distribution of LNG to support energy needs in the electricity sector [32]. In its implementation, this regulation focuses on collaboration between the government and private parties to build the required infrastructure, such as LNG terminals and pipeline networks. This is important to reduce dependence on other fossil fuels and increase the use of renewable energy in power generation.
   - Decree of the Minister of Energy and Mineral Resources NO 08 of 2017 regulates the Domestic Market Obligation (DMO) policy set at 25% of gas production, aimed at ensuring domestic gas supply for national energy needs. This policy requires gas producers to sell a portion of their production in the domestic market [33]. With the implementation of this policy, it is expected to enhance national energy resilience by ensuring that domestic gas needs are met before being exported. However, the challenge faced is the high interest of producers to export gas due to higher international prices, which often results in a lack of gas supply for domestic needs.
   - Decree of the Minister of Energy and Mineral Resources NO 10 of 2020 regulates the utilization of natural gas for power generation and sets a price cap for natural gas at the plant gate at 14.5% of the Indonesia Crude Price (ICP) Index. This aims to maintain price stability for gas for industries and consumers [34]. With this regulation, it is hoped that the operational costs of power plants can be reduced, so that electricity tariffs for consumers remain affordable. However, its implementation faces challenges in terms of appropriate price adjustments and oversight of gas producer compliance.

Through these regulations, the Indonesian government strives to create a more sustainable and independent energy system, although challenges in implementation remain, especially related to market compliance and international price dynamics.

5.4 Carbon Economic Incentive

Indonesia also implements the Carbon Economic Value (CEV) policy through Presidential Regulation Number 98 of 2021 [6–8]. This policy aims to provide incentives for companies that successfully reduce greenhouse gas emissions. Through a carbon trading system and carbon tax, the government hopes to encourage Green Innovation and Green Technology innovation [35]. Investment in Green Technology is key to the implementation of CEV. Green Technology, such as renewable energy and electric vehicles, is a concrete example of innovation that can help reduce carbon footprints. According to research, Green Innovation not only contributes to emission reductions but also offers new economic opportunities for companies and communities. By leveraging this technology, companies can improve operational efficiency and reduce long-term costs, while also meeting the demands of increasingly environmentally conscious consumers. Therefore, collaboration between the government and the private sector is crucial to creating an ecosystem that supports Green Innovation and achieving national emission reduction targets [36, 37].

5.5 Implementation in the Maritime Sector

The implementation of sustainable practices in Indonesia's maritime sector encompasses a multifaceted approach, integrating sustainable port management, green technology in maritime transportation, policy challenges, and stakeholder participation. Sustainable port management focuses on reducing port emissions, optimizing resource use, and ensuring efficient waste management, which helps ports operate with minimal environmental impact. Simultaneously, the adoption of green technology such as energy-efficient vessels, alternative fuels, and emission control systems drives cleaner maritime transportation, supporting Indonesia's commitment to reducing greenhouse gas emissions. However, implementing these initiatives poses significant challenges, including high upfront costs, technology gaps, and the need for regulatory clarity and consistency. Overcoming these challenges requires strong participation from stakeholders, including government bodies, port authorities, maritime companies, and local communities, who play crucial roles in supporting policies, investing in cleaner technology, and fostering a collaborative approach to sustainable maritime practices.

1. **Sustainable Port Management**

   To achieve net zero emissions by 2060, sustainable port management has become one of the priorities. The government has implemented Green and Smart Port certification, focusing on waste management and the utilization of renewable energy [20]. A total of 27 ports have received this certification as an initial step toward sustainability. Green and Smart Port is expected to serve as a model for port development in Indonesia, focusing on the balance between economic and ecological aspects. This concept includes the application of information technology and digitalization to improve efficiency and reduce carbon footprint. In this context, the development of smart ports also becomes important, as it integrates modern technology to optimize business processes and minimize unnecessary physical interactions. This aligns with the statement of the Secretary of the Directorate General of Sea Transportation, emphasizing the need for collaboration among agencies to ensure that activities in ports prioritize not only economic benefits but also environmental sustainability [38].

2. **Green Technology in Maritime Transportation**
   The use of Green Technology in maritime transportation is increasingly gaining attention in Indonesia, especially through the application of LNG-fueled ships that are known to be cleaner compared to conventional fuels. This initiative aligns with the commitment of the International Maritime Organization (IMO) which targets the reduction of carbon emissions from ships. The Indonesian government, as a member of the IMO, has issued several regulations to support the implementation of green shipping, including the obligation to use low sulfur fuels and exhaust gas cleaning systems (scrubbers) on ships. These steps aim to reduce marine environmental pollution due to high consumption of fossil fuel consumption in the maritime transport sector and encourage the use of environmentally friendly energy [39]. In addition, the government also encourages the use of electric vehicles for land transportation that is integrated with ports, creating a more sustainable transportation system. These efforts not only focus on reducing emissions in the maritime sector but also encompass the entire transportation ecosystem. By building environmentally friendly infrastructure and raising awareness among stakeholders, it is hoped that the shipping industry can contribute more significantly to the Net Zero Emission (NZE) 2060 goals. In this context, collaboration between the government, the shipping industry, and technology providers becomes key to realizing innovations that support green shipping and the protection of the maritime environment [25, 26, 40].
3. **Challenges in Policy Implementation**
   Despite various policies and initiatives, significant challenges remain in their implementation. One of the main challenges is the lack of public understanding regarding government policies related to climate change and emission reduction. Therefore, intensive socialization is needed to raise public awareness about the importance of climate change mitigation actions [19].
4. **Stakeholder Participation**
   The participation of all stakeholders is crucial in achieving emission reduction targets. The involvement of local governments, the private sector, and civil society needs to be enhanced so that the policies implemented can run effectively and gain broad support [41]. Maritime energy management and emission reduction in Indonesia are important steps towards environmental sustainability. With various policies such as NDC, NEK, and the development of Green Technology, Indonesia shows a strong commitment to facing the challenges of climate change. However, the success of the implementation of these policies greatly depends on the active participation of all parties and a good understanding of the importance

of mitigation actions. Through these strategic steps, Indonesia not only strives to meet its commitments to climate change but also to harness the economic potential of sustainable energy transition.

6 Policy Initiatives and Industry Innovation

Incentives for clean technology, many regulatory frameworks promote the adoption of energy-efficient technologies and alternative fuels (e.g., LNG, hydrogen, biofuels) through subsidies, tax breaks, or grants. Research and Development (R&D), governments may fund R&D initiatives to develop advanced energy management systems and emissions reduction technologies. Stricter regulations may increase operational costs for shipping companies, prompting them to adopt cleaner technologies. Such changes could also lead to shifts in market dynamics, with a potential increase in demand for green shipping services. The complexity of international regulations can create compliance challenges, particularly for smaller operators who may lack the resources to adapt to changing standards. Regulatory frameworks that include emissions trading systems or carbon pricing can incentivize the shipping industry to innovate in energy management and emissions reduction strategies.

Decarbonization initiatives are expected to intensify, with increasing calls for the shipping industry to transition to zero-emission vessels by mid-century. There is a growing emphasis on collaboration between governments, industry stakeholders, and research institutions to share knowledge and best practices in emissions reduction. The regulatory landscape for maritime energy management and emissions reduction is evolving rapidly, influenced by international agreements, regional policies, and technological advancements. The implications of these regulations are profound, affecting operational practices, investment strategies, and the overall sustainability of the maritime industry. As the world moves towards stricter environmental standards, ongoing dialogue among stakeholders will be crucial to achieving effective and equitable solutions. In Indonesia's maritime industry, the implementation of the Energy Efficiency Design Index (EEDI), Energy Efficiency Existing Ship Index (EEXI), and Carbon Intensity Indicator (CII) represents a significant commitment to reducing emissions and increasing energy efficiency for vessels, aligning with IMO's goals.

1. **Energy Efficiency Design Index (EEDI)**

   EEDI sets minimum energy efficiency standards for new ships, requiring them to be designed with fuel efficiency in mind, thus reducing CO_2 emissions. For new ship constructions, EEDI compliance is integrated within national shipbuilding standards, monitored by authorities such as BKI (Biro Klasifikasi Indonesia). Shipyards and designers work to meet these standards by optimizing hull designs, reducing ship weight, and integrating efficient propulsion technologies. Limited access to advanced technologies and infrastructure constraints in Indonesian shipyards can make it difficult to meet high EEDI standards, especially for smaller, local shipbuilders.

2. **Energy Efficiency Existing Ship Index (EEXI)**
   EEXI applies to existing ships, providing a metric to assess energy efficiency and determine if older vessels need retrofitting to improve fuel consumption and lower emissions. The Indonesian government is encouraging vessel owners to evaluate their fleet's compliance with EEXI standards, especially for tankers and bulk carriers. Ships not meeting the EEXI may be required to undergo modifications, such as engine power limitation (EPL) or the installation of energy-saving devices. The cost of retrofitting older ships is a significant obstacle, as many Indonesian shipping companies operate on limited budgets and may lack access to financing or subsidies for such upgrades.
3. **Carbon Intensity Indicator (CII)**
   CII measures the operational carbon intensity of ships, calculated based on the fuel consumed per ton-mile, effectively targeting emissions reduction in daily operations. CII reporting is integrated with the operational management of ships, requiring shipping companies to monitor fuel use and optimize operations. Indonesian shipping companies, particularly those operating internationally, are encouraged to follow IMO's DCS (Data Collection System) and SEEMP (Ship Energy Efficiency Management Plan) for CII compliance. Operational monitoring requires advanced data collection and analysis systems that smaller operators in Indonesia may not yet possess. Additionally, consistent implementation of CII standards is challenging in the face of limited technical know-how and potential increased operational costs.

Indonesia's Ministry of Transportation and related maritime regulatory bodies are working to align the national regulatory framework with IMO standards. This effort includes incentives for adopting energy-efficient technologies, providing training programs on operational efficiency, and exploring financing models for retrofitting ships. The government is also engaging in capacity-building initiatives to strengthen local expertise in emission management. By adopting these measures, Indonesia aims to enhance sustainability in its maritime sector, contributing to the global effort to reduce greenhouse gas emissions and achieve IMO's 2030 and 2050 emissions reduction goals. In the context of Indonesia's maritime energy management and emissions reduction, industry innovation plays a crucial role in shaping policies that align with global sustainability goals. The maritime sector in Indonesia, including shipping and energy distribution (e.g., LNG), is vital for its archipelagic nature. Here are several innovations and their policy implications:

1. **Cleaner Fuels for Shipping**

   - Innovation, adoption of LNG as a marine fuel is gaining momentum as an alternative to conventional heavy fuel oils. LNG significantly reduces sulfur oxides (SOx), nitrogen oxides (NOx), and CO_2 emissions, making it a cleaner option.
   - Policy Implications, Indonesia could enhance regulations mandating the transition to LNG-powered ships, especially for domestic routes. Incentives for

retrofitting existing ships with LNG-powered engines and building new LNG-fueled vessels are key. Aligning with International Maritime Organization (IMO) emissions targets could also drive regional standards.

2. **Hybrid and Electrification Technologies**
    - Innovation, the development of hybrid propulsion systems (combining LNG or low-sulfur fuels with electric batteries) and the use of fully electric vessels for shorter routes is advancing. Hybrid vessels can optimize fuel consumption and reduce emissions.
    - Policy Implications, Indonesian authorities could introduce subsidies or tax incentives for the adoption of hybrid or electric maritime technologies. This can be coupled with the development of necessary charging infrastructure at key ports. Policy frameworks could focus on supporting research and development (R&D) in maritime electrification and local shipbuilding capabilities.
3. **Digitalization and Smart Shipping**
    - Innovation, digital technologies like AI, blockchain, and IoT (Internet of Things) are being applied to optimize vessel routing, reduce fuel consumption, and enhance predictive maintenance, thus minimizing downtime and emissions.
    - Policy Implications, Indonesian maritime policies should encourage the adoption of digital solutions through public–private partnerships, improving operational efficiency and reducing emissions. Regulations around data sharing, cybersecurity, and the standardization of smart technologies across the shipping industry could be established to promote widespread adoption.
4. **Port Infrastructure Development**
    - Innovation: Ports are critical in managing the energy transition for maritime logistics. Innovations include the use of renewable energy sources for port operations (e.g., solar-powered terminals), cold ironing (shore-to-ship power supply to reduce emissions while docked), and waste-to-energy systems.
    - Policy Implications: Indonesia could revise its port development plans to focus on sustainability and energy efficiency. Policies that promote investments in green port infrastructure and renewable energy utilization are necessary to support maritime decarbonization. Regulations on mandatory cold ironing for certain classes of vessels and investment in renewable energy grids at major ports could also drive progress.
5. **Carbon Capture and Storage (CCS) for Shipping**
    - Innovation, emerging technologies like onboard carbon capture and storage (CCS) systems could drastically reduce CO_2 emissions from ships. Although still in early development stages, CCS for ships holds long-term potential.

- Policy Implications, R&D funding could be allocated toward developing and piloting CCS technologies for maritime applications. Policies incentivizing early adoption and partnerships with global research institutions could accelerate Indonesia's leadership in this field [42, 43].

6. **Green Shipbuilding**

- Innovation, the design of energy-efficient ships with advanced hull forms, air lubrication systems, and the use of lightweight materials is reducing the energy consumption of vessels.
- Policy Implications, Indonesia could establish shipbuilding standards that mandate energy efficiency features and eco-friendly materials. Financial support for shipbuilders to meet these standards could be implemented, driving innovation in Indonesia's maritime industry.

7. **Maritime Emissions Trading System (ETS)**

- Innovation, implementing a maritime emissions trading system (ETS) would set caps on the emissions ships can produce, with the option for ships to buy and sell emissions allowances.
- Policy Implications, Indonesia could adopt a regional ETS, incentivizing shipping companies to reduce emissions or invest in cleaner technologies. This would need to be harmonized with regional and international markets for emissions trading.

7 Conclusion

Indonesia's journey toward sustainable maritime energy management and emissions reduction showcases its commitment to aligning with global environmental standards while addressing national economic and ecological goals. By adopting international frameworks such as the Paris Agreement and the UNFCCC, Indonesia has demonstrated its dedication to mitigating climate change through robust regulations, innovative policies, and cross-sector collaboration.

The integration of renewable energy, biofuels, and advanced technologies like carbon capture and storage (CCS) underscores the nation's ambition to balance economic growth with environmental sustainability. However, challenges such as infrastructure gaps, technological limitations, and public awareness remain barriers that require continued effort from policymakers, industry leaders, and communities.

Indonesia's focus on incentivizing green innovation, fostering public–private partnerships, and enhancing regulatory compliance ensures a pathway toward achieving net-zero emissions by 2060. By leveraging its strategic position as a maritime nation and fostering advancements in clean energy, Indonesia contributes to global climate goals and sets a precedent for sustainable development in the maritime sector.

References

1. The Ministry of Energy and Mineral Resources, Regulation 50 of 2017; amended by The Ministry of Energy and Mineral Resources, Regulation 4 of 2020
2. Manalu YSH, Wahyuningtyas P, Nugroho TDH, Akadiati VAP (2024) Determining carbon emissions disclosure in Indonesian companies. Governors 3(1):10–20.
3. Firdaus I (2022) Dukungan Kebijakan Dan Peraturan Perundang-Undangan Untuk Mengakselarasi Aktivitas Riset Energi Baru Terbarukan di Indone-sia. Jurnal RechtsVinding 11(3):411–434
4. Hardiana I, Trixie B (2014) Menuju Indonesia sebagai Negara Poros Mari-tim. https://setkab.go.id/menuju-indonesia-sebagai-negara-poros-maritim/
5. Peraturan Presiden Nomor 16 Tahun 2017 Tentang Kebijakan Kelautan Indonesia
6. Peraturan Presiden (Perpres) Nomor 98 Tahun 2021 tentang Penyelenggaraan Nilai Ekonomi Karbon untuk Pencapaian Target Kontribusi yang Ditetapkan Secara Nasional dan Pengendalian Emisi Gas Rumah Kaca dalam Pembangunan Nasional. Indonesia (2021)
7. PERPRES NO. 98 TAHUN 2021, Presiden Republik Indonesia, Jakarta
8. Regulation of The President of The Republic of Indonesia Number 98 of 2021 on the Implementation of Carbon Pricing to Achieve The Nationally Determined Contribution Target and Control Over Greenhouse Gas Emissions in the National Development (2021)
9. Peraturan Menteri Lingkungan Hidup dan Kehutanan Republik Indonesia No-mor 7 Tahun 2023 Tentang Tata Cara Perdagangan Karbon Sektor Kehu-tanan
10. Peraturan Menteri Energi dan Sumber Daya Mineral Republik Indonesia Nomor 2 Tahun 2023 Tentang Penyelenggaraan Penangkapan dan Penyim-panan Karbon, Serta Penangkapan, Pemanfaatan, dan Penyimpanan Karbon Pada Kegiatan Usaha Hulu Minyak Dan Gas Bumi
11. Paris Agreement (United Nations Framework Convention on Climate Change) (2015)
12. Barus EB, Wijaya S (2022) Penerapan Pajak Karbon di Swedia dan Finlandia Serta Perbandingannya dengan Indonesia. J PAJAK Indonesian Indonesian Tax Rev 5(2):256–279. https://doi.org/10.31092/jpi.v5i2.1653
13. Furusaki T, Asmussen M, Centre for nature and climate. https://www.weforum.org/stories/2024/04/why-reducing-barriers-for-maritime-fuel-projects-is-key-to-progressing-on-decarbonization/#:~:text=NatureandClimate-,Maritimetransportac-countsfor2–3%25ofglobalgreenhousegas,thechallenge—butbarriersremain
14. Peraturan Presiden Republik Indonesia Nomor 61 Tahun 2011 tentang Rencana Aksi Nasional Penurunan Emisi Gas Rumah Kaca
15. Dewantara EA, Yulinawati H, Suswantoro E (2020) Emisi Co2 Kapal Peti Kemas Di Jakarta International Terminal Container, Pelabuhan Tanjung Priok Saat Pandemi Covid-19, J. Tek. Lingkung. Fak. Arsit. Lanskap Teknol. Lingkung. Univ. Trisakti 1(1):1–7
16. Centre KDI (2022) Indonesia carbon trading. Kata Data Insight Cent, Agustus, pp 1–77
17. Wirawan J, Setijaningsih HT (2022) Analisis Pengungkapan Emisi Karbon Di Indonesia. Jurnal Muara Ilmu Ekonomi dan Bisnis 6(1):235. https://doi.org/10.24912/jmieb.v6i1.18398
18. Bae Choi B, Lee D, Psaros J (2013) An analysis of Australian company carbon emission disclosures. Pac. Account. Rev. 25(1):58–79. https://doi.org/10.1108/01140581311318968
19. Santoso R (2024) Upaya Pengurangan Emisi Karbon, 1 February 2024, pp 1–2
20. Kementerian ESDM Ditjen Migas (2022) Tekan Emisi Karbon, Indonesia Naikkan Target ENDC Jadi 32 Persen, 14 November 2022, p 1
21. Climate Action Tracker, 2023. Indonesia Main Climate Targets. 4 December, p. 1.
22. Enhanced Nationally Determined Contribution Republic of Indonesia. United Nations Framework Convention on Climate Change (UNFCCC), Rio de Janeiro & New York
23. Indonesia Blue Economy Roadmap (2023) The Ministry of National Development Planning
24. Aziz M (2024) Opportunities and challenges toward Indonesia net zero emission in 2060. Seminar Nasional Master
25. Regulation of The Ministry of Environment and Forestry of The Republic of Indonesia Number SK.168/MENLHK/PKTL/PLA.1/2022 on the Indonesia's Forestry and Other Land Use (FOLU) Net Sink 2020 for Climate Change Control (2022)

26. Kementerian Lingkungan Hidup dan Kehutanan (2024) Kemajuan Aksi Peru-bahan Iklim Indonesia: leading by examples, 12 January 2024, p 1
27. Palebangan H, Yanuar (2019) Analisa Kapal Berbahan Bakar LNG sebagai Marine Fuel dalam Mengurangi Emisi Gas Buang Terhadap Lalu Lintas Kapal di Pelabuhan Bitung. Warta Penelitian Perhubungan 31(1):25–34
28. PLN EPI Gandeng Konsorsium Indokorea Gas Kembangkan Infrastruktur Mid-stream LNG di Nusa Tenggara, 28 March, p 1
29. Rencana Usaha Penyediaan Tenaga Listrik Listrik (RUPTL) (2021) Perusahaan Listrik Negara (PLN)
30. PERPRES No. 22 Tahun 2017, Presiden Republik Indonesia, Jakarta
31. Law No. 30/2007 on Energy in 2007 and Presidential Regulation No. 79/2014 in 2014
32. KEPMEN ESDM No. 2K/TL.01/MEM.L (2022) Menteri Energi dan Sumber Daya Mineral Republik Indonesia, Jakarta
33. KEPMEN ESDM (2017) KEPMEN ESDM No. 08 Tahun 2017 Menteri Energi dan Sumber Daya Mineral Republik Indonesia, Jakarta
34. KEPMEN ESDM No. 10 Tahun (2020) Menteri Energi dan Sumber Daya Mineral Republik Indonesia, Jakarta
35. Kementerian Koordinator Bidang Kemaritiman dan Investigasi (2023) Pemerintah Indonesia Tegaskan Komitmen Wujudkan Nol Emisi Karbon di Tahun 2060, 3 December 2023, p 1
36. Nuurhasanat AP, Haq A (2024) Pengaruh Pengungkapan Emisi Karbon, Inovasi Hijau, dan Investasi Hijau Terhadap Nilai Perusahaan. EKOMA: Jurnal Ekonomi, Manajemen, Akuntansi 3(5):2287–2297
37. Chayaningtyas SR, Isnaini Z, Ramadhani RS (2022) Green corporate social responsibility: green innovation Dan Nilai Perusahaan. Jurnal Ap-likasi Akutansi 6(2):87–108
38. PELINDO (2023) Dua Pelabuhan Regional 2 Menerima Penghargaan Green and Smart Port dari Kemenkomarves, 14 December, pp 1–2
39. Dirjen Hubla (2024) Kemenhub Dukung Penerapan Green Shipping untuk Lin-dungi Lingkungan Maritim, 11 January 2004, p 1
40. Regulasi Seputar Emisi di Indonesia (2023) https://transisienergi.id/regulasi-seputar-emisi-di-indonesia/
41. Peraturan Menteri Koordinator Bidang Kemaritiman dan Investasi Republik In-donesia Nomor 1 Tahun 2023 Tentang Pedoman Pelaksanaan Peman-tauan, Evaluasi, Pelaporan, dan Penyesuaian Rencana Aksi Kebijakan Kelautan Indonesia Tahun 2021–2025.
42. Regulation of The President of The Republic Indonesia Number 14 of 2024 on The Organization of Carbon Capture and Storage Activities (2024)
43. Regulation of The Indonesian Minister of Energy and Mineral Resources ("MEMR") Number 2 of 2023 on The Implementation of Carbon Capture and Storage ("CCS") and Carbon Capture, Utilization, and Storage ("CCUS") in Upstream Oil and Gas Business Activities (2023)

Chapter 11
Future Trends and Innovations in Maritime Energy Management

Budianto, Parman, and Muhammad Al Hazman

Abstract The maritime sector, which accounts for more than 80% of worldwide merchandise trade, is grappling with escalating environmental and energy-related issues. Rising emissions from fossil fuel reliance, stringent International Maritime Organization (IMO) regulations, and the urgent need for decarbonization drive innovation in maritime energy management. This paper explores future trends, including alternative fuels like methanol, hybrid power systems, energy-efficient technologies, and renewable energy integration. Methanol is gaining recognition as a practical fuel for marine use because of its minimal environmental footprint and compatibility with existing infrastructure. Hybrid systems combining batteries, fuel cells, and renewable energy sources demonstrate potential for significant emissions reductions. Advanced energy-efficient technologies, such as optimized hull designs and predictive fuel models, further enhance operational efficiency. These innovations, supported by digitalization and sustainable energy strategies, offer pathways for achieving the IMO's decarbonization targets while ensuring economic viability and operational reliability in maritime transportation.

Keywords Renewable energy · Decarbonization · Energy-efficient technologies · Sustainable shipping · IMO regulations · Green marine transportation

Budianto (✉) · Parman · M. Al Hazman
Department of Shipbuilding Engineering, Shipbuilding Institute of Polytechnic Surabaya, Surabaya, Indonesia
e-mail: budianto@ppns.ac.id

Parman
e-mail: parman@ppns.ac.id

M. Al Hazman
e-mail: al.hazman@ppns.ac.id

M. Elsisi et al. (eds.), *Maritime Infrastructure for Energy Management and Emission Reduction Using Digital Transformation*, Studies in Infrastructure and Control,
https://doi.org/10.1007/978-981-96-4438-4_11

1 Introduction

Maritime transportation is a vital component of the global economy, managing over 80% of worldwide merchandise trade by volume. This significance is emphasized by multiple studies, such as Hu et al. [1], who highlights the industry's critical role in global trade, and Mou et al. [2], who confirm that maritime trade accounts for a substantial portion of total merchandise movement. Additionally, Kontovas [3] discusses the broader environmental impact of shipping, while Verschuur et al. [4] underscore its economic importance through an analysis of shipping data. Collectively, these references affirm that the maritime industry is a cornerstone of global trade, significantly influencing economic activities and international commerce. This growth has driven substantial changes in fleet diversity and size, resulting in numerous challenges around energy consumption, environmental impact, and noise pollution. The maritime industry, a major consumer of natural gas, petroleum, and other fossil fuels, now faces increasing pressures to reduce energy usage and environmental footprint in response to a growing global energy crisis [5] Shipping's reliance on bunker fuel, a high-carbon fossil fuel, directly influences the industry's significant emissions profile, releasing considerable amounts of CO_2, nitrogen oxides (NO_x), sulfur oxides (SO_x), and particulate matter into the atmosphere [6].

In 2018, for example, the global shipping industry emitted approximately 1.03 billion tons of CO_2, representing roughly 3% of the world's total CO_2 emissions [7]. If substantial decarbonization measures are not adopted, projections indicate that shipping CO_2 emissions could account for as much as 18% of global anthropogenic CO_2 emissions by 2050, placing the industry among the highest emitters worldwide. Furthermore, the sector remains a significant source of nitrogen oxide emissions, contributing around 15% to the global total. These emissions, particularly black carbon, NO_x, and SO_x, degrade air quality in coastal areas and pose serious health hazards, especially for populations in densely populated port cities [5]. Efforts to control emissions have led to new regulatory frameworks, including the International Maritime Organization's (IMO) 2023 Strategy on greenhouse gas emissions. The strategy emphasizes the need for operational and technical measures to improve energy efficiency and align the maritime sector with global decarbonization goals (IMO, 2023). However, the effectiveness of these regulatory measures depends heavily on how consistently they are enforced across global maritime operations.

The maritime industry has responded by exploring innovative technologies and energy management solutions to comply with regulations and reduce operational costs. Hybrid power systems that combine electrical energy with traditional fossil fuels are emerging as promising solutions, providing flexibility and significant economic benefits by enabling ships to adapt their energy sources to operational needs and environmental conditions [8]. However, managing hybrid systems presents its own challenges due to the need to integrate these diverse energy sources with conventional diesel engines. Successful implementation will require balancing operational reliability with the complexities of multi-source energy management. The

2022 Maritime Decarbonization Strategy highlights the critical need for substantial investment in infrastructure to support alternative fuels, such as methanol [8]. As the sector undergoes a profound transformation, this chapter explores emerging trends in maritime energy management, with a focus on alternative fuels, energy-efficient technologies, and energy management innovations. These advancements are not only necessary for meeting stringent regulatory standards but also for ensuring the industry's sustainability in a rapidly evolving global energy landscape.

2 Alternative Fuels: Methanol as a Marine Fuel

When renewable methanol is blended into conventional methanol in increasing proportions, the carbon dioxide (CO_2) emissions associated with its use as a fuel can be substantially reduced. These advantages have made methanol an attractive option for operators of ocean-going vessels, short-sea shippers, ferries, cruise ships, and inland watercraft.

With a specific energy of 19,700 kJ/kg, methanol possesses a significantly lower energy density compared to liquefied natural gas (LNG). To achieve the same energy content, methanol requires approximately 2.54 times the storage volume of conventional fuels. However, methanol avoids some of the major volumetric efficiency losses seen in LNG, which arise from factors such as tank shape, insulation requirements, boil-off gas, and transfer losses.

A life-cycle analysis (LCA) of fuels, as illustrated in Fig. 1, encompasses all stages from raw material extraction, fuel production, and transportation to storage, bunkering, and combustion aboard a vessel. Future regulatory frameworks, including stricter emissions standards and potential carbon taxes, are likely to discourage carbon-intensive production methods. This shift could make energy- and carbon-efficient production pathways more economically viable over time.

The selection of raw materials as feedstock plays a pivotal role in minimizing greenhouse gas (GHG) emissions throughout the life cycle of methanol production. Among the available feedstocks, natural gas is the most commonly utilized. The production process, which is energy-intensive, consists of three primary stages:

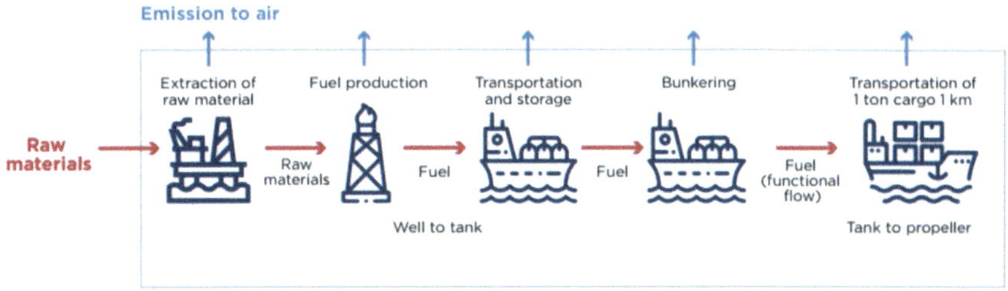

Fig. 1 Analysis of the maritime fuel life cycle from well to propeller

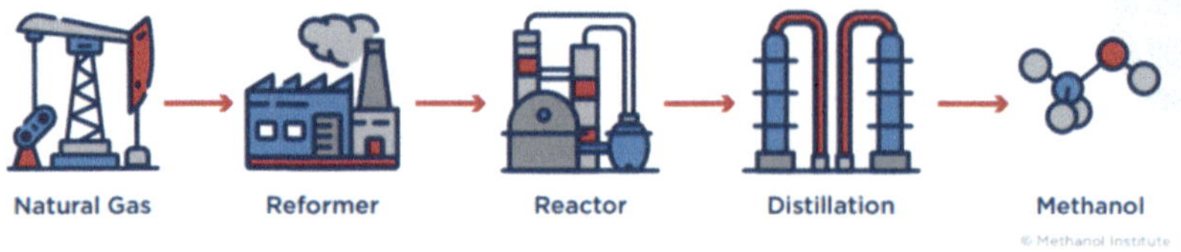

Fig. 2 Methanol production from natural gas

synthesis gas (syngas) preparation, methanol synthesis, and methanol purification or distillation.

In the syngas preparation stage, natural gas undergoes partial oxidation with steam, converting it into syngas—a mixture of carbon monoxide, carbon dioxide (CO_2), and hydrogen. The syngas is subsequently processed to synthesize methanol, with hydrogen and water generated as by-products. During the distillation stage, water is separated from methanol through reboiling, which can be tailored depending on the desired water content, potentially simplifying the production process.

The heat required for reboiling can be efficiently obtained by cooling the syngas. Additionally, the process produces surplus heat, which can be utilized to generate electricity. This self-sustaining system allows the production facility to rely on internally generated energy, thereby reducing the need for external energy inputs (Fig. 2).

Well-to-tank emissions refer to greenhouse gas (GHG) emissions generated during the conversion of natural gas into methanol. When natural gas serves as the feedstock, these emissions are relatively high, leading to well-to-propeller emissions that are slightly greater than those of conventional fuels.

Alternative feedstocks, such as biomass and biogas from sources like wastewater treatment plants and landfills, offer a more sustainable pathway. Methanol produced from these feedstocks is referred to as biomethanol. The process of syngas production from biomass or biogas closely mirrors that of natural gas feedstock, involving heating the material to specific temperatures and pressures. However, emissions from power generation equipment supplying the necessary energy remain a primary source of environmental impact. The type of energy used significantly influences the life cycle GHG emissions.

An additional low-carbon methanol production method involves carbon dioxide recovery (CDR), where surplus CO_2 from syngas (generated through steam methane reforming) is converted into methanol. Incorporating renewable electricity into natural gas-based methanol production can further decrease the carbon footprint by partially replacing conventional energy sources.

Methanol's combustion properties inherently reduce sulfur oxides (SOx) and particulate matter (PM) emissions, as its chemical structure (CH_3OH) lacks sulfur and carbon-to-carbon bonds, which are primary contributors to PM formation. Moreover, methanol's lower adiabatic flame temperature compared to diesel can help mitigate peak cylinder temperatures, thereby reducing nitrogen oxides (NOx) formation

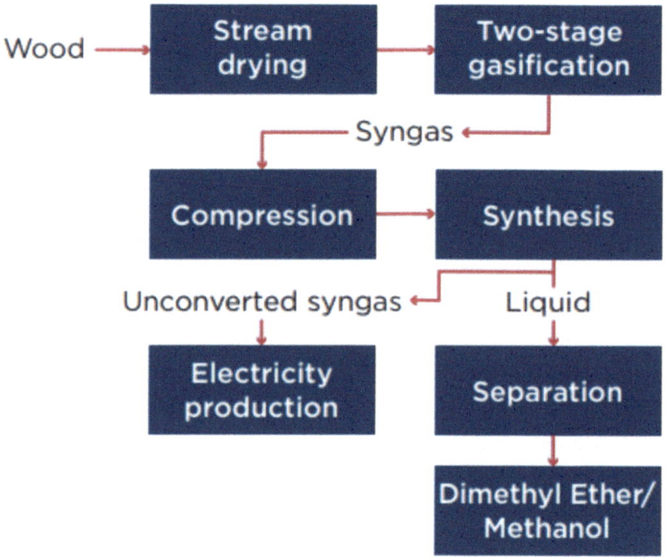

Fig. 3 Biomass (wood) to methanol process [9]

during combustion. Although after treatment systems, such as exhaust gas recirculation (EGR) or selective catalytic reduction (SCR), are often required to meet IMO Tier III NOx emission standards, adding water to methanol has been proposed as a viable alternative to control NOx emissions without relying on such systems.

Life-cycle analyses indicate that methanol produces approximately 45% lower NOx emissions and 8% lower SOx emissions per unit of energy compared to conventional fuels, highlighting its potential as a cleaner marine fuel (Fig. 3).

2.1 *Methanol as Marine Fuel*

Methanol's inclusion in the IMO's Interim Guidelines for Low Flash Point Fuels, approved in November 2020, marked the beginning of its adoption as a marine fuel. Its applications on vessels range from powering internal combustion engines to serving as a fuel source for fuel cells. Containing methanol onboard is simpler than containing LNG, as it requires only minor adjustments to systems designed for conventional marine fuels, mainly to address methanol's low flashpoint. A low-pressure system, operating at around 10 bar, can supply fuel to the engine.

2.2 Characteristics of Methanol

Methanol fires nearly invisible flames at low temperatures, making them challenging to detect until they spread. Methanol vapors are flammable within a concentration range of 6–36.5% in air, and even mixtures with 25% methanol can ignite, requiring alcohol-resistant foams for extinguishment.

Methanol's corrosive nature also demands attention to materials used in tank coatings, piping, and other fuel-handling components.

2.3 Toxicity Environmental Effect

Despite these challenges, methanol has a significantly lower environmental impact in case of spills compared to traditional hydrocarbon fuels. Methanol readily dissolves in water, posing minimal risk to marine ecosystems at low concentrations. It is naturally produced by phytoplankton and consumed by bacteria, integrating into the ocean's food chain. Table 1 lists the properties of methanol.

Table 1 Properties of methanol

Methanol property	Value
Energy density (MJ/L)	15.7
Heat of vaporization (kJ/kg)	1098
Autoignition temperature (°C)	450
Liquid density (kg/m3)	798
Adiabatic flame temperature at 1 bar (°C)	1980
Molecular weight (g/mol)	32.04
Melting point (°C)	−97.8
Boiling point at 1 bar (°C)	65
Critical temperature (°C)	239.4
Critical pressure (bar)	80.48
Flammable range in dry air (%)	6–36.5
Cetane number	<5
Octane number	109
Flash point (°C)	12
Heavy Fuel Oil (HFO) equivalent volume	2.54

2.4 Regulatory Compliance Considerations International Standards

The IMO has introduced various standards and safety codes to support the adoption of low-flashpoint fuels like methanol. The Maritime Safety Committee (MSC) adopted MSC.1/Circ.1621. Additionally, the ABS Guide for Gas and Other Low-Flashpoint Fuel Ready Vessels includes a "Methanol Fuel Ready" notation applicable to both new builds and retrofitted vessels, ensuring compliance with the International Gas Carrier (IGC) Code and other regulatory standards.

2.5 ABS Rules and Guide

ABS incorporates two IMO safety codes directly into its regulations to govern the transportation and utilization of natural gas and other low-flashpoint fuels. Sections 5C-8 and 5C-13 of these rules also include additional ABS-specific requirements and interpretations, along with unified standards and interpretations established by the International Association of Classification Societies (IACS).

This notation is applicable to newly constructed vessels as well as conversions of existing ships designed for gases or other low-flashpoint fuels, including methanol, irrespective of the vessel's size.

The guide also extends to ships currently operating on conventional fuels but equipped with features that enable future transitions to low-flashpoint fuel systems. Vessels that comply with the standards outlined in the International Gas Carrier (IGC) Code can receive recognition and notations based on their individual assessments. These assessments require adherence to IGC Code standards, compliance with Part 5C-8 of the Marine Vessel Rules, and approval from the relevant Flag Administration.

2.6 The Anticipated Function of Methanol as a Marine Fuel

The environmental positive effect of decreased emissions from methanol combustion might play a pivotal role in lowering greenhouse gas emissions within the maritime sector. Methanol's established trade infrastructure may also offer cost and supply advantages over other alternative fuels.

A key challenge for vessel owners during this fuel transition is selecting a viable alternative fuel for compliance by 2050. The feasibility of early adoption depends on the demand and supply dynamics for such fuels. Although methanol has a well-developed trade network, its high demand for manufacturing limits the surplus available for maritime use unless incentives or synthetic production are implemented. Synthetic methanol production, however, can introduce additional costs. For up-to-date methanol pricing, supply information, and safe handling practices, Methanex

Corporation, a leading methanol producer, serves as a valuable resource, as does the Methanol Institute, which actively promotes methanol as a fuel. More insights can be found in the *Methanol as a Marine Fuel Report* by FC Business Intelligence Ltd. and the Methanol Institute's *Methanol Safe Handling Manual*.

Methanol-powered vessels, especially cargo carriers with methanol propulsion systems, are increasingly being built, and if methanol is renewably sourced, these vessels could significantly cut life-cycle emissions while bolstering renewable methanol supplies for broader use.

The adaptability of methanol-fueled dual-fuel marine engines could also facilitate operations with a variety of alternative greener fuels in the coming years. Such engines might utilize methanol if it's available, with the flexibility to switch with other fuels depending on availability and cost.

Methanol's liquid state offers practical advantages over LNG and other gas fuels, as it simplifies storage and retrofitting on vessels. Methanol can be stored on board more easily and economically than gaseous fuels. Additionally, retrofitting conventional fuel tanks for methanol storage is simpler than modifying them for LNG.

Ongoing research aims to expand methanol's availability and infrastructure while advancing its onboard applications. Figure 4 illustrates predicted fuel of marine trends until 2050 as the stakeholders work to achieve the IMO's GHG emissions-reduction targets.

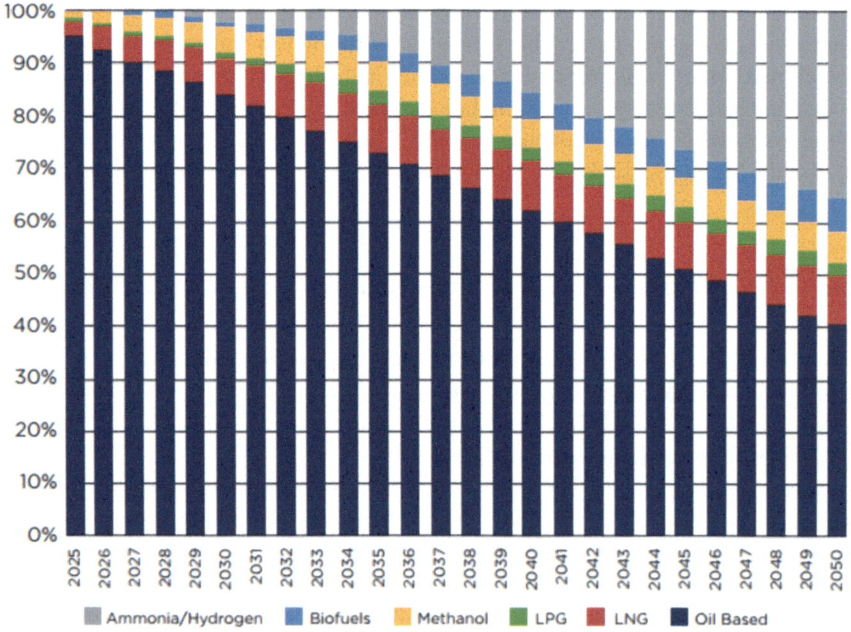

Fig. 4 Projected marine fuel use to 2050

3 Hybrid Power Systems in Maritime Application

Hybrid power systems are increasingly recognized as a transformative solution for improving efficiency of energy and decreasing emissions in maritime operations. These systems typically integrate conventional diesel engines with alternative sources of power, such as batteries, fuel cells, and renewable energy technologies like wind and solar power. The combination of different energy sources allows for a more flexible and adaptive approach to power generation, addressing the industry's need for sustainability. Unlike traditional single-fuel systems, hybrid configurations can switch seamlessly between energy sources, maximizing fuel efficiency and minimizing environmental impact, especially in variable operating conditions. This integration is becoming essential as the shipping industry faces growing regulatory pressure to reduce its carbon footprint and adhere to international emissions standards.

The use of batteries as part of hybrid power systems is a key innovation that offers substantial benefits in terms of fuel savings and emission reductions. Batteries can store excess energy produced by diesel engines during periods of low demand, which can then be used when additional power is needed, reducing the load on the engine. This approach not only decreases fuel consumption but also minimizes the wear and tear on diesel engines, potentially extending their lifespan. Battery-assisted hybrid systems are particularly effective for vessels operating on short sea routes, ferries, and inland waterways, where frequent stops and varying power demands make them an ideal choice. Studies indicate that implementing battery hybrids can reduce fuel usage by up to 20%, providing a significant advantage for ship operators looking to cut operational costs and meet emissions targets [10].

Fuel cells represent another promising component of hybrid power systems, offering a low-emission alternative to diesel engines. Fuel cells generate electricity through an electrochemical reaction using hydrogen or other low-carbon fuels, emitting only water and heat as by-products. When integrated into a hybrid system, fuel cells can complement battery storage by providing consistent, long-duration power, which is particularly useful for auxiliary systems and long voyages. The flexibility of fuel cells enables ships to reduce their reliance on fossil fuels, significantly lowering their carbon footprint. Additionally, ongoing advancements in fuel cell technology are making them more efficient and viable for large-scale maritime applications, positioning them as a critical component in the transition toward zero-emission shipping [8].

Incorporating renewable energy sources like solar and wind power into hybrid systems further enhances their potential for reducing emissions and improving energy efficiency. Solar panels can be installed on the deck to harness sunlight and generate supplementary power for onboard electrical systems, reducing the load on diesel engines. Wind-assisted propulsion technologies, such as rotor sails and kite sails, can harness wind energy to aid propulsion, decreasing the overall fuel consumption. These renewable integrations are particularly beneficial on long voyages with stable weather conditions, where they can significantly reduce a ship's environmental impact. The adoption of renewable energy in hybrid systems not only aligns with the industry's sustainability goals but also offers a pathway to compliance with stricter environmental regulations, paving the way for a greener future in maritime energy management.

3.1 Diesel-Battery Hybrid Systems

Hybrid power transmission systems are advanced, purely electrical systems that have overcome many of the traditional drawbacks associated with purely electric designs. Additionally, they have mitigated some of the limitations of internal combustion engines. A major advantage of these systems over conventional diesel engines is their ability to operate at a fixed load and consistent RPM, allowing the engine to function at its optimal efficiency. This enhances overall engine performance while reducing both emissions and fuel consumption. Furthermore, during braking or stopping, energy is captured and stored electronically in batteries through a regenerative braking system. This process reduces the workload on the combustion engine, leading to lower emissions and improved fuel efficiency. Hybrid boats feature various configurations but typically consist of an energy storage system, a power generation unit, and a power transmission system (Fig. 5).

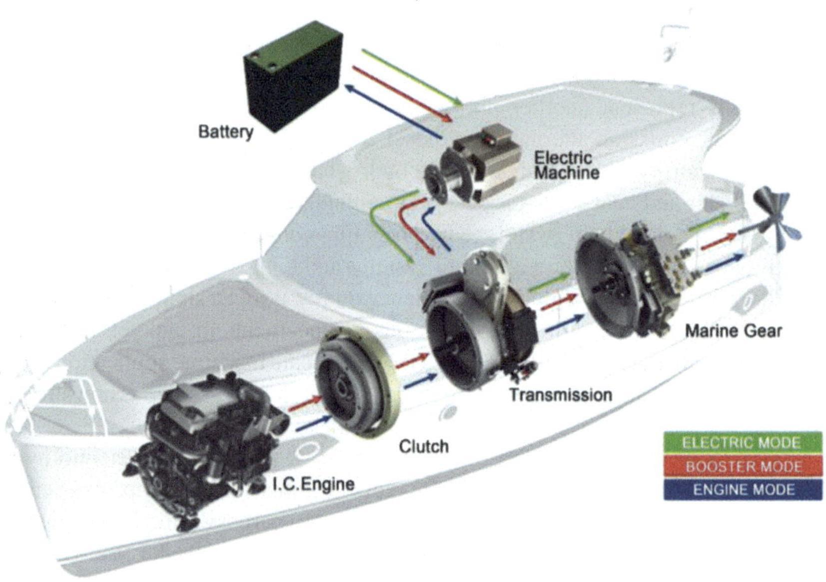

Fig. 5 Schematic of hybrid boat [8]

(a) **Diesel Engine**: The diesel engine used in hybrid boats resembles the engines found in conventional cars. However, it is typically smaller and incorporates advanced technologies that enhance efficiency and reduce emissions compared to traditional vehicle engines. These innovations improve overall performance while minimizing environmental impact. The design can also be categorized based on the degree of hybridization, referring to the ratio of electrical power to the total power output.

(b) **Electric Motors**: In hybrid boats, electric motors have the dual capability of functioning as both a motor and a generator. When additional power is needed, the motor can draw energy from the batteries to provide propulsion. Conversely, when the boat is coasting or moving downhill, the electric motor switches to generator mode, feeding power back into the battery system.

(c) **Energy Storage System**: Hybrid boats utilize various energy storage systems, including batteries, flywheels, ultra-capacitors, and hydrogen fuel cells. Among these, batteries are the most commonly used. The key attribute of a battery is its capacity, typically measured in ampere-hours (Ah). Another crucial feature is the stored energy, calculated as the product of capacity and the average discharge voltage, and expressed in watt-hours (WH). Additionally, the battery's performance and selection are influenced by its state of charge (SoC), which indicates the remaining energy as a percentage.

3.2 Fuel Cell Integration with Ships

The integration of fuel cells in the maritime industry is gaining traction as a viable solution to reduce emissions and enhance energy efficiency in marine applications. Fuel cells, particularly Proton Exchange Membrane Fuel Cells (PEMFCs) and Solid Oxide Fuel Cells (SOFCs), are being explored for various applications including main propulsion, auxiliary power, and emergency power supply on ships. These technologies are recognized for their potential to significantly decrease greenhouse gas emissions and improve overall energy efficiency compared to traditional marine diesel engines [11].

The adoption of fuel cells in maritime applications is driven by stringent environmental regulations aimed at reducing carbon emissions from shipping. The International Maritime Organization (IMO) has set ambitious targets for emission reductions, prompting the industry to seek alternative power sources that can comply with these regulations [12]. Studies have shown that fuel cells can provide a zero-emission solution when powered by hydrogen, which can be produced through renewable sources, thus aligning with global decarbonization efforts [13]. Moreover, the hybridization of fuel cells with conventional power sources, such as diesel generators, has been proposed to optimize performance and reliability while transitioning to cleaner technologies [14].

Research indicates that fuel cells can offer significant advantages in terms of operational efficiency and environmental impact. For instance, PEMFCs are noted for their high efficiency and low noise levels, making them suitable for various marine applications, including passenger ferries and submarines [15]. Additionally, SOFCs are recognized for their fuel flexibility and high conversion efficiency, which can be particularly beneficial for deep-sea [11]. However, the integration of these systems is not without challenges. Issues such as the need for specialized regulations, the establishment of hydrogen bunkering infrastructure, and the technical complexities of integrating fuel cells into existing ship designs must be addressed [16].

Furthermore, experimental studies have demonstrated the feasibility of using fuel cells in marine environments, with ongoing research focusing on optimizing their design and operational parameters under maritime conditions [17]. The development of energy management strategies for hybrid systems is also crucial to ensure efficient power distribution among various energy sources onboard. As the maritime industry continues to explore and implement fuel cell technologies, it is essential to foster collaboration between stakeholders, including shipbuilders, regulatory bodies, and energy providers, to facilitate the transition toward sustainable maritime operations.

Table 2 compares different types of fuel cells based on their main fuel sources, electrolytes, operating temperatures, technological maturity, and application areas. Each fuel cell type has distinct characteristics that determine its suitability for specific uses. For instance, Proton Exchange Membrane Fuel Cells (PEMFCs), which utilize hydrogen or methanol, operate at relatively low temperatures (up to 100 °C) and are known for their quick startup capabilities. These properties make PEMFCs ideal for small power sources and automotive applications, and they are currently in the development and demonstration phase. Molten Carbonate Fuel Cells (MCFCs), which use natural gas or coal gas, function at higher temperatures (600–700 °C), allowing for efficient hybrid power generation. MCFCs are still under development, focusing on scaling up their applications.

Table 2 Fuel cell types [14]

Fuel cell types	Main fuel	Electrolyte	Temperature of generation (°C)	Level of technology	Subject of application
PEMFC	Hydrogen, methanol	Ion conductive polymer film	Ordinary temperature ~100 °C	Development and demonstration phase	Small power source, Automobile
MCFC	Natural gas, coal gas	Molten carbonate	600–700 °C	Development phase	Hybrid power generation
PAFC	Natural gas, methanol	Phosphoric acid	150–200 °C	Commercialization phase	Distributed power system
SOFC	Natural gas, coal gas	Solid oxide	700–1000 °C	Development phase	Hybrid power generation
AFC	Hydrogen	Potassium hydroxide	80–120 °C	Commercialization phase	Space missions

Phosphoric Acid Fuel Cells (PAFCs), employing natural gas or methanol, operate at moderate temperatures (150–200 °C) and have reached the commercialization stage. PAFCs are typically used in distributed power systems due to their reliability and efficiency. Solid Oxide Fuel Cells (SOFCs), which use natural gas or coal gas with a solid oxide electrolyte, run at very high temperatures (700–1000 °C). These high temperatures enable greater efficiency, making SOFCs suitable for hybrid power generation, though they remain in the development phase. Finally, Alkaline Fuel Cells (AFCs), powered by hydrogen and using potassium hydroxide as the electrolyte, function at lower temperatures (80–120 °C) and have achieved commercialization, mainly serving space missions due to their ability to operate effectively with pure hydrogen fuel.

In conclusion, the diverse landscape of hydrogen fuel cell technologies offers a transformative pathway for reducing the environmental impact of the maritime industry. The choice of fuel cell technology depends on factors such as operating temperature, efficiency, and specific application needs. For instance, Proton Exchange Membrane Fuel Cells (PEMFCs) and Alkaline Fuel Cells (AFCs) are well-suited for lower temperature applications and portable uses, while Molten Carbonate Fuel Cells (MCFCs) and Solid Oxide Fuel Cells (SOFCs) are more appropriate for large-scale, high-efficiency power generation due to their higher operating temperatures. The technological maturity of these fuel cells varies, with some, like Phosphoric Acid Fuel Cells (PAFCs) and AFCs, already commercialized, while others, such as PEMFCs and SOFCs, remain under active development. Despite the challenges, ongoing research and development, coupled with supportive regulatory frameworks and investments in renewable hydrogen production, position hydrogen fuel cells as a critical component of future maritime energy strategies. With continued innovation and collaboration across the industry, these technologies have the potential to meet a wide range of energy demands and become a cornerstone of sustainable maritime transportation, helping the sector achieve its ambitious goals for decarbonization and environmental protection.

3.3 Renewable Energy Integration

The integration of renewable energy sources, particularly solar and wind, into hybrid power systems for maritime applications is gaining traction due to its potential to enhance fuel efficiency and reduce emissions. Solar panels installed on vessels can generate supplementary power for onboard systems, while wind-assisted propulsion technologies, such as Flettner rotors and sails, can significantly alleviate the load on conventional engines. This dual approach not only optimizes energy consumption but also contributes to the decarbonization of the shipping industry.

Recent studies indicate that wind-assisted ship propulsion (WASP) technologies can lead to substantial fuel savings. For instance, the energy savings from various wind propulsion systems can range from 0.4 to 60%, depending on the technology employed and the operational conditions of the ship [18]. Specifically, rigid sails can achieve fuel reductions of 5–60%, while soft sails can save between 4.2 and 35% [19]. The economic viability of these technologies is further supported by analyses that demonstrate their potential to significantly lower greenhouse gas (GHG) emissions in maritime operations, with estimates suggesting that wind energy could contribute to a reduction of GHG emissions by up to 50% by 2050 [20].

Moreover, the implementation of hybrid systems that combine solar and wind energy has been shown to enhance the reliability and quality of power supplied to maritime vessels. Such systems are capable of functioning effectively even in the absence of one energy source, thereby ensuring continuous operation. The flexibility of hybrid renewable energy systems is critical, especially considering the intermittent nature of wind and solar power. For instance, the integration of fuzzy logic controllers in managing battery systems within these hybrid setups has been highlighted as a means to optimize energy usage and improve overall efficiency [21]. In addition to fuel efficiency, the operational benefits of utilizing renewable energy sources are evident in the context of long voyages. Stable weather conditions can maximize the effectiveness of wind-assisted technologies, allowing vessels to harness wind energy more efficiently, thus further reducing reliance on fossil fuels [18]. The systematic overview of emerging technologies in the green maritime sector emphasizes the importance of digitalization and hull drag reduction technologies in enhancing the efficiency of hybrid systems [22].

Figure 6 illustrates a hybrid power system configuration for a maritime vessel, integrating various energy sources and components to enhance efficiency and reduce emissions. At the core of the system are two main generators that supply power for both propulsion and onboard electrical needs. The energy flow, represented by blue arrows, shows how power from the generators is directed toward the propulsion system and service loads, which include essential ship operations. In addition to the main generators, the system incorporates photovoltaic (solar) modules that capture solar energy, converting it into electrical power. This renewable energy is stored in batteries, which can supply power to the propulsion system or service loads when needed, reducing reliance on traditional fuel and lowering emissions.

The communication and control signal flow, represented by green arrows, highlights the role of the shipboard energy management system (EMS) in coordinating the various components. The EMS monitors and controls the distribution of energy, optimizing the use of renewable sources like solar power and managing the battery storage efficiently. It ensures that power is supplied to service loads and propulsion based on real-time demand while maintaining optimal performance. The integration of batteries and photovoltaic modules provides flexibility, allowing the ship to operate efficiently under varying conditions by utilizing stored energy and renewable inputs. This system design reflects a shift toward sustainable energy management in maritime applications, aiming to reduce fuel consumption and the environmental footprint of shipping operations.

The diagram depicts a hybrid energy management system for maritime vessels, illustrating two case examples: the "Viking Lady OSV" with a power output of 330 kW for auxiliary systems and 8040 kW for propulsion, and the first 500 kW ship fuel cell project in China developed by the 712 Institute. These examples highlight the

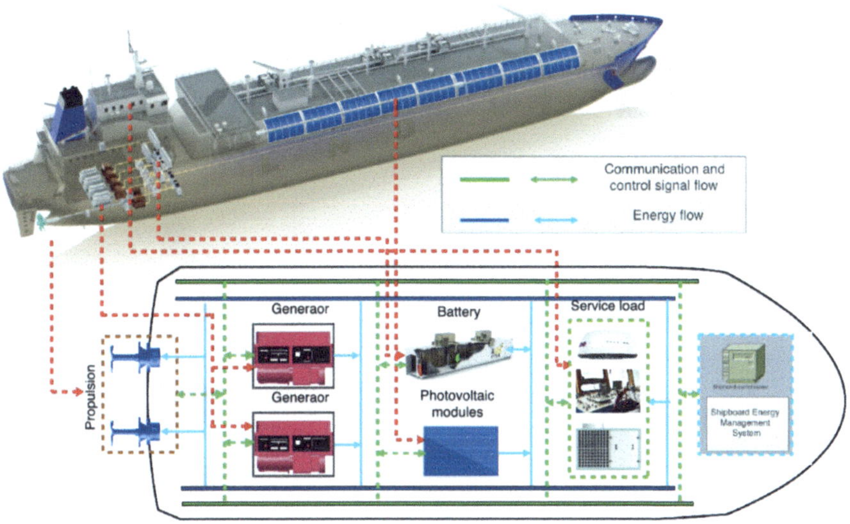

Fig. 6 Typical topology of future all-electric ships [23]

integration of multiple energy sources within a vessel's power distribution framework, aiming to enhance efficiency and sustainability. The hybrid energy system includes renewable sources like photovoltaic (PV) panels, fuel cells, supercapacitors, and batteries, all connected to a common power distribution bus (Bus A and Bus B). The integration of diverse energy sources allows the system to dynamically balance power supply based on demand and operational conditions (Fig. 7).

The diagram shows the connection of different energy storage and generation units to the buses that supply low-voltage hotel loads, which include essential onboard systems such as lighting, communication, and auxiliary services. Photovoltaic panels (PV) provide renewable solar energy, contributing to the power supply, especially in daylight conditions. Fuel cells, as another key component, generate electricity through the electrochemical reaction of hydrogen, offering a clean and efficient energy source. The use of supercapacitors provides rapid energy storage and release, making them ideal for handling peak power demands and transient loads. Batteries, known for their energy storage capacity, provide a stable and sustained power source, complementing the other components in the system.

The hybrid system design, as illustrated, emphasizes the importance of integrating various energy sources to optimize power management on ships. By combining renewable energy from PV panels with the efficient output of fuel cells and the rapid response capabilities of supercapacitors, the system can maintain reliable power for hotel loads while reducing fuel consumption and emissions. This approach allows for flexibility in energy sourcing, where the ship can utilize stored energy during peak demand periods or when renewable sources are limited. Overall, the integration of these technologies supports the goal of achieving lower emissions, enhanced energy efficiency, and improved operational performance in modern maritime vessels.

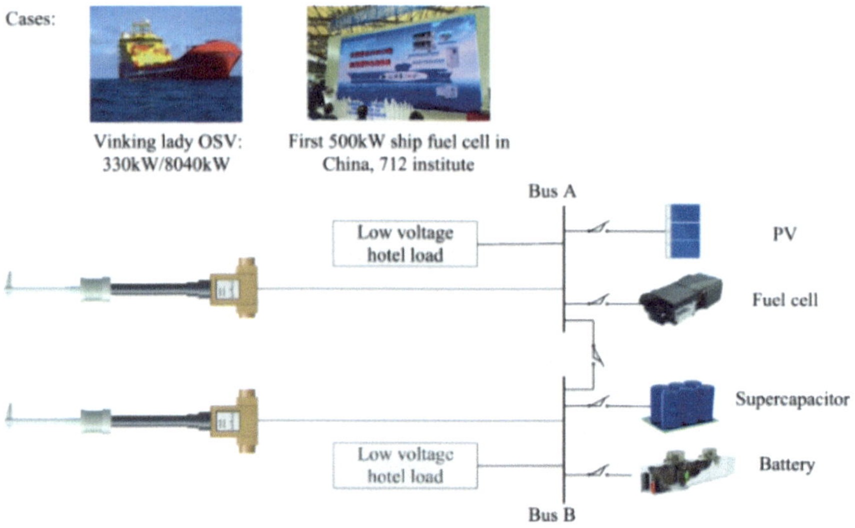

Fig. 7 Integration of fuel cell into all-electric ships [24]

In conclusion, the integration of renewable energy sources such as solar and wind power into hybrid power systems for maritime applications represents a significant step toward enhancing energy efficiency and reducing the carbon footprint of the shipping industry. The use of solar panels to generate supplementary power and wind-assisted propulsion technologies like Flettner rotors and sails alleviates the load on conventional engines, thereby optimizing fuel consumption. Recent studies indicate substantial potential for fuel savings and greenhouse gas (GHG) emissions reductions through the implementation of wind-assisted ship propulsion (WASP) technologies, with energy savings ranging from 0.4 to 60% depending on the specific technology and operational conditions [18]. These advancements underscore the economic viability of hybrid renewable energy systems, supporting the broader industry goal of achieving decarbonization targets by 2050.

Hybrid power systems that combine solar and wind energy have proven particularly effective in enhancing the reliability and quality of power supplied to maritime vessels. The integration of energy storage solutions such as batteries and supercapacitors, managed by advanced energy management systems (EMS), ensures that power demand is met even when renewable sources are intermittent. These hybrid configurations not only reduce reliance on fossil fuels but also optimize overall energy usage, making them adaptable to varying operational conditions. The flexibility offered by such systems is crucial for maintaining efficient power distribution and maximizing the effectiveness of renewable energy inputs, especially during long voyages with stable weather conditions.

The hybrid system design, illustrated through examples like the "Viking Lady OSV" and the first 500 kW ship fuel cell project in China, demonstrates the potential of combining diverse energy sources within a vessel's power distribution framework

[24]. By integrating photovoltaic panels, fuel cells, supercapacitors, and batteries, the system can dynamically balance the power supply, catering to both propulsion and essential onboard loads. This approach not only supports sustainable energy management but also enhances the operational performance of maritime vessels by reducing fuel consumption and emissions. The use of advanced control strategies within the EMS further optimizes the distribution of energy, ensuring efficient utilization of stored and renewable power.

Overall, the shift toward hybrid renewable energy systems in maritime applications highlights the industry's commitment to sustainability and innovation. By leveraging multiple energy sources and advanced power management techniques, these systems provide a viable pathway toward reducing the environmental impact of shipping operations. The continued development and adoption of such technologies are key to achieving long-term energy efficiency and decarbonization goals, paving the way for a greener and more sustainable future in maritime transportation.

4 Energy-Efficient Technologies in Maritime Operations

The maritime industry has witnessed significant advancements in energy-efficient technologies aimed at optimizing fuel usage and reducing emissions. These technologies encompass a broad spectrum of applications, including improved hull designs, advanced propulsion systems, and innovative digital solutions. This literature review synthesizes recent findings on these technologies, highlighting their contributions to enhancing energy efficiency in maritime operations. One of the primary areas of focus has been the optimization of fuel consumption through predictive modeling and operational adjustments. Su et al. [25] emphasize the importance of fuel consumption prediction models that assist ship operators in identifying anomalies and optimizing routes and speeds [25]. This predictive capability is complemented by operational measures such as speed optimization, which has been shown to significantly impact fuel efficiency due to the non-linear relationship between fuel consumption and sailing speed [26, 27]. Moreover, the integration of machine learning techniques for real-time fuel consumption prediction has emerged as a valuable tool for enhancing operational efficiency [28, 29]. Hull design modifications also play a crucial role in improving energy efficiency. Research by Ariani et al. [30] indicates that factors such as hull shape, material selection, and coatings can significantly reduce resistance and, consequently, fuel consumption [30]. The Energy Efficiency Design Index (EEDI) and the Ship Energy Efficiency Management Plan (SEEMP) are frameworks that guide the implementation of these design measures, promoting the adoption of technologies that enhance energy efficiency in both new and existing vessels [31, 32].

The Fig. 8 illustrates an integrated energy management system for a ship, utilizing a combination of renewable energy sources, thermal management technologies, and advanced power generation units to optimize efficiency. The system harnesses wind power and photovoltaic (solar) power, which are converted into electricity and stored

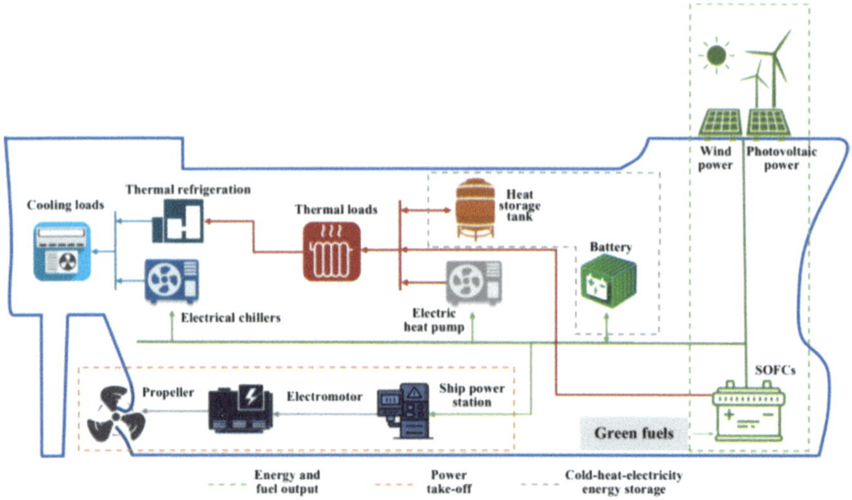

Fig. 8 Diagram of ship multi-energy coupling energy supply [33]

in batteries or used directly by Solid Oxide Fuel Cells (SOFCs). These green fuels contribute to powering the ship's main components, including the electro generator, electromotor, and propeller. The energy flow is managed to balance various ship operations, such as propulsion, cooling, and thermal loads. For cooling, electrical chillers provide thermal refrigeration, while an electric heat pump is used to meet thermal heating requirements, with excess heat stored in a dedicated heat storage tank. The ship power station coordinates energy distribution across the system, ensuring optimal utilization of renewable inputs. Additionally, the system integrates cold-heat-electricity storage capabilities, enhancing overall energy efficiency and flexibility. This comprehensive approach not only reduces the reliance on fossil fuels but also minimizes emissions, paving the way for sustainable maritime operations by leveraging multiple renewable sources and advanced thermal management solutions.

Furthermore, trim optimization, which involves adjusting the ship's balance to minimize resistance, has gained traction as an effective operational strategy that does not require structural modifications [32, 34]. The exploration of alternative fuels is another critical aspect of enhancing energy efficiency in maritime operations. Elgohary et al. highlight the potential of liquefied natural gas (LNG) as a cleaner alternative to conventional marine fuels, aligning with the International Maritime Organization's (IMO) stringent emission regulations [35]. The transition to alternative fuels, alongside advancements in hybrid power systems, represents a significant shift toward sustainable maritime practices [27]. Additionally, the use of multi-energy systems that integrate renewable energy sources, such as wind and solar, is being explored to further reduce reliance on traditional fuels and enhance overall energy efficiency [33]. Operational strategies, including fleet deployment and scheduling, are also pivotal in optimizing fuel consumption. Studies have shown that effective route planning, considering environmental factors such as ocean currents and weather conditions, can lead to substantial fuel savings [36]. The implementation of advanced algorithms for navigation optimization has demonstrated the potential to reduce fuel consumption by strategically planning sailing routes [37]. Moreover, the economic implications of fuel consumption optimization are significant, as reducing bunker fuel costs directly impacts the operational expenses of shipping companies [38]. In conclusion, the maritime industry's pursuit of energy-efficient technologies encompasses a multifaceted approach that integrates predictive modeling, hull design innovations, alternative fuels, and operational optimizations. The ongoing research and development in these areas are crucial for achieving the IMO's targets for reducing greenhouse gas emissions and enhancing the sustainability of maritime operations.

5 Future Outlook and Challenges Renewable Energy

The future of renewable energy sources exhibits a discernible upward trajectory, and it is a significant factor in the effort to mitigate environmental impacts and combat climate change. One of the most significant trends is the growing utilization of renewable energy sources, including solar, wind, and hydroelectric power, which are being embraced on a large scale by countries and industries. Solar and wind energy are preferred in the context of the energy transition due to their abundance and increasing cost-efficiency. Moreover, battery-based energy storage technologies are also developing rapidly, thereby enabling a more stable supply of electricity from weather-dependent energy sources [8]. A considerable number of countries are making significant investments in infrastructure for clean energy, with the expectation that this will result in a gradual replacement of fossil-based power generation. In conjunction with this progress, it is projected that renewable energy will become the foundation of the global energy system in the coming decades. The forthcoming trend in clean energy is the utilization of green hydrogen as a low-carbon alternative fuel with considerable potential across a range of sectors, particularly in the transport and heavy industry sectors. Green hydrogen is produced through the electrolysis of water using electrical energy derived from renewable sources, thereby eliminating the emission of carbon dioxide during the production process. It is anticipated that green hydrogen will supersede fossil fuels in a number of industrial applications that are challenging to transition to electrical energy, including the steel, cement and chemical industries, which typically necessitate exceedingly high temperatures during production. In the field of transportation, hydrogen also presents a promising prospect as a fuel for heavy vehicles and ships that have high energy requirements and need to travel long distances. Moreover, the concept of the smart grid represents another trend that is conducive to the advancement of clean energy. These grids utilize digital technology to facilitate the real-time management and optimization of energy distribution, enhance system efficiency and facilitate the integration of renewable energy into the grid. The implementation of smart grids enables more effective management of fluctuations in energy supply from renewable sources, such as solar and wind, thereby enhancing the stability of the energy supply. Furthermore, smart grids facilitate the administration of energy consumption at the consumer level through the utilization of smart home technology and energy storage systems. These technologies enable consumers to manage their energy use more efficiently, thereby contributing directly to the reduction of fossil-based energy consumption on the grid.

6 Digitalization

The process of digitalization represents one of the most significant and pervasive trends of the future, exerting a profound influence on a multitude of domains, including industry and social interactions. One of the most significant developments in the field of digitalization is the growing utilization of artificial intelligence (AI) and machine learning (ML). These technologies not only automate processes but also facilitate intelligent decision-making. AI and ML are currently being applied in a number of sectors, including healthcare, finance, manufacturing, and retail. To illustrate, in the healthcare sector, AI is employed to facilitate the diagnosis and early detection of diseases through the analysis of medical images, as well as to enhance the precision of predictions in patient care. In the financial sector, AI is employed for the purposes of fraud detection, risk analysis, and the provision of a more personalized customer service. It is anticipated that this trend will continue to expand, with AI becoming an indispensable element of everyday life, offering more profound insights and accelerating business processes. Another trend in the digitalization of business is the use of big data and data analytics to facilitate more informed, evidence-based decision-making. In the contemporary business environment, organizations are able to amass substantial quantities of data from a multitude of sources, including consumer behavior, production patterns and online interactions. The utilization of big data enables organizations to gain a more profound comprehension of consumer requirements, anticipate market developments, and enhance the efficacy of their operational processes. To illustrate, in the retail sector, data can be employed to forecast product demand, whereas in the transport industry, data facilitates traffic management and more effective route planning. This trend demonstrates that an increasing number of strategic decisions are being informed by in-depth data analysis, thereby enhancing efficiency and reducing risk. Furthermore, the Internet of Things (IoT) is emerging as a pivotal aspect of the forthcoming trajectory of digitalization, with a growing number of interconnected devices forming intricate and intelligent networks. The Internet of Things (IoT) is not solely utilized in the context of smart homes; it is also employed in industry (Industrial IoT), agriculture, and transport. In the industrial sector, the use of IoT sensors facilitates the real-time monitoring of machinery, thereby enabling the prevention of breakdowns and the optimization of production processes. In the agricultural sector, the Internet of Things (IoT) is employed for the monitoring of soil, moisture, and weather conditions, thereby facilitating more efficient and environmentally conscious agricultural practices. These trends indicate that the Internet of Things (IoT) will become increasingly integrated into various aspects of our lives, leading to the creation of a more intelligent, automated, and interconnected environment.

The field of cybersecurity represents a significant and integral aspect of the broader digitalization trend. As the quantity of data in circulation on the internet continues to grow, so too does the potential for cybersecurity threats. The prevalence of cybercrime, including activities such as hacking, data theft, and ransomware, represents a significant concern for organizations and governments across the globe. Consequently, there has been a notable increase in investment in cybersecurity, with a particular emphasis on encryption, firewalls, intrusion detection, and AI-based security technologies. Furthermore, there is a growing emphasis on the importance of cybersecurity education and awareness, both among professionals and the general public [1]. In the future, cybersecurity will extend beyond the protection of data to encompass the assurance of the integrity and reliability of the digital infrastructure that underpins modern life. Furthermore, developments in digitalization are also reinforcing the trend of digital transformation within organizations and governments. Organizations across a range of sectors are now engaged in a competitive process of adopting digital technologies with the objective of improving operational efficiency, delivering enhanced services, and improving customer engagement. In addition, governments are adapting their practices by implementing digital public services that facilitate public access to a range of administrative services, including health and education. As a result of digital transformation, public services have become more efficient, transparent, and accessible. This digital transformation demonstrates how technology is transforming the manner in which organizations and governments operate, thereby conferring considerable added value to society.

7 Conclusion

The diagram depicts a hybrid energy management system for maritime vessels, illustrating two case examples: the "Viking Lady OSV" with a power output of 330 kW for auxiliary systems and 8040 kW for propulsion, and the first 500 kW ship fuel cell project in China developed by the 712 Institute. These examples highlight the integration of multiple energy sources within a vessel's power distribution framework, aiming to enhance efficiency and sustainability. The hybrid energy system includes renewable sources like photovoltaic (PV) panels, fuel cells, supercapacitors, and batteries, all connected to a common power distribution bus (Bus A and Bus B). The integration of diverse energy sources allows the system to dynamically balance power supply based on demand and operational conditions.

References

1. Hu X, Hu K, Tao D, Zhong Y, Han Y (2023) GIS-data-driven efficient and safe path planning for autonomous ships in maritime transportation. Electronics (Basel) 12(10):2206. https://doi.org/10.3390/electronics12102206
2. Mou N et al (2021) Traffic inequality and relations in maritime silk road: a network flow analysis. ISPRS Int J Geoinf 10(1):40. https://doi.org/10.3390/ijgi10010040
3. Kontovas CA (2020) Integration of air quality and climate change policies in shipping: the case of sulphur emissions regulation. Mar Policy 113:103815. https://doi.org/10.1016/j.marpol.2020.103815
4. Verschuur J, Koks EE, Hall JW (2021) Global economic impacts of COVID-19 lockdown measures stand out in high-frequency shipping data. PLoS ONE 16(4):e0248818. https://doi.org/10.1371/journal.pone.0248818
5. Yan R, Wang S, Zhen L, Laporte G (2021) Emerging approaches applied to maritime transport research: past and future. Commun Transp Res 1:100011. https://doi.org/10.1016/j.commtr.2021.100011
6. Wan Z, Zhu M, Chen S, Sperling D (2016) Pollution: three steps to a green shipping industry. Nature 530(7590):275–277. https://doi.org/10.1038/530275a
7. Aakko-Saksa PT et al (2023) Reduction in greenhouse gas and other emissions from ship engines: current trends and future options. Prog Energy Combust Sci 94:101055. https://doi.org/10.1016/j.pecs.2022.101055
8. Moshiul AM, Mohammad R, Hira FA (2023) Alternative fuel selection framework toward decarbonizing maritime deep-sea shipping. Sustainability 15(6):5571. https://doi.org/10.3390/su15065571
9. Sikarwar VS, Zhao M, Fennell PS, Shah N, Anthony EJ (2017) Progress in biofuel production from gasification. Prog Energy Combust Sci 61:189–248. https://doi.org/10.1016/j.pecs.2017.04.001
10. Karimpour R, Karimpour M (2016) Development of hybrid propulsion system for energy management and emission reduction in maritime transport system. Open J Mar Sci 06(04):482–497. https://doi.org/10.4236/ojms.2016.64040
11. van Veldhuizen B, van Biert L, Aravind PV, Visser K (2023) Solid oxide fuel cells for marine applications. Int J Energy Res 2023:1–35. https://doi.org/10.1155/2023/5163448
12. Wang X, Zhu J, Han M (2023) Significant development potential of the solid oxide fuel cell for the technical progress of the marine main propulsion plant in the context of energy conservation and emission reduction. New Energy Exploit Appl 2(1):1–7. https://doi.org/10.54963/neea.v2i1.120
13. Rattazzi D, Rivarolo M, Massardo AF (2021) An innovative tool for the evaluation and comparison of different fuels and technologies onboard ships. E3S Web Conf 238:08001. https://doi.org/10.1051/e3sconf/202123808001
14. Roh G, Kim H, Jeon H, Yoon K (2019) Fuel consumption and CO2 emission reductions of ships powered by a fuel-cell-based hybrid power source. J Mar Sci Eng 7(7):230. https://doi.org/10.3390/jmse7070230
15. Xing H, Stuart C, Spence S, Chen H (2021) Fuel cell power systems for maritime applications: progress and perspectives. Sustainability 13(3):1213. https://doi.org/10.3390/su13031213
16. Gadducci E et al (2022) Characterization and experimental comparison of commercial PEMFC stacks for marine applications. E3S Web Conf 334:04003. https://doi.org/10.1051/e3sconf/202233404003
17. Pietra A, Gianni M, Zuliani N, Malabotti S, Taccani R (2022) Experimental characterization of a PEM fuel cell for marine power generation. E3S Web Conf 334:05002. https://doi.org/10.1051/e3sconf/202233405002
18. Chou T, Kosmas V, Acciaro M, Renken K (2021) A comeback of wind power in shipping: an economic and operational review on the wind-assisted ship propulsion technology. Sustainability 13(4):1880. https://doi.org/10.3390/su13041880

19. Grosso M, Marques dos Santos FL, Gkoumas K, Stępniak M, Pekár F (2021) The role of research and innovation in Europe for the decarbonisation of waterborne transport. Sustainability 13(18):10447. https://doi.org/10.3390/su131810447
20. Talluri L, Nalianda DK, Giuliani E (2018) Techno economic and environmental assessment of Flettner rotors for marine propulsion. Ocean Eng 154:1–15. https://doi.org/10.1016/j.oceaneng.2018.02.020
21. Venkateshkumar M, Raghavan R (2017) Hybrid photovoltaic and wind power system with battery management system using fuzzy logic controller. Int J Adv Appl Sci 6(3):268. https://doi.org/10.11591/ijaas.v6.i3.pp268-273
22. Vidović T, Šimunović J, Radica G, Penga Ž (2023) Systematic overview of newly available technologies in the green maritime sector. Energies (Basel) 16(2):641. https://doi.org/10.3390/en16020641
23. Fang S, Wang Y, Gou B, Xu Y (2020) Toward future green maritime transportation: an overview of seaport microgrids and all-electric ships. IEEE Trans Veh Technol 69(1):207–219. https://doi.org/10.1109/TVT.2019.2950538
24. Tashakori Abkenar A, Nazari A, Jayasinghe SDG, Kapoor A, Negnevitsky M (2017) Fuel cell power management using genetic expression programming in all-electric ships. IEEE Trans Energy Convers 32(2):779–787. https://doi.org/10.1109/TEC.2017.2693275
25. Su M, Su Z, Cao S, Park K-S, Bae S-H (2023) Fuel consumption prediction and optimization model for pure car/truck transport ships. J Mar Sci Eng 11(6):1231. https://doi.org/10.3390/jmse11061231
26. Yang L, Chen G, Zhao J, Rytter NGM (2020) Ship speed optimization considering ocean currents to enhance environmental sustainability in maritime shipping. Sustainability 12(9):3649. https://doi.org/10.3390/su12093649
27. Gao C-F, Hu Z-H (2021) Speed optimization for container ship fleet deployment considering fuel consumption. Sustainability 13(9):5242. https://doi.org/10.3390/su13095242
28. Ma Y, Zhao Y, Yu J, Zhou J, Kuang H (2023) An interpretable gray box model for ship fuel consumption prediction based on the SHAP framework. J Mar Sci Eng 11(5):1059. https://doi.org/10.3390/jmse11051059
29. Hu Z, Jin Y, Hu Q, Sen S, Zhou T, Osman MT (2019) Prediction of fuel consumption for enroute ship based on machine learning. IEEE Access 7:119497–119505. https://doi.org/10.1109/ACCESS.2019.2933630
30. Ariani B, Ariesta RC, Prasetya R, Oktaviani M, Hakim MH (2022) Investigation of hull design modifications on fuel consumption and energy efficiency design index (EEDI). Kapal: Jurnal Ilmu Pengetahuan dan Teknologi Kelautan 19(3):133–140. https://doi.org/10.14710/kapal.v19i3.47029
31. Beşikçi EB, Kececi T, Arslan O, Turan O (2016) An application of fuzzy-AHP to ship operational energy efficiency measures. Ocean Eng 121:392–402. https://doi.org/10.1016/j.oceaneng.2016.05.031
32. Elkafas AG (2022) Advanced operational measure for reducing fuel consumption onboard ships. Environ Sci Pollut Res 29(60):90509–90519. https://doi.org/10.1007/s11356-022-22116-7
33. Wang Z et al (2023) Optimizing energy management and case study of multi-energy coupled supply for green ships. J Mar Sci Eng 11(7):1286. https://doi.org/10.3390/jmse11071286
34. Sherbaz S, Duan W (2014) Ship trim optimization: assessment of influence of trim on resistance of MOERI container ship. Sci World J 2014:1–6. https://doi.org/10.1155/2014/603695
35. Elgohary MM, Seddiek IS, Salem AM (2015) Overview of alternative fuels with emphasis on the potential of liquefied natural gas as future marine fuel. Proc Inst Mech Eng Part M: J Eng Marit Environ 229(4):365–375. https://doi.org/10.1177/1475090214522778
36. Li S, Zhang P, Sun Z (2023) Optimization of fuel supply strategy for LNG dual-fuel ships. J Phys Conf Ser 2491(1):012014. https://doi.org/10.1088/1742-6596/2491/1/012014
37. Zhang W, Feng Y, Gu C (2023) Multi-objective navigation optimization of sailing time and fuel consumption under the influence of wind and waves. In: Ding X, Loskot P (eds) Third international conference on advanced algorithms and neural networks (AANN 2023), SPIE, p 8, October 2023. https://doi.org/10.1117/12.3004659

38. Gu Y, Wang Y, Zhang J (2022) Fleet deployment and speed optimization of container ships considering bunker fuel consumption heterogeneity. Manag Syst Eng 1(1):3. https://doi.org/10.1007/s44176-022-00003-2

The manufacturer's authorised representative in the EU is Springer
Nature Customer Service Centre GmbH, Europaplatz 3, 69115 Heidelberg,
Germany. If you have any concerns regarding our products, please
contact ProductSafety@springernature.com

Printed and bound by CPI Group (UK) Ltd, Croydon, CR0 4YY
01/10/2026
02225929-0004